40 INQUIRY EXERCISES for the COLLEGE BIOLOGY LAB

40 INQUIRY EXERCISES for the COLLEGE BIOLOGY LAB

A. Daniel Johnson

Arlington, Virginia

Claire Reinburg, Director
Jennifer Horak, Managing Editor
Judy Cusick, Senior Editor
Andrew Cocke, Associate Editor
Betty Smith, Associate Editor

Art and Design
Will Thomas, Jr., Director—Cover and Interior Design

Printing and Production
Catherine Lorrain, Director
Nguyet Tran, Assistant Production Manager

National Science Teachers Association
Francis Q. Eberle, PhD, Executive Director
David Beacom, Publisher

11 10 09 4 3 2 1

Library of Congress Cataloging-in-Publication Data
Johnson, A. Daniel.
40 inquiry exercises for the college biology lab / by A. Daniel Johnson.
p. cm.
Includes bibliographical references and index.
ISBN 978-1-933531-39-7
1. Biology--Examinations, questions, etc. 2. Biology--Examinations--Study guides. I. Title.
II. Title: Forty inquiry exercises for the college biology lab.
QH316.J64 2009
570.78--dc22
2008051246

CONTENTS

PART I

PART II

Acknowledgments

In writing this book, I drew heavily from my own work as a lab developer, coordinator, and instructor. However, one person cannot accomplish a project this complex alone. It requires help and support from many others. I would like to thank the 70-plus graduate teaching assistants who led introductory laboratory courses in the Department of Biology at Wake Forest University from 1998 to 2008. They have provided an enormous amount of feedback about the organization and flow of units and individual exercises. Many of them made suggestions for revisions and improvements, and provided ideas for new exercises or novel ways to use existing ones that ultimately made it into this book. I also would like to thank the undergraduate students of Wake Forest University for putting these units to the ultimate test.

I also wish to thank several faculty members of the Department of Biology at Wake Forest University for their significant contributions to specific lab units and for their support of inquiry-based lab instruction overall: Pat Lord, Brian Tague, Carole Browne, and Pete Weigl. I am especially indebted to Herman Eure, former chair of biology and currently dean for faculty development, for his continuous support and encouragement during our program's transition to inquiry–based laboratories. My thanks also go to our preparatory staffers—Allen Emory, Gant Hewitt, Shannon Mallison, and Mary Tietjen—all of whom helped develop the preparatory notes that accompany each unit and provided insights for the instructors' notes.

In preparing and sharing parts of this book I have received innumerable helpful comments from reviewers and from instructors at other institutions who have adapted these exercises to their own classes. Their insights made this a better book, and I sincerely thank them for their contribution. I would also like to thank Judy Cusick, Claire Reinburg, and the rest of the editorial staff at NSTA Press for their insight and suggestions for improvements and especially for their support of an atypical book such as this.

Most of all, I want to thank my long-suffering wife, Bev Nesbit, who has patiently endured the many late nights and working weekends needed to make this book a reality.

Preface

In 1998, the Howard Hughes Medical Institute (HHMI) issued *Beyond Bio101: The Transformation of Undergraduate Education*. The report outlined several strategies used by faculty at various institutions that are changing the way undergraduate science is taught. The section on laboratory experiences begins with a snapshot of the current situation on many campuses:

> *Biology students approach teaching laboratories with mixed emotions. For some, laboratory courses are windows on the world of science, allowing them to gain experience with the techniques, concepts, and emotions that go with real research. For others, laboratories are exercises in preordination, a tedious derivation of answers that are already known to questions that do not seem important.*
>
> *Often the best laboratory experience is one in which students pursue their own research under faculty guidance. In fact, given the success of undergraduate research, more and more faculty members have begun asking: Why not make teaching laboratories more like research projects? Instead of just showing students what it is like to do science, why not confront them with real problems and ask them to come up with their own solutions?* (Olson et al. 1998, p. 30; © Howard Hughes Memorial Institute. Used with permission.)

HHMI, the Carnegie Foundation for the Advancement of Teaching, the National Research Council (NRC), and numerous other groups have repeatedly called for fundamental changes in how undergraduate biology is taught (Kenny et al. 1998; NRC 2003). Each set of recommendations is based on the same theme: Students learn more, retain knowledge longer, and are better able to apply it if they are taught using active, inquiry-based strategies that let them participate in the discovery of knowledge. The advantages of inquiry-based instructional methods are not just a matter of expert opinion; numerous studies have shown they lead to significantly greater gains in student learning outcomes (for reviews and examples, see Arce and Betancourt 1997; Bain 2004; Chickering and Gamson 1987; Coppola, Ege, and Lawton 1997; Gardiner 1994; Hofstein and Lunetta 1982; National Institute of Education 1984; NRC 2003; National Survey of Student Engagement 2000).

Unfortunately, most undergraduates continue to be taught by traditional didactic exposition. They listen passively to lectures, and then in lab perform exercises that reiterate principles they supposedly just learned. Given the evidence that active, inquiry-based instructional practices increase learning gains, why do faculty continue to lecture? Why do demonstration-oriented exercises hang on as the dominant style of lab instruction? Some insight can be gained by looking at high school teachers, who have experienced similar calls for reform.

Efforts to incorporate more inquiry into high school science curricula go back more than a century. In 1892 the National Education Association (NEA) asked a panel of education leaders to examine the structure, content, and organization of the high school curriculum. The "Committee of Ten" (as it came to be called) made numerous recommendations in its final report. The following excerpt from the report shows that many of its recommendations reflect an inquiry-based approach to teaching science (emphases are this author's):

> *The Conference on Natural History unanimously agreed that the study of botany and zoology ought to be introduced into the primary schools at the very beginning of the school course, and be pursued steadily.... In the next place they agreed that in these early lessons in natural science* ***no textbook should be used****; but that the study should constantly be associated with the study of literature, language, and drawing.... Like the report on Physics, Chemistry, and Astronomy, the report on Natural History* ***emphasizes the absolute necessity of laboratory work by the pupils on plants and animals****, and would have careful drawing insisted on from the beginning of the instruction.... [T]he Conference on Natural History recommends that the pupils should be made to* ***express themselves clearly and exactly in words, or by drawings, in describing the objects which they observe....*** (NEA 1893)

Efforts to incorporate more inquiry-based instruction continued through the following century. John Dewey (author of *Democracy and Education*) called for more inquiry-based instruction in a 1916 speech to the National Education Association (Dewey 1916). By the 1950s and 1960s, the National Science Foundation had made progress toward actually implementing some changes; it sponsored several inquiry-based K–12 curriculum improvement projects, including the Biological Sciences Curriculum Study (BSCS), Chemical Education Materials (CHEM) Study, and the Science Curriculum Improvement Study (SCIS). Unfortunately the changes were short-lived and fairly localized. By 1980, exposition had returned as the dominant mode of instruction at all grade levels (Hurd et al. 1980).

During the 1980s, concerns grew that K–12 students in the United States lagged behind students in the rest of the world in science and math proficiency (NCEE 1982), and from 1989 to 2003, the American Association for the Advancement of Science and National Research Council repeatedly challenged science educators from kindergarten through college to integrate active, inquiry-based instructional methods into their classrooms (AAAS 1989, 1993; NRC 1996, 1999, 2003). Yet in 2005, a survey of state science curriculum standards found that most of the standards still emphasized content coverage; few integrated inquiry-based teaching in any systematic way (Gross et al. 2005). In short, after more than a century of work by education leaders, passive exposition remains the dominant method of instruction for most K–12 programs.

Why is there so much resistance to inquiry-based methods? In *Science Teaching and Development of Thinking*, Lawson (2002) lists 10 reasons precollege instructors commonly gave for not using inquiry-based teaching methods; most could easily apply to undergraduate instructors as well.

- Development and implementation take too much time (the most-cited reason).
- There is insufficient content coverage.
- The reading level required is too high/difficult for my students.
- The risk is too high; I do not know how the instructional units will turn out, and administration will not understand what I am doing and will penalize me.
- There are no strong students in the regular biology class (or in the nonmajors course for undergraduates).
- Students are too immature and waste too much time to use it successfully.
- I have been teaching this way for too long to change now.
- The textbook/manual/ancillary materials restrict the order in which we cover topics.
- The students and I are too uncomfortable with it.
- It is too expensive. My teaching lab is not equipped properly, and there is no budget to buy necessary materials.

It has been further suggested that undergraduate instructors are less likely than high school teachers to adopt inquiry-based instruction because there is less central oversight of the curriculum on the college level (NRC 2003 [especially see references]; Tanner and Allen 2006). A central administrative directive for K–12 reforms can help ensure that inquiry-based methods will be used systematically in several courses or across an entire curriculum. The college-level instructor, on the other hand, must make changes with minimal guidance or support.

Because there is no central source for materials and resources that meet these various needs, college instructors must cobble together solutions from a variety of sources. Simply modifying materials originally designed for high schools is not a solution because the two audiences are fundamentally different. Students' maturity, intellectual skills, and attention span differ, as do instructional goals.

The number of students served also is a factor. College instructors routinely have dozens to hundreds of students enrolled in a single large introductory course. The amount of time available for lab-related work also differs greatly.

The relative scarcity of resources designed specifically for college instructors is another source of resistance. There is an immense body of validated curricula, lesson plans, instructional resource guides, textbooks, and lab exercises available to K–12 instructors. The amount of material developed specifically for college instructors is orders of magnitude smaller.

PURPOSE AND STRUCTURE OF THIS BOOK

To incorporate more inquiry at the undergraduate level, faculty need a practical introduction to the general theory and best practices of inquiry-based teaching; tools to help them create, administer, and evaluate new inquiry-based courses and update existing courses so they can follow a more inquiry-based approach; instructional supporting materials designed specifically for their audience; and laboratory exercises that can be adapted to a variety of inquiry-based teaching and learning strategies.

This book was designed to be a general resource guide for college faculty who want to add inquiry-based methods to their biology laboratory courses. It focuses mostly on the laboratory setting, but many of the principles and methods described in Chapters 1–4 can be applied to a variety of course situations. For those who are new to this topic, Chapters 1, 2, and 4 provide a basic introduction to inquiry as an instructional practice and offer guidelines for developing an inquiry-based course using an outcomes-oriented approach. Chapter 3 reviews assessment methods and provides guidance in how to teach novice instructors (such as teaching assistants, for example) to use inquiry. Those who are already familiar with these topic areas can skip the early chapters will little loss of continuity.

Some readers may want to supplement an existing lab course with just one or two of these lab units. Others may decide (as the author's department did) to reorganize their entire lab program around inquiry. Chapter 2 discusses strategies for making this transition. Again, I have only provided an overview; details are in references at the end of each chapter and unit.

The main body of this book consists of 16 modular lab units; some are inquiry-oriented adaptations of well-established exercises, while others are entirely new. An introduction preceding the units (p. 81) summarizes the differences between them and explains the conventions used. The units are self-contained as much as possible, so they may be arranged in any order. Each unit was developed for students in either a nonmajor or introductory majors biology course. They span a range of topics and vary in length, overall structure, and difficulty. **The Summary of Units on page 86 provides information on each unit's intended audience, model and questions, major concepts, prior skills and knowledge needed, and degree of difficulty.**

The units have undergone extensive testing with undergraduates and been revised repeatedly, so they are very robust. It is important to remember, though, that every student group is unique; what one particular audience finds difficult, another may find straightforward. Inquiry requires instructors to be flexible and responsive to students' needs. Readers should not be afraid to experiment and modify exercises to fit their particular instructional goals.

Each unit contains classroom-ready student exercises, plus Instructors' Notes and Preparatory Notes. The student exercises are at the level at which they are used at the author's institution. They range in complexity from exercises that are best suited for nonmajors just starting in biology, to exercises designed for majors in sophomore-level courses. However, most can be revised to fit a different target audience. The background material that precedes the student exercises has been written so it can be revised, rearranged, or stripped down as needed to fit the target audience and the level of inquiry and difficulty desired by the instructor.

The Instructors' Notes provide supplemental background information and instructional goals and typical outcomes. They indicate where students may struggle with a unit and suggest how the instructor can guide them. Suggestions for modifying units to fit a range of course formats and audiences are included, as are suggested modifications if a small lab section (fewer than 12 students) would have difficulty completing a unit.

The Preparatory Notes for each unit list specific equipment and quantities of supplies needed for a single lab section of 20 students. Quantities are simply scaled up for larger or multiple lab sections. Detailed instructions are provided for obtaining and maintaining model organisms (if any) and for preparing and storing reagents.

Key terms and concepts appear in **boldface type**.

References

American Association for the Advancement of Science (AAAS). 1989. *Project 2061: Science for all Americans.* New York: Oxford University Press.

American Association for the Advancement of Science. 1993. *Project 2061: Benchmarks for science literacy.* New York: Oxford University Press.

Arce, J., and R. Betancourt. 1997. Student-designed experiments in scientific lab instruction. *Journal of College Science Teaching* 27: 114–118.

Bain, K. 2004. *What the best college teachers do.* Cambridge, MA: Harvard University Press.

Chickering, A. W., and Z. F. Gamson. 1987. Seven principles for good practice in undergraduate education. *AAHE Bulletin* 39: 3–7.

Coppola, B. P., S. N. Ege, and R. G. Lawton. 1997. The University of Michigan undergraduate chemistry curriculum: Instructional strategies and assessment. *Journal of Chemical Education* 74: 84–94.

Dewey, J. 1916. Method in science teaching. *General Science Quarterly* 1: 3.

Gardiner, L. F. 1994. *Redesigning higher education: Producing dramatic gains in student learning.* San Francisco: Jossey-Bass.

Gross, P. R., U. Goodenough, L. S. Lerner, S. Haack, M. Schwartz, R. Schwartz, and C. E. Finn Jr. 2005. *The state of state science standards 2005.* Washington, DC: Thomas B. Fordham Institute.

Hofstein, A., and V. N. Lunetta. 1982. The role of the laboratory in science teaching: Neglected aspects of research. *Review of Educational Research* 52: 201–217.

Hurd, P. D., R. W. Bybee, J. B. Kahle, and R. E. Yeager. 1980. Biology education in secondary schools of the United States. *American Biology Teacher* 42: 388–410.

Kenny, S. S., B. Alberts, W. C. Booth, M. Glaser, C. E. Glassick, S. O. Ikenberry, K. H. Jamieson, R. M. O'Neil, C. Reid-Wallace, C.-L. Tien, and C. N. Yang. 1998. *Reinventing undergraduate education: A blueprint for America's research universities.* Stony Brook, NY: Boyer Commission on Educating Undergraduates in the Research University.

Lawson, A. E. 2002. *Science teaching and development of thinking*. Belmont, CA: Wadsworth/ Thompson Learning.

National Commission on Excellence in Education (NCEE). 1982. *A nation at risk: The imperative for educational reform*. Washington, DC: NCEE.

National Education Association (NEA). 1893. *Report of the commission on secondary school studies*. Washington, DC: NEA.

National Institute of Education. 1984. *Involvement in learning: Realizing the potential of American higher education.* Washington, DC: National Institute of Education.

National Research Council (NRC). 1996. *National science education standards.* Washington, DC: National Academy Press.

National Research Council (NRC). 1999. *Transforming undergraduate education in science, mathematics, engineering, and technology.* Washington, DC: National Academy Press.

National Research Council (NRC). 2003. *Bio2010: Transforming undergraduate education for future research biologists.* Washington, DC: National Academies Press.

National Survey of Student Engagement. 2000. *National benchmarks of effective educational practice.* Bloomington, IN: Indiana University Center for Post-Secondary Research and Planning.

Olson, S., J. Richardson, J. B. Verringia, and D. Jarmul, eds. 1998. *Beyond Bio101: The transformation of undergraduate biology education.* Chevy Chase, MD: Howard Hughes Medical Institute.

Tanner, K., and D. Allen. 2006. Approaches to biology teaching and learning: On integrating pedagogical training into the graduate experiences of future science faculty. *CBE—Life Sciences Education* 5: 1–6.

About the Author

A. Daniel (Dan) Johnson is a North Carolina native who obtained his BS in biology from the University of North Carolina at Charlotte. After three years in the pharmaceutical industry he entered Wake Forest University School of Medicine where he earned a PhD in cell biology in 1992. He subsequently completed postdoctoral fellowships at the Texas Heart Institute in Houston and the University of Virginia in Charlottesville.

In 1998 he returned to Wake Forest University, this time to the Department of Biology as core curriculum coordinator, where he currently holds the rank of senior lecturer. He teaches general biology for nonmajors; introductory cell biology and physiology courses for premajors; and graduate courses in instructional methods, professional skills development, and bioethics on both the undergraduate and medical school campuses. He leads faculty development workshops through the campus Teaching and Learning Center and has served as an instructional methods consultant on several awarded national grants. Dr. Johnson is an active member and regular workshop leader for the Association of Biology Laboratory Education (ABLE), and in 2008 he was voted to its governing board.

Dr. Johnson and his wife volunteer their free time to Historic Bethabara Park in Winston–Salem, North Carolina. They are the principal caretakers for the Hortus Medicus, the park's historically accurate restoration of its circa 1761 Moravian medical garden.

Part I

CHAPTER 1

A Brief Introduction to Inquiry

When the term *inquiry* comes up in conversations about science curriculum reform and improvement, it usually is shorthand for **inquiry-based learning (IBL)** and, by extension, inquiry-based instruction. But what exactly is meant by inquiry-based learning? How does it differ from "traditional" learning? What is the difference between the way most of us teach (and were taught) and inquiry-based instruction?

First and foremost, inquiry is more than a collection of teaching techniques and classroom principles; it is a mind-set. The instructor focuses on developing the abilities and skills of the learner to use knowledge effectively. In contrast, traditional, didactic instruction focuses mainly on accumulation of content knowledge; it is highly fact- and content-oriented. Halonen, Brown-Anderson, and McKeachie (2002) describe the two philosophies this way:

> *Content-centered teachers tend to define their primary objective as sharing important facts and concepts with students, with limited attention to the process of learning itself and the thinking that learning requires. Many content-centered teachers believe that merely providing exposure to the ideas of the discipline will*

cause students' thinking to evolve naturally over time. Some believe that the capacity to think is innate, and that spending valuable class time promoting changes in thinking seems unnecessary or even misguided.

In contrast, learner-centered teaching elevates the process of learning by requiring students to grapple with ideas, not just passively receive them. Teachers with this pedagogical philosophy accept and relish their responsibility for fostering changes in how students think by emphasizing active learning strategies. Cognitive scientists report that underlying brain structures change to support enduring learning when students think about the course material in more meaningful ways.... Knowledge about how memory functions bolsters the viewpoint that students can improve their thinking skills through well-designed college courses." (2002, pp. 284–285. Copyright 2002 Cengage Learning, Inc. Reproduced with permission.)

The central objectives of inquiry are to (a) encourage students to be active participants in discovering knowledge for themselves and (b) provide them with legitimate opportunities to do so. The following scenarios illustrate these objectives. Two different teaching assistants are leading two general biology laboratory sections. The topic of the day is enzyme function in both classes, but the teaching assistants proceed very differently. One follows a more traditional approach, while the other uses inquiry-centered instructional methods.

Scenario 1: The instructor starts the lab with a 30-minute lecture on biological functions of enzymesand reviews some ideas covered in lecture the previous day. At the end of the lecture, the instructor tells students that their goal for the day is to demonstrate the correlation between enzyme activity of purified beta-galactosidase and temperature. Students have a detailed assay protocol to follow and so are turned loose to complete the exercise. The worksheet that summarizes students' results for the exercise is due next week. Two hours after lab began, the last student leaves for the day.

Scenario 2: The instructor starts class with a question: "What are some common methods we use to prevent food spoilage, and why do you think each one works?" Students are given one minute to come up with their own answers, then they turn to a lab partner and share their answers. After another two minutes, the instructor asks pairs to share their explanations with the class while he or she collates the explanations on a whiteboard. From time to time, the instructor asks quieter students for their ideas and probes others with follow-up questions. For example, when one pair suggests that refrigeration slows microbial growth, the instructor asks, "How specifically does the cold slow growth? What is cold doing to the organisms?" The class fumbles with this question for a few minutes until the quiet pair at the back table suggests that cold somehow inhibits enzymes that microbes use for metabolism and energy production.

After the class has compiled a list of ideas to test, the instructor points out that they cannot test them all but that the lab does have materials available to measure the speed at which bacterial enzymes break down sugar. Then—about 45 minutes after lab started—the instructor gives the class its challenge for the day: Working in groups of three to four, students are to find out whether cold could slow down enzyme activity enough to prevent spoilage. The instructor gives the students some purified beta-galactosidase (one of several enzymes that bacteria use to hydrolyze sugar for energy), a substrate solution and a one-page handout outlining the general steps for measuring enzyme activity. Students must devise the specific procedures they will use and include proper controls. A two-page written summary of their methods and observations is due next week. Three days after the lab has ended, this instructor still is answering questions by e-mail.

These two scenarios are at different points on the continuum that runs between traditional, purely didactic methods at one extreme and purely open-ended, inquiry-based learning and instruction at the other. Scenario 1 exemplifies a demonstration-type lab, in which most of the time students assume a passive role in learning. Note the following elements of Scenario 1:

1. Lab begins with a review of basic course content that many students likely understand already from the earlier lecture and that all students are personally responsible for knowing.
2. The goal of the lab is to demonstrate/confirm (yet again) a well-established piece of general knowledge (hence the pejorative label, "cookbook lab"). Students know what to expect from the exercise and have well-defined procedures to follow. Little thought is necessary to finish the exercise, so students never become cognitively engaged in it.
3. Students work in isolation. They are not required to demonstrate their thinking proce sses openly to peers or to defend their thinking against challenges. They have no opportunity to test and revise their ideas against the thinking of others.
4. The lab provides no applicable context for the content knowledge. Why should students care that enzyme activity changes with temperature?
5. The worksheet eliminates the need for students to think about how to communicate their results effectively.
6. There is no in-progress assessment of learning. Students receive no feedback regarding their knowledge and skills before the final graded assignment is handed in.

The lab structure in Scenario 2 requires students to actively participate in their learning processes. It begins with an obvious intellectual challenge that builds continuously. To succeed, each student must be deeply engaged with the topic at hand. Some other features worth noting are the following:

1. The instructor asks follow-up questions that uncover students' ongoing thinking processes. This strategy keeps thinking processes out in the open, so students see and learn to model successful patterns. The strategy also helps students move forward when their thought processes have stalled or branched off in unproductive directions.
2. The instructor encourages students to stay actively engaged anytime interest wanes and to think beyond their initial responses.
3. Multiple testable hypotheses are shared and discussed with peers. The "right" ideas are not the only ones considered.
4. Students are required to devise their own procedures and to communicate their rationales. This forces students to think about *how* their new knowledge will be obtained.
5. There are no predetermined results that students must come to; making their own observations and interpreting them is the priority. Moreover, students will not be able to completely answer the challenge question. There is room for interpreting experimental data as well as for further experimentation.
6. Students' understanding is assessed informally several times before the final grade is given.

Neither of these two approaches (didacticism versus inquiry) is fundamentally better than the other. Used properly, both have their place in the classroom. However, they are not interchangeable; at certain times one approach meets instructional goals and students' needs better than the other. To understand why requires looking at how humans learn.

CONSTRUCTIVISM PREDICTS MANY STUDENT LEARNING PATTERNS

Constructivism is a model of human learning that emerged from the work of John Dewey in the early 1900s, Jean Piaget in the 1950s, and David Ausubel in the 1960s and 1970s. Recent research in the fields of cognitive neuroscience and human behavior has confirmed most of the basic tenets of constructivism. According to the constructivist model, thinking patterns and knowledge cannot be transferred unchanged from one person to another because a learner is not a blank slate. As information (in the form of content knowledge or thinking-process skills) is transmitted, the receiving individuals construct their own mental models with it; the models reflect their unique life experiences and past learning. An individual's constructed knowledge exists as two major elements:

- A series of compartmentalized mental models that consist of both content knowledge (i.e., factual information) and related thinking process skills
- A larger-scale mental scaffold that links together the various mental models and determines which models are used most often

When individuals are challenged with learning new content knowledge or process skills, they will attempt to do so by using one of their preexisting mental models. Learners strongly resist developing new mental models as long as an existing model can solve the challenge. If an extant model is used successfully, any new knowledge or skills gained become closely associated with that particular mental model only. In the future, learners will tend to use their newly acquired content information only in the context of that particular mental model, using just the cognitive processing skills associated with it. This process of associating new content and basic process skills with existing mental models occurs routinely and is a necessary component of learning. However, new knowledge and process skills gained this way tend *not* to be applied to other situations, leading to what is often called "shallow learning."

Deeper learning occurs when a learner faces a question, problem, or situation that his or her current mental models fail completely to resolve. Once all prior mental models fail, the individual begins (usually unconsciously) to assemble one or more new "provisional" mental models and to test them against the current unsolved problem. While provisional models are in play, the person is particularly receptive to learning new content knowledge and process skills. In addition, links to potentially relevant content knowledge and process skills from the person's other preexisting mental models are established as part of the new provisional mental models. Once a new mental model has been constructed that appears to solve the current problem satisfactorily, it is reinforced and becomes stronger. The new mental model is placed within the larger mental scaffold, and other provisional models are abandoned. If the new mental model is not used regularly after it is created, it fades and is lost. Conversely, a new model that is used subsequently becomes even more stable and grows as additional content knowledge and skills become associated with it. Formation of these new but highly stable mental models is referred to as "deep learning."

Within the constructivist model, the mental scaffold is a manifestation of the underlying principles that guide a person's thinking processes *in toto*. This scaffold largely determines which mental models will be used first and how frequently. It is also the mechanism by which connections between mental models are made and by which multiple mental models are brought to bear on a problem simultaneously. Thinking patterns that make up the scaffold include learned priorities, early developmental and educational experiences, and habitual behavior; the remainder of the scaffold is linked to basic personality traits, fundamental belief systems, and one's sense of self. Because it is so deeply ingrained, the mental scaffold is the component of learning that is most resistant to change and requires the most concentrated effort to do so.

When traditional didacticism and inquiry are compared from the constructivist perspective, the inherent advantages and disadvantages of each approach to teaching become more apparent. Certain student behaviors that regularly frustrate instructors also begin to make more sense. For example, many general education courses are disciplinary surveys that are taught in a strictly didactic, content-centered style. The emphasis is almost entirely on acquisition, memorization, and direct recall of the central facts and content knowledge that underlie the discipline.

To an instructor, this task may seem to be simple and straightforward. After all, many students have already developed several mental models and process skills with which to place that mass of factual information in proper context. However, some percentage of students at the college level will not yet have developed a mental model that allows them to accomplish this apparently simple task. The instructor is unlikely to ever model the thinking-process skills that students should use to accomplish the required task. These students never have an opportunity to learn the necessary skills or to develop a successful mental model that can accomplish the goals of the course. As a result these students take required general education courses two, three, or more times without ever receiving a passing grade.

Even when students pass their required survey courses successfully, they often do not develop the process skills that are intended. Try this experiment: Ask a large group of students to explain a moderately complex concept from a lecture in a didactic survey course taken the preceding semester (or even material covered by a prior exam); then have the students apply it to a novel situation. For the majority, the relevant content knowledge remains highly compartmentalized and is unavailable for recall. Others may recall the information erroneously but have significant misunderstanding of the details or misapply it. A few students will be able to apply their prior knowledge to the novel problem, but reluctantly or with great difficulty.

These outcomes are disheartening, but should not be surprising. A typical general education class is structured to ensure they will occur. Remember that according to the constructivist model, an individual challenged with learning new knowledge or skills will try to accomplish the task using a preexisting mental model. Unlike the students who cannot complete their general education requirements, a majority of traditional-aged college students have a robust mental model that they use to identify relevant facts and retain them for a short time, then recognize correct and incorrect statements relating to those facts on a multiple-choice test. Most are drawing on the same mental models they used in high school to get into college initially. These students were never challenged, so now their existing mental models fail to accomplish the required tasks at the college level.

The outcome would not be different if the same application challenge were issued to students in a typical teaching laboratory instead of a lecture hall. As stated in the Preface, the laboratory experience of most undergraduates is highly scripted and content-centered. Despite claims that the teaching lab is where students get to "learn by doing," in reality most students do not gain as much as they could from laboratory experiences because their preexisting mental models are not challenged in a way that fosters deep, meaningful learning. In short, they are not being taught using inquiry methods.

INQUIRY ENCOURAGES DEEPER, MORE FUNCTIONAL LEARNING

In the constructivist view, the ultimate responsibility for learning rests with the learner. The responsibility of the instructor is to serve as a facilitator of the *process* of learning, not as the final source of authority or information. This includes providing learners with achievable challenges, that is, challenges that are beyond their current mental abilities and skill sets but that the learner still has a reasonable chance of accomplishing. When learners succeed in meeting challenges, it builds their confidence and increases their motivation to take on more advanced challenges.

Let's return to Scenario 2, where the instructor used an inquiry-based approach in a general biology lab, and see how learning is being facilitated. Scenario 2 depicts an idealized situation and not every student will respond equally well. Compared to the traditional demonstration lab in Scenario 1, however, even highly resistant students are likely to show significantly greater learning gains.

Students were asked several leading questions:

- *What are some common methods we use to prevent food spoilage?*
- *How does cold slow growth? What is cold doing to microbes?*
- *Will cold slow down enzyme activity enough to prevent spoilage?*

Most students in a general biology course already know that enzyme activity usually rises with temperature; this fact is part of an existing mental model. Using their existing mental models, students also can make reasonable predictions about how cold will affect enzyme activity. However, the final answer to the last question cannot be determined with certainty. From the constructivist viewpoint, students are faced with a challenge that (for most of them) their current mental models cannot solve. These are a few questions students might raise in response to this challenge, for which they cannot provide answers:

- *How cold does it have to be to stop enzyme activity entirely?*
- *Does cold affect all enzymes exactly the same way?*
- *If just one metabolic enzyme is inhibited by cold, is that enough to stop microbial metabolism completely?*

At this point, students will begin unconsciously assembling provisional mental models that might allow them to answer this question. Normally this activity happens within each individual, but because learners are particularly receptive to new content knowledge and process skills at this stage, the instructor requires them to share their thinking processes with peers, then the class, rather than allowing them to work alone. Multiple ideas are considered, and students are asked to look beyond their initial responses and evaluate their provisional mental models more carefully. Students see multiple thinking processes modeled and learn which ones are most successful. There also is the opportunity to build links to potentially relevant content knowledge and process skills that their peers discovered, but they themselves did not.

As they conduct their experiments, students must cooperate in designing an enzyme assay that includes the proper experimental variables and controls. They also must discuss how to interpret their results and present their data. Each of these behaviors reinforces new content and knowledge-processing skills within students' new provisional mental models. As a result, the new provisional model becomes more stable and is linked through the mental scaffold to other existing models.

Inquiry-based instruction does much more than just promote formation of new mental models though. If well executed, inquiry encourages students to revisit and test connections among their preexisting mental models, thus strengthening and consolidating prior knowledge as well. Advanced students can be encouraged to engage in **metacognition**, that is, thinking consciously about how they are thinking and learning. For most students, what happens during learning occurs at an unconscious level. If an instructor poses the appropriate questions, receptive students can be trained to follow their own learning process on a conscious level. Once students become aware of their own learning processes, they can be introduced to formal metacognitive strategies that give them direct access to their mental scaffolds and that heighten the ability to deploy their mental models more flexibly. In the general population, this level of cognitive self-regulation is uncommon and develops well after the undergraduate years. Using inquiry, though, it can be developed much earlier. Metacognitive teaching methods are beyond the scope of this discussion. Those who are interested in knowing more should consult Pintrich, Brown, and Weinstein (1994) and Weinstein (2000, 2002) as starting points for further information.

A FEW WORDS IN DEFENSE OF DIDACTICISM

The preceding discussion may leave readers with the mistaken idea that content-centered instruction should be abandoned. That is not the case; neither didacticism nor inquiry is fundamentally better. They represent two different sets of teaching tools, and each set can achieve certain instructional goals very effectively. The key is to employ each one at the appropriate time. A validated strategy for choosing instructional methods is discussed in Chapter 2.

A traditional lecture remains a good choice if an instructor's goal is content dissemination only. For instance, a lecture is probably the better approach when teaching students the phylogenetic relationships among classes of invertebrates. The amount of time needed for students to develop their own mental models would be considerable, and it is unlikely they would complete the challenge successfully. Similarly, many programs provide undergraduates with straightforward lab safety training. There simply is no need to use inquiry to explain the established procedures for chemical waste disposal, handling and disposing of sharps, use of safety glasses, or similar routine procedures. In the same vein, demonstration laboratories are the best way to train students to perform technically difficult operations or assays. One good example would be teaching students sterile technique in a microbiology laboratory. On the other hand, a more inquiry-oriented approach is likely to be better for introducing students to the concept and goals of phylogenetics and for exploring how that information could be applied.

In the author's experience, didacticism and inquiry can be mixed in a single course (or even a single lab session) very effectively. Imagine that a new two-semester introductory biology course is being developed for training undergraduates to be K–12 teachers. State licensing regulations mandate that these students know a predefined set of content, but the faculty know from experience that these teacher trainees need assistance in building thinking-process skills. The new course might start by laying out the conflict as a constructivist-style challenge to the students:

- *Why is the content mandated by the state considered vital for all students?*
- *How can that content be presented to K–12 students using an inquiry-based approach?*

The challenge to the student teachers would be to develop a general strategy and specific methods for providing state-mandated content to K–12 students while using inquiry-based methods appropriate for each age group. As the teacher trainees identify areas where their current content knowledge of biology is below that mandated by the state standards, faculty instructors would provide didactic minilectures that review the relevant content.

WHY IS DIDACTIC TEACHING SO PREDOMINANT?

Very few instructors operate entirely at either of the two extremes shown in the earlier scenarios. Most use both didactic and inquiry-based instructional methods in their classrooms. That said, college science instructors overwhelmingly rely on didactic methods. Most faculty claim that they encourage deep learning and critical thinking in their classroom. Yet data provided by an observation protocol or other objective evaluation tool usually show otherwise. When pressed to explain why, faculty give reasons that tend to fall into three categories: personal history, predictability, and ease of development and assessment.

Personal History. Biology as a discipline has long emphasized breadth of content knowledge, with the assumption that thinking skills develop naturally. Most current faculty learned successfully in this environment, so they tend to assume others do as well. Moreover, novice teachers tend to emulate techniques and methods they experienced personally.

Predictability. In a content-centered classroom the instructor is the central figure. One individual controls the pace, so the quantity of content delivered can be predicted accurately. Frequently an instructor has been teaching long enough to predict which topics will be most difficult for students and what questions will be asked. Contrast this with inquiry, where the instructor must change and adapt to students' needs and questions. Instructional outcomes are less predictable than they are with didacticism. Inquiry-based teaching requires instructors to be more flexible in their expectations of students. Furthermore, students are already comfortable with their passive role in learning; active learning makes them uncomfortable (especially when it is first introduced), and they are more likely to complain.

Ease of Development and Assessment. Most textbooks and laboratory manuals for undergraduate biology are designed for a content-centered curriculum. Question banks (either from publishers or from the faculty member's old exams) make the process of assessing student learning fairly straightforward. In contrast, inquiry-based teaching often means an instructor must develop new lab exercises, find alternative textbooks, and write new homework assignments. Assessment becomes a more significant problem as well: If the course emphasis is not on gain in content knowledge, how can students be assessed for grades?

Collectively, these elements produce a sort of pedagogical "natural selection" that strongly favors continued use of traditional didactic teaching methods.

WHAT METHODS AND PRACTICES QUALIFY AS INQUIRY?

Fortunately, building an inquiry-based teaching practice is not as arduous as it first looks. Remember that inquiry is defined as any teaching method that encourages students to construct or discover knowledge for themselves, as practicing scientists do. Any teaching practice or exercise that mimics or models the behavior of a scientist in the process of discovery qualifies as inquiry. Most instructors already use some inquiry methods and can learn to incorporate additional techniques quite easily.

General approaches to laboratory instruction can be categorized based on (1) the extent of instructor involvement and (2) the level of challenge students face. Terminology varies among authors; for clarity, the subsequent discussion will use the categories summarized in Table 1, which is adapted from Herron (1971).

Table 1

Features of each category of class exercise

Type of Exercise	Instructor Provides Problem	Instructor Provides Procedure	Solution or Outcome Is Known	Instructor's Involvement
Demonstration	Yes	Yes	Yes	Varies
Structured	Yes	Yes	Instructor only	High
Guided	Yes	No	No	Moderate
Collaborative	Yes	Shared	No	As peer
Open	Part or none	No	No	Minimal

Structured Inquiry. The instructor guides students through an investigation or project, asking them focused questions, giving them suggestions and ideas, and acting as a supervisor of students' work. In labs, the instructor provides a general procedure, but the expected outcome is unknown. This method is particularly suited for large-enrollment courses and for introducing groups of scientifically naive students to inquiry.

Guided Inquiry. The organization is the same as a structured inquiry. However, the procedure for conducting the investigation is developed by the students. Lab courses in which students design and conduct their own experiments usually fall into this category.

Collaborative Inquiry. Students and the instructor work side by side in an authentic investigation of a novel question. Neither party knows the precise outcome of the project. Procedures are developed in collaboration. This form of inquiry is best suited to advanced lab courses and is difficult to use in a nonlab setting.

Open Inquiry. Students investigate questions they themselves have formulated; use procedures of their own design; and conduct, modify, and report on their own experiments. The instructor facilitates this approach by providing physical resources, but for the most part students work on their own. This style of inquiry is, for all intents, the same work pattern that graduate students are expected to follow.

To illustrate the differences between types of inquiry, let's return to the scenarios at the beginning of this chapter. Scenario 1 clearly is a demonstration exercise (alternatively called a confirmation exercise). Students know they should find a linear correlation between temperature and enzyme activity for beta-galactosidase. As a stand-alone lab, this demonstration exercise has little pedagogical value. Suppose though that students are conducting a lengthy inquiry exercise that requires them to measure enzyme activity in extracts of *E. coli* grown under different metabolic stresses. If students cannot perform the basic assay accurately, it is a waste of reagents and their time to conduct the bacterial metabolic stress trials. So the basic assay might be presented as a mastery test; students cannot proceed with experiments until they achieve a predetermined level of competence. This creates a greater incentive to master the procedure quickly and perform it accurately.

In Scenario 1, students were given a worksheet on which to summarize their results. The worksheet can be amended to include a short informal assessment of students' thinking processes. The worksheet also could include one or more application scenarios. Here is a possible scenario:

> *Suppose your results for the enzyme assay show there is no enzyme activity in a sample collected from stressed bacteria. What are three reasonable explanations for why there is no activity? If you observe this outcome, how will you distinguish between these three possibilities?*

This simple informal assessment gives the instructor insight into students' overall ability to interpret their results before they conduct the actual trials. If students cannot interpret their results properly, they may need additional instruction before proceeding.

Scenario 2 is inquiry-based, but what type of inquiry is it? The instructor asks several questions that lead students to their challenge for the day: Can cold slow down enzyme activity enough to prevent spoilage? Students are given tubes of enzyme and substrate and a one-page handout outlining the general steps for measuring enzyme activity. They work with their peers to devise the specific procedures they will follow. This scenario potentially could be classified as either structured or guided inquiry, because, in both types, the problem or challenge originates with the instructor, not the students. Moreover, the final outcome is (likely) known to the instructor but not the students. The major difference is how much procedural guidance the instructor provides and how involved he or she is with students' actual work. Since the instructor provided students with a basic protocol for the enzyme assay, this scenario seems to fit most closely the description of a structured inquiry. However it can be argued that, since the instructor gave no further guidance beyond that, it fits the description of a guided inquiry instead.

In practice, it is not essential to know exactly what category of inquiry is being used. However, thinking about what type of inquiry is being employed can help instructors know how much written and oral guidance to give students. These categories can be useful for planning and discussing exercises under development or for communicating their approximate level of difficulty. Finally, the categories help guide developers in sequencing labs. Students with no prior experience with structured inquiry are going to struggle with guided or collaborative inquiry. They may succeed in meeting the challenges posed, but they are likely to be so demoralized by the experience that they actively resist subsequent inquiry-based efforts. It is vital to introduce students to inquiry in measured steps, rather than simply throwing them into it without considering their current abilities.

Going back to Scenario 2, how could it be modified to provide the more advanced types of inquiry-based instruction? In a true guided inquiry, the procedures should be devised almost entirely by the student. Rather than providing even a skeleton of a protocol, students would be required to find their beta-galactosidase enzyme assay protocol in the primary literature. Alternatively, students could be given just the enzyme assay, but have to devise their own method for extracting the active enzyme from the live *E. coli*. Both strategies leave an essential piece of the procedure up to the students to develop.

In general, collaborative and open inquiries are not practical in large enrollment labs. However, it is helpful to understand how they fit into the overall scheme. For a collaborative inquiry, a group of two to three students working with a faculty member might be challenged to adapt a standard beta-galactosidase assay to measure metabolic activity of marine archaebacteria from thermal vents. In an open inquiry, one or two undergraduates who have learned about the beta-galactosidase assay in lecture may want to determine whether human lactose intolerance can be diagnosed by measuring beta-galactosidase enzyme levels in human saliva. They are provided space to work and access to reagents but mostly work alone to answer the question they have posed. Students who reach this stage are functioning essentially as independent investigators.

BENEFITS OF MIXING INQUIRY STYLES

It should be no surprise that teaching with inquiry requires the instructor to be more flexible and responsive to individual students' needs compared with didactic methods. This seems to be a fundamental obstacle to using inquiry in large-enrollment laboratory courses with multiple sections and instructors. However, in developing the units presented in this book, the author has learned two very important lessons about inquiry-based teaching. First, inquiry is not only possible in such a setting; it actually makes it easier to manage and maintain consistency between instructors and sections. As later chapters will explain, developing a good inquiry-based course starts with establishing clear content and performance goals for each stage and for the course overall. These goals provide each instructor with a clear road map to the course and benchmarks for assessing how well students are meeting those goals. Second, mixing inquiry styles makes it possible to engage and challenge a large number of students across a broad range of ability levels.

To understand how mixed inquiry works, the reader should look at the organizational structure of one of the lab units. Every unit contains one or more exercises organized as structured inquiries. Most units also have an option for students to design and conduct their own experiments, that is, to engage in a guided inquiry. Unless otherwise noted, the units are designed so that students complete the structured inquiry as a class in one week, then work in smaller groups to design and conduct their own experiment during the following week.

The structured phase of each unit has been designed so that students discover basic content knowledge for themselves. To do so, students must apply certain process skills that the instructor is seeking to build. (For convenience, the primary and secondary learning goals, and the process skills being developed, are outlined in the Instructors' Notes for each unit.) Since each unit is designed to lead students fairly autonomously through the discovery process, differences in group facilitation skills of instructors is not a major issue in the structured phase. Thus the lab coordinator or faculty supervisor does not need to spend as much time ensuring that every instructor is conducting the lab in exactly the same way.

For the guided phase of each unit, student groups of two to four outline their experiment in advance on an experimental outline form (see Sample Form for Students' Experimental Outline in Appendix B) that they submit to their instructor for approval; experiments that have been proposed frequently are listed in the Instructors' Notes. Often small groups will devise experiments that are simple extensions of the procedures given in the first part of the unit (such as testing a higher temperature or a broader pH range than was used the first week). For many students—particularly nonmajors and students new to inquiry—this will be sufficiently challenging. As students gain confidence, they design more complex experiments that depart more from the procedures of the first week. Since small groups work independently to conduct experiments, each is free to design an experiment that interests and challenges its group members but still is within their ability to accomplish. The instructor can encourage students to push a little beyond their current abilities and knowledge each time they design and execute a new experiment. As students' skills improve, they frequently ask novel questions that are well outside the experience of the instructor. With a little creativity, though, it is still usually possible to accommodate their experimental designs.

Combining structured and guided phases offers other advantages. Small groups often evolve into informal learning teams. Individuals begin to share knowledge (peer instruction) and test each other's thinking and understanding in much the same way that the structured inquiry is designed to do. More advanced students also serve as peer and near-peer role models for students whose thinking skills have not progressed as far yet. Instructors gain more time during lab to work with students individually or in small groups. There are fewer model systems that students must master each semester, so the instructor can conduct more in-depth explorations of key principles and help students refine their thinking-process skills. For lab coordinators, structured and guided inquiry reduce the number of model organisms and equipment that must be obtained and maintained. Time once spent managing materials becomes available to assist and coach novice instructors in inquiry-based instruction.

GOING BEYOND THE BASICS

This chapter provided only a very brief outline of the fundamental differences between inquiry and traditional didactic teaching and the key features of the major inquiry-based instructional styles. Those wanting to know more about the general theory and practice of inquiry should consult Bell, Smetna, and Binns (2005), Eick, Meadows, and Balkcom (2005), Gardiner (1994), Lawson (2002), and Mintzes and Leonard (2006). For a discussion of constructivism as it relates to teaching and learning, readers should consult Gardiner (1994) and Taylor, Gilmer, and Tobin (2002).

References

Bell, R. L., L. Smetana, and I. Binns. 2005. Simplifying inquiry instruction. *The Science Teacher* 72: 30–33.

Colburn, A. 2000. An inquiry primer. *Science Scope* 23: 42–44.

Eick, C., L. Meadows, and R. Balkcom. 2005. Breaking into inquiry: Scaffolding supports beginning efforts to implement inquiry in the classroom. *The Science Teacher* 72: 49–53.

Gardiner, L. F. 1994. *Redesigning higher education: Producing dramatic gains in student learning.* San Francisco, CA: Jossey-Bass.

Halonen, J. S., F. Brown-Anderson, and W. J. McKeachie. 2002. Teaching thinking. In *McKeachie's Teaching Tips.* 11th ed., ed. W. J. McKeachie, 284–290. Stamford, CT: Cengage Learning.

Herron, M. D. 1971. The nature of scientific enquiry. *School Review* 79: 171–212.

Lawson, A. E. 2002. *Science teaching and development of thinking.* Belmont, CA: Wadsworth/Thompson Learning.

Mintzes, J. J., and W.H. Leonard, eds. 2006. *Handbook of college science teaching.* Arlington, VA: NSTA Press.

Pintrich, P. R., D. R. Brown, and C. E. Weinstein. 1994. *Student motivation, cognition, and learning: Essays in honor of Wilbert J. McKeachie.* Hillsdale, NJ: Lawrence Erlbaum.

Taylor, P. C., P. J. Gilmer, and K. Tobin. 2002. *Transforming undergraduate science teaching: Social constructivist perspectives.* New York: Peter Lang.

Uno, G. E. 1999. *Handbook on teaching undergraduate science courses: A survival training manual.* Fort Worth, TX: Saunders College Publishing.

Weinstein, C. E. 2000. Self-regulation interventions with a focus on learning strategies. In *Handbook of Self-Regulation*, eds. M. Boekaerts, P. R. Pintrich, and M. Zeidner, 727–747. San Diego: Academic Press.

Weinstein, C. E. 2002. Teaching students how to learn. In *McKeachie's Teaching Tips*. 11th ed., ed. W. J. McKeachie, 270–283. Stamford, CT: Cengage Learning.

CHAPTER 2

An Outcomes-Oriented Approach to Implementing Inquiry

Moving from a traditional lab curriculum to one that uses inquiry as its main instructional method requires careful planning. Whether the goal is a minor update to individual courses or the overhauling of an entire program, changes made without conscientious forethought are likely to fail. Creating a robust, coherent curriculum plan involves more than just choosing a set of well-designed individual lab exercises; it requires writing down clearly defined student learning goals that provide a logical context for the lab exercises. Good curriculum plans also take into account the needs and expectations of the particular student population being served as well as institutional circumstances and limitations. Finally, curriculum planning can help instructors decide when a didactic approach really is more appropriate.

This chapter provides a short overview of the process of curricular planning, with special emphasis on inquiry-based instruction. It outlines a validated method that curriculum designers and coordinators can use to establish program goals, select instructional methods and strategies, and develop a rational assessment strategy. This approach is not just for developing lab courses; it can be used to design any type of course of instruction.

A "SCIENTIFIC" APPROACH TO TEACHING

Scientists are trained to look for evidence, evaluate it critically, then make conclusions that fit the available data. Yet surprisingly few instructors use an evidence-based approach to developing undergraduate curricula. It is not for lack of data; as Derek Bok (2006) aptly wrote, "[E]mpirical papers on teaching undergraduates lie thick upon the ground" (p. 50). Furthermore, there are a number of useful tools that can make the development process itself easier.

In 2006, Handelsman, Miller, and Pfund, at the Wisconsin Program for Scientific Teaching (supported by the Howard Hughes Medical Institute), published *Scientific Teaching,* which summarized the current understanding of best instructional practices. The authors encouraged faculty curriculum designers to adopt a "backward design" approach, originally described by Wiggins and McTighe (2001). To align curricular practices with learning goals, Wiggins and McTighe recommend

- focusing on learning outcomes (rather than teaching outcomes and content from cover to cover),
- using validated tools and models for development, and
- basing decisions on evidence of effectiveness.

In the mid-1960s, Robert M. Diamond, former assistant vice chancellor for instructional development at Syracuse University, introduced a similar outcomes-oriented approach that is still valuable today (Diamond 1998). Diamond recommends that instructors

- start planning at the end; work from "ideal to real";
- define the general learning outcomes (aka, teaching or learning goals) first;
- based on these outcomes, set specific measurable performance goals; and
- choose course formats, instructional methods, and assessment instruments that will meet those goals as efficiently as possible.

While their terminology differs, the sequence of steps in these two approaches is essentially the same:

1. Decide on the general learning outcomes for students first.
2. Set specific performance goals by which students' progress toward the general learning outcomes can be measured.
3. Choose and implement teaching methods that have been demonstrated to help students achieve the established performance goals.
4. Regularly assess progress toward achieving those goals.
5. Revise the performance goals, teaching methods, and other instructional elements when necessary to facilitate learning.

Figure 1 is a visual outline of the process. Learning outcomes, performance goals, and instructional methods will be discussed in this chapter. Assessment is such a critical feature of the process that it is discussed separately in Chapter 3.

Figure 1

The outcomes-centered approach to course design
Once general learning goals have been established, there is a continuing cycle of goal-setting, development, and assessment. This iterative process helps ensure that course quality remains high and meets students' learning needs.

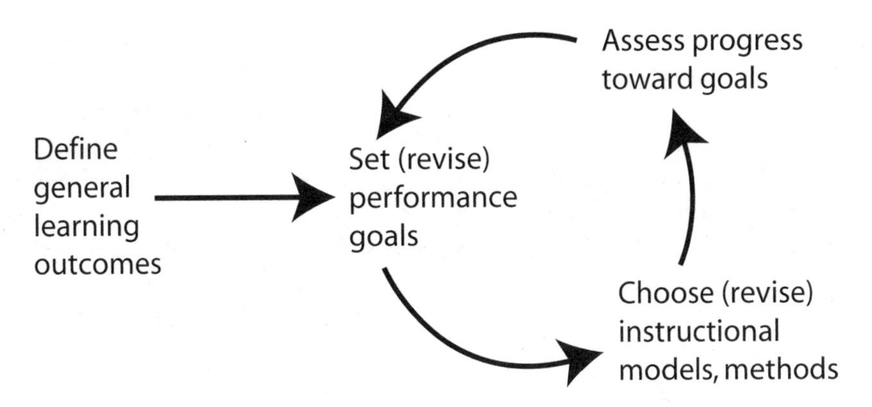

DECIDING GENERAL LEARNING OUTCOMES

The outcomes-centered approach is based on a simple yet powerful premise: Planning starts with a decision of what final endpoints students are expected to achieve. An efficient plan subsequently is developed for helping students to reach those goals. Various authors use different names for these final curricular endpoints: teaching goals, learning goals, outcome goals, and others. This discussion will use the term *general learning outcomes* (GLOs). These statements describe content knowledge, thinking-process skills, mental habits, and attitudes that students are expected to gain from a course or curricular sequence.

To understand why choosing end goals is the best place to start planning, consider the following hypothetical example. Ambry Bryce College is designing a new lab course in biology for nonmajors. The Curriculum Committee starts by choosing five GLOs for the new course:

1. Develop an ability to draw reasonable inferences from observations of the natural world
2. Develop an ability to apply principles of biology to social issues
3. Gain an understanding of the role of technology in human health
4. Learn major theories and principles of biology
5. Attract more students to biology as a major

Most faculties, when meeting to discuss a new or revised course (or curriculum), spend time hashing out specific content details before ever establishing general learning outcomes. So it is no surprise that the Ambry Bryce Curriculum Committee spends considerable time arguing about whether tuberculosis or HIV infection should be the topic for introducing disease as a social issue; furthermore, the committee members never even consider what realistic learning outcomes for the course might be. Since the majority of the Curriculum Committee's time is spent debating topics to be covered, they will probably create a course that focuses entirely on factual content. Worse, the course likely will consist of isolated concepts and will lack any unifying learning goals. Establishing general learning outcomes first provides guidance and boundaries for subsequent planning.

There are several instruments available that can help faculty identify and prioritize general learning outcomes. Probably the most widely used tool is the Teaching Goals Inventory, developed by Angelo and Cross (1993). Their inventory was designed with three purposes: (1) to help faculty choose informal classroom assessment methods, (2) to help individual faculty members become aware of course goals, and (3) to provide a starting point for discussion among colleagues. The inventory also has proven extremely useful for developing new courses and for analyzing connections between courses in a curricular sequence. Its creators have given permission for the inventory to be used freely, and it is widely available online (e.g., *http://campus.mst.edu/asess/tgi/tgi/html*).

The inventory consists of 51 general goal statements, each of which can be rated from 1 to 5. The statements are divided into six major clusters, depending on the general skills each goal fosters. All individuals involved with developing a new course complete the inventory, basing their ratings for each statement on their *own* visions for the new course. The individual results are tallied, then combined to create an overview of the group's collective vision for the course. The group's five to seven highest ranked goals on the inventory become the working list of general learning outcomes for the new course. Completing the inventory takes five to ten minutes, and one individual can tally the results for five to six respondents in less than an hour.

It is essential to understand that the Teaching Goals Inventory is a *starting point* for discussion. It cannot uncover every possible learning outcome, and the language of the general learning outcomes should be modified to reflect the specific course or curriculum. That said, the inventory is a powerful tool for guiding curricular discussion and encouraging faculty to think in general terms before addressing specific details of a single course or course sequence.

The Curriculum Committee members at Ambry Bryce College decided to take the Teaching Goals Inventory. Let's look at the learning outcomes for the nonmajors lab course to which the the inventory led the committee.

- *Learning Outcome #1: Develop an ability to draw reasonable inferences from observations of the natural world.* The wording of this outcome is identical to Item #4 in the actual Teaching Goals Inventory, except for the added phrase "of the natural world." The five members of the Curriculum Committee were unanimous in rating this as their most important objective in the new course.
- *Learning Outcome #2: Develop an ability to apply principles of biology to social issues.* This outcome originated in Item #28 of the inventory: "Develop an informed concern about contemporary social issues." Four of five committee members rated this goal as essential, but the fifth felt the wording in the inventory did not really fit with the empirical spirit of biology. After some debate, the group settled on revised wording.
- *Learning Outcome #3: Gain an understanding of the role of technology in human health.* This outcome was inspired by Item #33 of the Teaching Goals Inventory: "Develop an informed understanding of the role of science and technology." The committee was split on this goal and could not reach a consensus. Eventually they decided the problem was that the wording in the inventory was too vague. Their students were extremely interested in how modern technological advances were going to affect their own health. So the committee drafted a new outcome statement that included the phrase "human health."
- *Learning Outcome #4: Learn major theories and principles of biology.* This wording is similar to Item #18 in the inventory: "Learn concepts and theories in this subject." The committee had originally been convened because the biology faculty felt their nonmajors course was too fact laden and was not teaching useful thinking skills. However, all five members of the committee felt that students still needed to learn at least some basic principles.
- *Learning Outcome #5: Attract more students to biology as a major.* This goal is not part of the Teaching Goals Inventory. Furthermore, it is not actually a learning outcome but rather a program goal. Although it may seem out of place in a list of course outcomes, this author would argue that it is a legitimate concern. A course designed to attract new majors presents fundamentally different design issues than does one intended for students who have already committed to biology or for students who have already decided not to major in biology and are only fulfilling divisional requirements.

SETTING SPECIFIC CONTENT AND PERFORMANCE GOALS

Once the general learning outcomes have been identified for a course, the next step is to decide what specific content should be included and what the course's specific performance goals are—that is, what students should be able to do by the end of the course. This discussion can be extremely contentious. Fortunately there are tools to help focus the debate.

Khodor, Halme and Walker (2004) developed the Biology Concept Framework (BCF) as a way to help course planners prioritize content material to enhance teaching and learning. In an article about the framework, they wrote:

> *Our BCF is hierarchical, places details in context, nests related concepts, and articulates concepts that are inherently obvious to experts but often difficult for novices to grasp. Our BCF is also cross-referenced, highlighting interconnections between concepts. We have found our BCF to be a versatile tool for design, evaluation, and revision of course goals and materials. There has been a call for creating Biology Concept Inventories, multiple-choice exams that test important biology concepts, analogous to those in physics, astronomy, and chemistry. We argue that the community of researchers and educators must first reach consensus about not only what concepts are important to test, but also how the concepts should be organized and how that organization might influence teaching and learning. We think that our BCF can serve as a catalyst for community-wide discussion on organizing the vast number of concepts in biology, as a model for others to formulate their own BCFs and as a contribution toward the creation of a comprehensive BCF.* (p. 111. Reprinted with permission from the American Society of Cell Biology.)

The BCF can be accessed online at *http://web.mit.edu/bioedgroup/HBCF/CBE-Summer2004.htm*. Developers can use the original version as a guide when discussing content priorities or as the basis for their own concept frameworks.

The performance goals in any such framework will consist of action statements that describe high but attainable challenges and tangible abilities that can be assessed in a clearly defined way. These statements do not use general verbs such as *know, understand*, or *appreciate.* Instead they use more specific verbs such as *diagram, explain in a short essay, predict, compare, contrast, prioritize*, and *classify*. Consider the five general learning outcomes discussed previously.

Learning Outcome #1: Develop an ability to draw reasonable inferences from observations of the natural world.

This is a reasonable outcome expectation, but how will instructors *actually measure students' progress toward achieving it?* The answer should be clear from performance goal statements created for each learning outcome. So, for Learning Outcome #1, the goals might be as follows.

Specific performance goals: By the end of the course, if students are given an unknown organism that they have not encountered previously, they can

- *classify* the unknown organism to domain, phylum, and probable class and provide reasons for their choices;
- *compare and contrast* the methods by which the unknown organism and a human transport nutrients, exchange gases, and regulate water balance. This includes describing any anatomic evidence to support their statements; and
- *predict* how the unknown organism might adapt to a 10°C change in normal temperature of its habitat and give reasons why their predictions would protect the unknown organism.

Well-crafted performance goals play a major role in making a course or curriculum learner centered, because they focus attention on what students *can and will do*, not what faculty will cover. In addition, they provide faculty with unambiguous criteria for assessing students' progress toward the general learning outcomes. Instructors also can use them as guidelines for conducting post-course self-evaluation (i.e., did my teaching methods meet these goals?). Performance goals help instructors choose the most effective instructional methods and communicate course expectations to students. For students, performance goals serve as a mechanism for self-assessing their progress.

Setting useful performance goals is the hardest step in the outcomes-oriented development process because it exposes individual instructors' biases and the tension between didactic and inquiry approaches to teaching. To facilitate the process, Diamond (1998) recommends the following group exercise.

> *Imagine a student has come to you saying, "I am sure I learned more in your course than my grade reflects. What do I have to physically do to convince you that I met your stated goals for this course?" Given that challenge, how would you respond to the student?* (pp. 134–135. Copyright John Wiley & Sons. Adapted with permission.)

Below are additional examples of performance goals for the other general learning outcomes listed on pages 23–24. (Note that some goal descriptions go further and define what is *not* expected of students.)

Learning Outcome #2: Develop an ability to apply principles of biology to social issues.

Specific performance goals: If a student is given a category of nonpoint source pollution (e.g., nitrogenous waste), the student can summarize and compare its effects on land, air, and water quality. The summary and comparison must include the length of time the pollutant remains active, points of accumulation, and the methods by which it is eliminated from the environment. Knowing specific chemical pathways and reactions is not required.

Learning Outcome #3: Gain an understanding of the role of technology in human health.

Specific performance goals: If given a newspaper story describing a technology-based treatment for human disease, a student can write a one-paragraph summary description of the technology that makes the new therapy possible. Independent research may be included, but is not required.

Learning Outcome #4: Learn major theories and principles of biology.

Specific performance goals: If given a common ecosystem, the student can draw a diagram or chart that summarizes at least three steps in energy flow through the system. The figure should indicate the relative amount of energy lost in each step and the types of organisms that occupy each step. Specific scientific names or exact energy values are not required.

Learning Outcome #5: Attract more students to biology as a major.

Specific performance goals: Two years after implementation, the number of majors in biology is at least 5% higher than the three-year average for academic years 2005–2008. In addition, at least 5% of newly declared biology majors will have taken the nonmajors course before declaring or changing majors. (Here the performance goals are not defined by student actions but rather by other measurable outcomes.)

Students will achieve some of the overarching performance goals over time. The goal listed under Learning Outcome #1 is a good example of this type of performance goal. No single experience in the lab course ensures that students can accomplish the goal. Instead, they must synthesize skills and content knowledge gained from several experiences. Other performance goals will be tied to individual experiences (e.g., the goal for Learning Outcome #2). During their discussions, the Curriculum Committee decided they wanted students to understand the differences in point and nonpoint source pollution and how each affects air, land, and water quality. The committee expects students to be able to meet this performance goal after completing a particular lab unit they already have in mind.

SPECIFIC CONTENT AND PERFORMANCE GOALS ARE FLEXIBLE

At this point, it is important to note that general learning outcomes tend to remain fixed once they are established for a course. In contrast, the specific content and performance goals are open to revision, as part of an iterative cycle. The cycle proceeds in this way: After the initial performance goals have been defined for a course (or curriculum), instructional methods and models are selected that will (in theory) help students achieve those goals. Students complete one full round of instruction (usually a semester or quarter); then they are assessed to determine their progress toward accomplishing the specific performance goals.

If the majority of students in a new course are not achieving the initial set of goals, the performance goals and instructional methods are re-evaluated. If the performance goals are deemed too unrealistic, they are revised to better fit the student population. If the performance goals appear to be reasonable, however, the instructional methods must not be achieving them effectively. The methods are changed and the performance goals are reworded to fit the new methods. Students are allowed to complete another round of instruction, and their progress toward the specific performance goals is reassessed. The cycle of revision and refinement continues until the performance goals, instructional methods, and assessment align and there is clear evidence that the course provides students with the thinking skills and content knowledge described by the general learning goals.

This iterative cycle of development and assessment may appear to be overly time-consuming, but in the author's experience it is time well spent, for several reasons. First, the goals, methods, and assessment usually align fairly quickly; frequently a course reaches a mature (i.e., meeting its learning goals) and stable form in as few as two semesters. Second, the time invested in planning is recaptured later. There is a systematic, data-driven process for evaluating course components, so less time is spent in random troubleshooting. Third, less time is required to train new instructors or teaching assistants. Each course has clear learning outcomes, performance goals, and assessment strategies that can be communicated quickly. From the perspective of the curriculum coordinator, less time needs to be spent ensuring that all lab sections are operating similarly.

CHOOSING INSTRUCTIONAL METHODS AND MODELS

The next decision to make in course development is, Which instructional methods will most effectively help students achieve the performance goals? It may seem odd to describe how to select instructional methods in a resource book on inquiry, but it is important to know at which point during planning the decisions about methods are best made. Moreover, the goals of inquiry (development of thinking-process and application skills) can be achieved by a variety of methods, not just labs. Finally, the outcomes-oriented approach to curricular design is useful in situations besides lab course development. For the sake of completeness, it is important to at least introduce other teaching methods.

Choosing instructional methods is a second point at which course development frequently derails. By default or habit, most instructors rely on didactic lectures, assigned readings, and perhaps some expository writing, with an accompanying lab usually consisting mostly of demonstration exercises. This way of proceeding is unfortunate because there are numerous other methods that better challenge students, help them become more engaged in the learning process, and encourage them to discover knowledge for themselves. The following six methods are just a sampling of possible alternatives. For further information about these and other instructional methods, the reader should consult Herreid (2007), McKeachie (2006), and Michaelsen, Knight, and Fink (2002).

Directed Discussion. This method ranges from simple think-pair-share exercises to the instructor leading class discussions. Discussion is easily adapted to the group's skill level but can be difficult to conduct and assess in large courses.

Case-Based Learning. Students analyze a fictional or true narrative story that illustrates scientific concepts. A large number of science teaching cases are available, and new ones can be developed or adapted quickly. Insufficient content coverage is a common instructor complaint, but even those instructors acknowledge that retention and ability to apply content knowledge to new problems is much greater with this method.

Team-Based Learning. This is a hybrid strategy devised by Larry Michaelsen (2002), in which students assigned to learning teams are jointly responsible for one another's progress. The method is suited to large and small classes. If well executed, it promotes high levels of engagement while continuing to provide content coverage. Team-based learning requires extensive planning, and necessary resources may not be readily available.

Problem-Based Learning. This is the dominant teaching method in law, business, and most medical schools. Instructors facilitate the learning process using interrupted cases but do not provide essential content. Students must identify learning issues, locate and master content knowledge needed to resolve case issues, then share that knowledge with peers. This is a potent method for building functional content knowledge, strong process skills, and collaborative skills, but the order and content of the cases requires extensive planning.

Directed Writing. Students complete a sequence of writing assignments crafted to reinforce course content, develop specific thinking-process skills, and encourage deeper reflection on course topics. Done well, a one-on-one dialog develops between the instructor and student that fosters rapid cognitive development. Crafting assignments and grading are very time-intensive.

Computer Simulations and Games. Students apply content knowledge and problem-solving skills within the context of a computer program or game. Simulations have grown in popularity as computing power has grown. They range from simple interactive graphical models of population genetics to complex re-creations of physiological recording equipment. Video games based on biological concepts or three-dimensional environments (like the inside of a cell) are coming online as well. Games and simulations reduce the need for expensive equipment (such as electrical stimulators, recorders, or monitoring equipment) and make difficult lab experiences more accessible. They may, however, reinforce the idea that there is a "right" answer that students should learn. Also, they reduce students' contact with actual organisms.

DIP IN A TOE OR DIVE RIGHT IN?

Assume for now that the decision is made to use inquiry methods in a new lab course. This creates another choice: whether to use inquiry-based instruction all at once or introduce it slowly, a few units at a time. There are advantages to each approach. Using inquiry for the entire course means students do not have to cope with two fundamentally different instructional methods simultaneously. It is also easier to develop an integrated sequence of lab units when the format and structure is repeated across all units. On the other hand, incorporating inquiry a unit or two at a time allows students and instructors alike to become familiar with it without being overwhelmed. It is also the best tactic when the curricular development goal is to make targeted revisions to an existing lab course.

The author and his colleagues have discovered that an outcomes-oriented approach works equally well whether inquiry-based instruction is introduced all at once or gradually. Imagine the following best-case scenario. The full faculty of a biology department is collaborating to create an entirely new introductory lab curriculum. They devise one master set of general learning outcomes that encompasses the entire program. Next they subdivide the general learning outcomes into four logical groups and use each group of outcomes to create new, separate lab courses. Performance goals are written for each set of general learning outcomes, which in turn are used to choose the instructional methods, lab units, and assessment tools. These four new lab courses embody a seamless progression of ideas and skills because they originated from a single unifying plan.

Obviously this level of collaboration is extremely hard to achieve. Yet it is still possible to create courses that fit together well and form a logical curricular sequence. The key is to use the same outcomes-oriented development method and planning instruments as a template for creating curricular "building blocks." If the same methods are consistently re-applied each time a course is created or revised, the underlying principles and structure of the courses will tend to be very similar, even if they are developed at different times. The same thinking applies to a single course: If each component block of the course is constructed following the same procedure, all blocks naturally tend to have complementary pedagogical structures. This building-blocks approach also can bind together a project that must be divided among several developers, so long as everyone involved works from the same set of general learning outcomes and communicates frequently about specific performance goals. One caveat: When a curriculum project is completed piecemeal, it is important to conduct some sort of post hoc assessment of the entire project to ensure that it meets the original outcomes and goals. Specific methods for conducting this type of assessment are described in Chapter 3.

CHOOSING LAB UNITS

For purposes of this discussion, let's assume that the Curriculum Committee for a hypothetical nonmajors' lab has decided to use structured inquiries as much as possible but is willing to supplement inquiry with more traditional exercises to accomplish its learning outcomes. How would the committee members go about choosing lab exercises to meet their general learning outcomes and specific performance goals? Let's return to the example on pages 24–25, which is repeated here.

Learning Outcome #1: Develop an ability to draw reasonable inferences from observations of the natural world.

Specific performance goals: By the end of the course, if students are given an unknown organism that they have not encountered previously, they can

- *classify* the unknown organism to domain, phylum, and probable class and provide reasons for their choices;
- *compare and contrast* the methods by which the unknown organism and a human transport nutrients, exchange gases, and regulate water balance. This includes describing any anatomic evidence to support their statements; and
- *predict* how the unknown organism might adapt to a 10°C change in normal temperature of its habitat and give reasons why their predictions would protect the unknown organism.

The first performance goal is straightforward, in that students must demonstrate mastery of a particular skill. What lab exercises would help students achieve that goal? Typically, classification is taught by surveying plant and animal taxa over the course of several weeks. But why is that skill important? Why should students retain that knowledge for any longer than it takes to pass the quiz or lab practical? One way to get around rote memorization is to introduce the topic with an inquiry exercise that actively engages students.

Unit 16 in this book, Measuring Biological Diversity, is particularly applicable to the first performance goal. Students must use a simple dichotomous key to identify aquatic invertebrates, but keying skills are not the primary focus. Rather, students learn to estimate diversity using differences in relative numbers and types of organisms present in two aquatic habitats (a natural creek and a stormwater stream). They learn by direct experience that the most prevalent taxa of invertebrates differ between high-quality and degraded aquatic habitats. By itself, the unit on biodiversity does not provide students with sufficient experience with identifying organisms to become proficient (i.e., able to achieve the stated performance goal). However, it *does* help students build a functional understanding of *why* classification skills are useful, which in turn will help them be more deeply engaged by a traditional didactic survey of taxa. To that end, the Curriculum Committee opts for a three-week sequence beginning with Measuring Biological Diversity, followed by a two-week survey of the physical features and lifestyles of major groups of invertebrates and vertebrates.

To achieve the second performance goal, students are expected to bring together content knowledge from multiple sources to synthesize a solution. Clearly, no single lab exercise can achieve this goal, but three different units provide relevant experience: Unit 8, Animal Hormones; Unit 10, Homeostasis and Heart Rate; and Unit 11, Metabolism and Oxygen Consumption. The unit on animal hormones explores how insect hormones regulate physiological processes; most of the central principles that students discover in this unit are applicable to hormones in other organisms, including humans. The unit on metabolism explores how gas transportation, consumption, and excretion are controlled, while the unit on heart rate emphasizes neurological controls of physiology. For the moment, the Curriculum Committee decides to include all three units in the new lab course. Additionally, at least one lab will include a guided inquiry, that is, a week where students design and execute their own experiments.

To meet the third performance goal under Learning Outcome #1, students must make predictions about physiological responses to temperature that they have not actually observed. Again, no one lab unit will ensure that every student can achieve this goal. Instead, students are expected to apply general principles and observations relating to temperature gained from multiple lab experiences. Two of the lab units specifically address the effects of temperature upon physiological processes as part of a structured inquiry—Unit 4, Properties of Enzymes, and Unit 11, Metabolism and Oxygen Consumption. In four other units, students often conduct experiments that relate to environmental temperature—Unit 6, Energetics and Photosynthesis; Unit 9, Neuromuscular Control; Unit 10, Homeostasis and Heart Rate; and Unit 13, Animal Behavior.

To keep track of the units, it is often useful to construct a matrix table that summarizes the specific performance goals and tells which lab units support each goal. Table 2, page 32, is an example matrix, using just the labs mentioned above.

Table 2

Matrix analysis of lab units

	General Learning Outcome #1		
Inquiry Lab Unit	**Classify an Unknown Organism**	**Compare Gas Exchange, Nutrient Transport, Water Balance Mechanisms**	**Predict Effects of Temperature Change on Physiology**
Properties of Enzymes			YES
Energetics and Photosynthesis			Possible
Animal Hormones		YES	
Neuromuscular Control			Possible
Homeostasis and Heart Rate		YES	Possible
Metabolism and Oxygen Consumption (w/one-week experimental lab)		YES	YES
Animal Behavior			Possible
Measuring Biological Diversity (incl. two weeks postlab)	YES		

The lab unit on metabolism can help students achieve more than one performance goal, so the Curriculum Committee revises its earlier choices and eliminates the units on animal hormones and regulation of heart rate. In their place, the committee adds a one-week guided inquiry to the metabolism exercise, so that students will have a chance to conduct experiments of their own.

At this point the lab under development consists of a two-week mixed inquiry on metabolism, a one-week structured inquiry on biodiversity, and a two-week follow-up sequence in which students conduct a survey of specific animal taxa. This process of addition and refinement continues until there is a rough collection of lab units that meet all of the established performance goals. Rather than follow along, this discussion will skip forward to the next step: sequencing and refining the rough collection of lab units.

SEQUENCING A NEW LAB COURSE

Typically the sequence of lab topics in a large introductory survey course follows the lecture sequence. At the introductory level, some repetition does help reinforce and prioritize key concepts. As students' integrative skills improve and they take more advanced classes, the sequence of lab and lecture topics can diverge, so long as the instructor explains why they differ.

Whether students should be introduced to new concepts in lab or lecture first is a different issue. In a didactic class, ideas usually are presented in lecture first, then confirmed in lab. Yet observations of the author's own program and conversations with other coordinators indicate that many students learn more deeply if they encounter topics in lab before lecture, not vice versa. From a constructivist perspective, this is not surprising. Encountering a concept in lecture first encourages students to be passive learners, that is, to think that there is a "correct" answer. A "lab-first" approach makes students wrestle with new concepts on their own before learning formal principles. They are more likely to have their existing mental models fail and to form new ones (Lawson 2002).

There is a downside to a "lab first" approach. Novice students comfortable with a demonstration approach to lab instruction often complain about having too little guidance the first time they face an inquiry-based lab; they complain louder still when key concepts are introduced in lab first. The instructor or coordinator must pay close attention to his or her measures of student performance (see Chapter 3, the section called Formal Assessment Methods.). If students are meeting performance goals, they merely are venting frustration at being challenged, but if they are not meeting performance goals, they may not have the cognitive maturity to manage the challenge yet. Consider splitting the difference: Introduce simpler ideas in lab first, but introduce more difficult concepts in lecture first.

What is the best order in which to present topics? Most instructors default to a sequence of increasing scale. However, other instructors have successfully started with physiology first, then moved outward toward ecosystems or inward to cells and molecules. Another option is to begin with an overview of biosphere-level processes, then examine in detail the individual systems that contribute to those processes. Finally, courses can be arranged around a single integrating theme (such as evolution), ecosystem (wetlands and aquatic habitats), or a particular group of organisms (humans, plants of economic importance) that connects topics and principles. None of these options is inherently better; what is most important is to consider the skill level and interests of your student population, then choose an arrangement that intellectually engages them as quickly as possible, while providing reasonable cognitive challenges.

REFINING A NEW COURSE

When a topic sequence is decided, individual units are refined to bring the course to its final form. So far, there has been little discussion of the realities of budget constraints, institutional resources, or available equipment and how they impact course development. This has not been an oversight by the author but a reflection of an intentional planning tactic that Robert Diamond terms "planning from ideal to real" (Diamond 1998). During development, the best instructional methods and labs often are eliminated from consideration for no better reason than "we can't do that." Designing the ideal course *first* creates an incentive to locate funds, equipment, expertise, and other resources needed to make it a reality. Even when the entire course cannot be conducted, usually much more can be managed than was originally thought possible. In addition, idealized course plans serve as blueprints for future improvements and help break down institutional resistance to curricular change.

For example, when the author began developing the first versions of the lab units presented in this book, inquiry accounted for less than 20% of all lab exercises. It was widely assumed that inquiry was not feasible in the context of large introductory lab courses and that routine student experiments would be impossible to manage. Ten years later, more than 90% of the laboratory exercises performed by students in the introductory biology courses at the author's institution are structured inquiries and, in over half, students (nonmajors and majors alike) conduct their own original experiments.

The initially selected set of lab units is refined by asking a number of questions that aim to uncover gaps in thinking-process skills development, identify discrepancies between the general learning outcomes and specific performance goals, and ensure that the proposed lab units can actually be conducted given the setting and resources. These questions do not need to be considered in order; again, a matrix table can assist with tracking. Based on this information, course developers can decide which units to expand, which to eliminate, and which to modify in order to create the course that fits their unique student populations. Clarifying questions fall into three general categories: instruction and pedagogy issues, technical issues, and course structure and administration.

Instruction and Pedagogy

1. What background knowledge and skills do students need for this lab? Will all students have them? If not, can students gain those skills and knowledge though online remedial tutorials, a supplemental lab meeting, or some other methods?
2. What thinking-process skills will this lab unit help develop? Do these skills fit the general learning outcomes and performance goals for the course?
3. What are the primary and secondary concepts that students will learn by completing this lab unit? Does this content knowledge fit the general learning outcomes and performance goals for the course?
4. How will students' progress toward achieving the performance goals and general learning outcomes be assessed? Is more than one assessment method used?
5. How does this unit connect back to and build on previous experiences? How does this unit connect to upcoming ones? How will the instructor ensure that students make these connections?

Technical Issues

1. Can we obtain, house, and care for the quantity of organisms required for this unit? If we cannot purchase the number that we need, can we rear them ourselves? Does staff have the necessary expertise to rear and handle the organisms?
2. Do we have the facilities to properly store, handle, prepare, and dispose of all chemical reagents required?

   This issue is increasingly important for many institutions. Teaching labs often employ hazardous chemicals, and authorities have been cracking down on their use and disposal. The units in this book contain instructions for proper handling and disposal of currently regulated chemicals. Older lab protocols may not indicate which chemicals are considered hazardous; check with the state office of the EPA or your Institutional Safety Office if you are ever unsure about a compound or waste product.
3. Is suitable equipment available?

   It is not necessary to have the highest quality research grade equipment. Do not hesitate to be creative; rather than buy a high-volume 37°C water bath, set up a 20 gal. aquarium and heat it with two aquarium heaters. Students are surprised at how much good science can be done with minimal equipment.
4. Are there sufficient backup resources?

   Students will use more organisms and reagents than necessary. They are also very hard on delicate equipment. Plan accordingly.

Course Administration

1. Is each lab unit reasonably challenging for the majority of students taking the course?
2. How will general learning outcomes and specific performance goals be communicated to students? To all instructors for the course? The simplest solution for instructors is shared, preprepared supplemental teaching notes.
3. What general techniques will instructors need to know in order to lead this unit? Will all instructors know how to conduct this unit as planned? Will they need supplemental training? How will they receive it?
4. If both didactic and inquiry-based instructional methods are being used, is there a reasonable balance between them? Does the unit encourage too much didactic instruction and discourage inquiry?
5. Should a supplemental guided inquiry be included as part of this unit?

The author believes that the benefits to students of conducting their own experiments far outweigh any disadvantages. Advantages include more rapid development of thinking process and application skills, deeper learning, and greater student engagement overall. The biggest disadvantage is the amount of time required. In the author's program, students typically complete a one-week (or longer) structured inquiry, then plan and execute their own experiments the following week. So at a minimum, two weeks are devoted to every topic. However, the depth of student learning makes up for the lack of additional breadth. In addition, lab units were selected that teach widely applicable skills and concepts, which compensates for some of the time spent conducting the guided inquiry.

Another potential disadvantage is the time required to prepare the additional materials that students request for their experiments. In practice, though, the time gained from having half as many labs to set up and pull down each semester more than compensates for the time used to prepare those materials. Before making this decision, read the Instructors' Notes for each selected unit. They include descriptions of experiments students have done previously and any specific concerns that should be considered.

Once the course developers are satisfied with the structure of the course and the necessary advance preparations are complete, it is time to test the course with students.

Roll-Out Strategies

When is the best time to introduce a new lab course? The fall semester or first quarter of the academic year is a frequent choice. The main disadvantage is that not everyone involved with planning the course may be available during summer, which pushes final preparations into the mad rush of late August. If a new course will eventually be offered in both fall and spring semesters or quarters, consider conducting its first trial at the end of the academic year instead. Course developers will have time to finalize details, and they will have a break during summer for making revisions before the course is offered the next time. If the new lab course is scheduled for the start of the academic year only, consider offering a single section during the spring semester or quarter. Summer school also is a good time to test new courses, when course enrollments are often lower.

Revisions Are Inevitable

The day students try a new lab unit or course for the first time is both exciting and anxiety-provoking. What if the lab fails to accomplish its goals? What if students do not like it? Innumerable "what ifs" go through the course designers' minds. As someone who has gone through this process many times, the author offers this advice: No matter how carefully the course was planned, the trial run will *not* go smoothly, and it will *not* be perfect. Students and lab instructors are going to point out problems and issues that the course designers never imagined. Do not be discouraged; the first version of the lab unit or course invariably needs revision. Remember, a robust lab program that is learner-oriented is continually changing and adapting to students' learning needs. Revisions are part of the ongoing cycle of assessment and refinement. To give the reader a frame of reference, the author is not particularly concerned if a course or lab unit requires as many as four to five revision cycles before reaching a stable, mature form that performs smoothly and consistently.

To make revisions easier, start collecting data immediately. Make notes of what instructors felt worked and what did not. Review the original planning questions; were any issues not addressed fully? Conduct a written or online survey of students to find out if they liked the model system, and ask specifically why or why not. Look at the assessment data; are there obvious gaps between the stated performance goals and what students accomplished? Based on the data that you collect and your observations, select the three to four most serious issues or problems with the current iteration of the lab unit or course. These should be the top priority when making revisions and preparing for the next iteration.

References

Angelo, T. A., and K. P. Cross. 1993. *Classroom assessment techniques: A handbook for college teachers.* San Francisco: Jossey-Bass.

Bok, D. 2006. *Our underachieving colleges.* Princeton, NJ: Princeton University Press.

Diamond, R. M. 1998. *Designing and assessing courses and curricula: A practical guide.* San Francisco: Jossey-Bass.

Handelsman, J., S. Miller, and C. Pfund. 2006. *Scientific teaching.* New York: W. H. Freeman.

Herreid, C. F. 2007. *Start with a story: The case study method of teaching college science.* Arlington, VA: NSTA Press.

Khodor, J., D. G. Halme, and G. C. Walker. 2004. A hierarchical biology concept framework: A tool for course design. *Cell Biology Education* 3: 111–121.

Lawson, A. E. 2002. *Science teaching and development of thinking.* Belmont, CA: Wadsworth/Thompson Learning.

McKeachie, W. J. 2006. *McKeachie's teaching tips,* 12th ed. Boston: Houghton Mifflin.

Michaelsen, L. K., A. B. Knight, and L. D. Fink. 2002. *Team-based learning: A transformative use of small groups.* Westport, CT: Praeger.

Wiggins, G., and J. McTighe. 2001. *Understanding by design.* Upper Saddle River, NJ: Prentice Hall.

CHAPTER 3
Assessing Inquiry-Based Instruction

Chapter 2 described how to set general learning outcomes for a course and specific performance goals that can be used to measure students' progress toward them. It examined various instructional methods for helping students accomplish the performance goals and, thereby, the learning outcomes. The next step in this process is to assess whether students are achieving the stated goals. Assessment encompasses more than just exams and papers; it also involves

- informally assessing student learning in progress,
- formally assessing students to determine their learning gains, and
- assessing whether the current instructional practices and course structure are effective.

Please refer back to Figure 1, page 21, to see a graphic representation of the place of assessment in the outcomes-centered approach to course design.

Summarizing the literature dedicated to assessment methods and strategies would require several books. The goals for this chapter are to

- provide developers and instructors with a framework for thinking about assessment;
- highlight some useful tools (at first, some of the methods may seem more appropriate for lecture courses than for laboratories. However, all of the methods described could reasonably be employed in the laboratory setting.); and
- identify selected sources of additional information.

GENERAL PRINCIPLES

A robust assessment plan is built on three principles. First, assessment is integral to the course (or curricular) development cycle, not an afterthought. Second, methods used to assess students should be carefully selected to match learning outcomes and performance goals. Third, assessment should look at both students and the course. These principles may seem obvious, but in practice they are often given little thought. Frequently the ways instructors measure students' abilities do not match the original instructional goals. Moreover, very few instructors conduct routine assessment of their teaching methods to ensure that students achieve the predetermined goals.

The process can be illustrated by revisiting the general learning outcomes and performance goals from Chapter 2 (see pp. 24–26). What follows is a model plan that reflects some fairly common ways for assessing whether students are meeting performance goals. Ignore for now the fact that few instructors actually write out detailed assessment plans this way. The goal is to demonstrate the thinking process.

Learning Outcome #1. Develop an ability to draw reasonable inferences from observations of the natural world.

Specific performance goals: By the end of the course, when students are given an unknown organism that they have not encountered previously, they can do the following:

a. *Classify* the unknown organism to domain, phylum, and probable class and provide reasons for their choices.

   Assessment: In a lab practical, students will be given an unknown organism and will be asked to identify it to proper taxa. Students will be graded based on whether they identified the organism correctly. Partial credit will be given if higher taxa are correct but more specific taxa are not.

b. *Compare and contrast* the methods by which the unknown organism and a human transport nutrients, exchange gases, and regulate water balance. This includes describing any anatomic evidence to support their statements.

   Assessment: On a written lab exam, students will be asked to create a table that contains the relevant information. Answers will be scored based on a preset rubric.

c. *Predict* how the unknown organism might adapt to a 10°C change in the normal temperature of its habitat. Give reasons why your prediction would protect the unknown organism.

   Assessment: Students will be asked this question as part of a written lab exam. Students will receive full credit if they can predict two independent responses to the change in environmental temperature and justify why they are protective.

While these assessment methods seem straightforward, there are two fundamental problems. For Goals (b) and (c), students demonstrate their skills on a written exam. This assessment method does not really measure whether students have gained functional skills they can carry into new situations, since it is unlikely they will encounter a similar situation outside of the classroom. The second, subtler problem is that these methods stress content knowledge associated with the specific performance goals. For Goal (a) the assessment is scored entirely on whether students can memorize relevant content knowledge (i.e., give the right answer to the classification problem). There is no measure of students' progress toward developing the thinking-process skills embodied in the general learning outcome.

How can this assessment plan be made more robust? The first priority is choosing assessment methods that measure content knowledge and process skills in about the same proportions as the course emphasizes. A second important change is to assess students more than one way. Just as not all students learn the same way, not all demonstrate new knowledge and skills in the same way. Third, robust assessment tests students in ways similar to how they will ultimately use their knowledge and skills. Finally, the plan should incorporate informal assessment, so that students can check their learning and self-correct before being assessed formally (i.e., for their grade).

An assessment plan that is more robust (and somewhat idealized) than the three assessesment goals listed above for a , b, and c might look like this:

Idealized Assessment Plan for Learning Outcome #1

1. In the final lab session, the instructor will take students to a field site on campus and give them dip nets, sweep nets, and other collection equipment. Using this equipment, students must collect five species that were not specifically studied during this course.
2. In lab, paired students will identify the 10 organisms they collected to Class. Together they will draw a phylogenetic tree for their ten specimens and write a one-paragraph summary of the evidence to support their classifications and tree. Before handing in the assignment, they must demonstrate their initial arrangement to at least one other pair of students; revisions will be allowed.

   Phylogenetic trees will be scored using a narrative rubric with branched choices. Deducted points will be added back onto the students' scores if the students' summaries provide reasonable evidence to support an incorrect classification.
3. In the second week of this lab, pairs of students will examine and dissect four of the preserved specimens they collected. Each pair will give a five-minute class presentation in which they describe or predict how each of their four organisms transports nutrients, exchanges gases, and regulates water balance, citing evidence from other organisms they have observed to support their conclusions.

   Presentations will be evaluated on three criteria. Factual accuracy of information about specimens, and strength of evidence for conclusion, will each be rated using a 5-point Likert scale. Factual evidence counts for 40% of total score, and strength of evidence for 40% of total score. Clarity of presentation will be evaluated by a peer review, counting for 20% of the total score.
4. After the presentations, each instructor will randomly select 10 organisms collected from the same general habitat by different students. Working individually, each student must rank the 10 organisms in order of increasing resistance to cold. When completed, groups of four students will collaborate to create a consensus rank order and write a shared one-page summary describing how each organism protects itself.

   Summary papers will be scored with a rubric based on a cumulative points strategy. Factual evidence counts for 30% of total score, strength of evidence for 40% of total score, and rationale for 30% of total score.

This idealized plan incorporates several improvements. Overall, there is equal emphasis on thinking processes, content knowledge, and ability to apply that knowledge. For instance, the rubric used to score students' phylogenetic trees includes the option to reinstate points that are taken off for incorrect tree arrangements if the students can support their alternative classifications with specific evidence.

Similarly, 60% of the grade for class presentations is based on presenting sound evidence and demonstrating clear thinking, while only 40% is based on factual information. The test of content knowledge is not just a recall test; rather, students are expected to demonstrate mastery by applying that prior knowledge to organisms they have not seen before and to provide support for their classifications.

This new plan incorporates multiple methods of assessment that resemble situations students are likely to encounter outside the classroom. Writing is used repeatedly as an assessment method; pairs write a one-paragraph summary of the evidence to support their classification and tree diagram and four students must write a one-page explanation of their thinking about how each organism protects itself from the cold. In both writing assignments the primary emphasis is on providing a well-reasoned argument. The plan also provides routine informal assessments through collaborative learning. Students work in pairs to classify their unknowns; then they share and defend their results with peers before handing in the first written assignment. The oral presentation is a collaborative effort, as is the final project in which four students must reach a consensus arrangement of 10 randomly selected organisms. Each time, students check their own understanding against that of their peers before they are graded. Even seemingly minor details make positive contributions to the assessment. In this case, students get to collect and select the organisms that they will study for two weeks, giving them a greater stake in the questions being asked.

This idealized plan provides a much better indication of how well students are progressing toward the general learning outcome, which is to develop an ability to draw reasonable inferences from observations of the natural world. It is also a powerful method for communicating course expectations to new instructors (both faculty and TAs) who are teaching an established course for the first time. Similarly, a well-written assessment plan can improve consistency between instructors who are teaching different sections of a large course; all those involved know what the rationale is behind each assessment element and exercise.

TIME FOR A REALITY CHECK

After reading the previous section, most instructors will likely throw up their hands in frustration. Planning an assessment this detailed for every learning outcome in a new course seems impossible, and rightly so. Developing an idealized plan such as this one can uncover unrealistic performance goals and help course planners think more broadly about assessment methods. However, even for a well-planned course, actually trying to implement an assessment plan this complex without supporting data is unrealistic and counterproductive.

Look back at the outcomes-centered approach to development shown in Figure 1, page 21. According to the model, assessment plans should evolve over time, with initial assessment data serving as the basis for revising and improving the performance goals and instructional methods. In turn, the assessment plan is updated to match the revised performance goals and teaching methods. These cycles create an assessment continuum, as shown in Figure 2. On the left side is a plan (like the first one described on pp. 40–41) that uses only a few simple, low-level assessment methods. On the right side lies an assessment plan that incorporates multiple high-level methods and approaches the idealized plan in its complexity. With each iteration the course assessment plan should move to the right along the continuum.

Figure 2

The assessment plan continuum

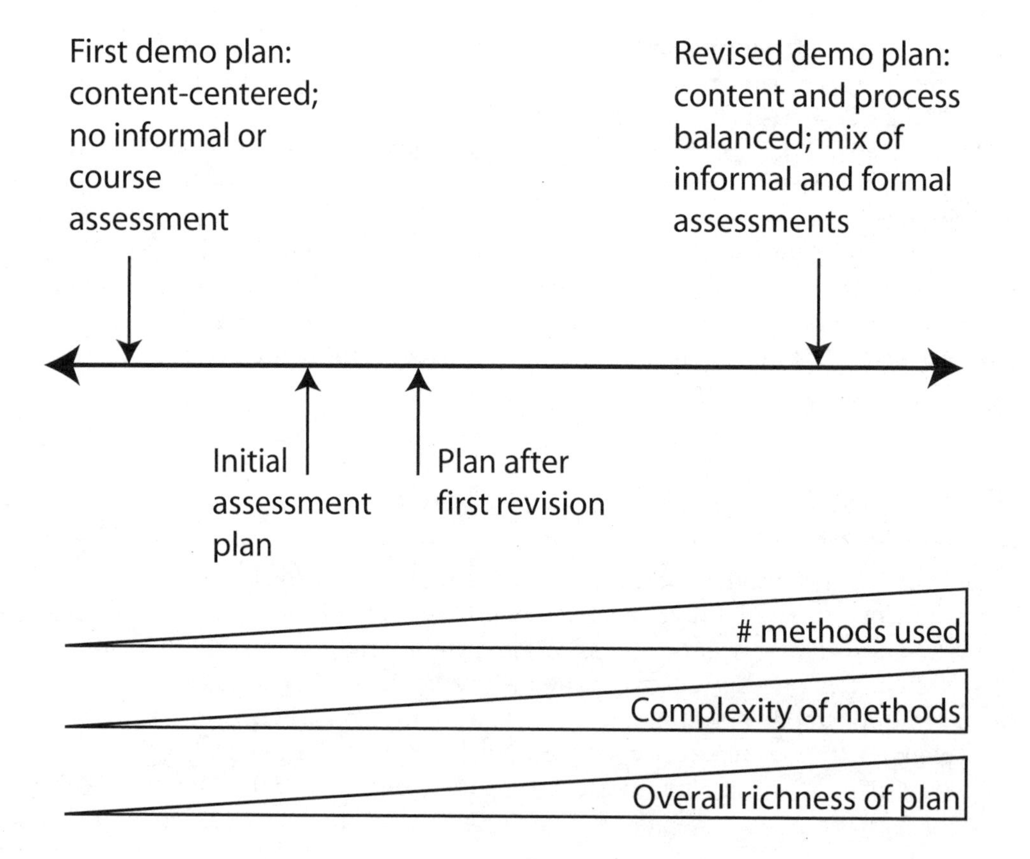

DEVELOPING AN INITIAL ASSESSMENT PLAN

Effective assessment plans include informal, formal, and course (or program) assessments. Validated instruments that course developers can use for each section are outlined below. The discussion is not exhaustive but is intended as a starting point for assessment planning. Methods that can be used for more than one type of assessment are described in the section where they are most often applied. None of the techniques is inherently better, but some do have broader applicability.

Informal Assessment Methods

Informal assessment methods help instructors gauge the accuracy and depth of what students are learning *while instruction still is in progress*. By definition, informal (also called formative) assessment is not used for grading. Its goal is to provide students with frequent constructive feedback about their learning so they can correct mistakes sooner rather than later. Formative assessment also tells instructors whether their teaching methods are accomplishing the intended goals.

Nearly all instructors use some kind of informal assessment; rarely does one lecture for an entire class period or allow a lab to continue for the entire period without breaking at least once to check comprehension. Informal assessment can be as simple as stopping in the middle of a lecture or lab on, for example, mitosis and asking students what would happen in metaphase if a cell had not previously replicated its DNA correctly during S-phase. This question is meant to determine whether students know what occurs in S-phase and why those events are important in mitosis. If students answer incorrectly, the instructor stops and reviews the events of S-phase before continuing. In one minute, this informal assessment lets students know whether or not they have mastered the content the instructor expects and lets the instructor know whether the students have brought their prior knowledge forward successfully.

There are significant drawbacks to relying entirely on *ad hoc* questions as the only type of informal assessment. The instructor gets a general idea whether students understand or not, but he or she does not know what proportion of them has the requisite knowledge. So long as one or two students answer correctly, instructors tend to move on rather than stopping to address the fact that the majority of the class might not know the answer. Fortunately, other informal assessment methods can provide this missing information. Other informal methods also help guide students' learning or increase their incentive to learn.

Angelo and Cross's classic text, *Classroom Assessment Techniques* (1993), is the leading reference on this topic. It describes 50 different classroom assessment techniques (CATs) that instructors can use to monitor student learning in progress. Some CATs are widely known. The one-minute paper (one of several names for this technique) probably is the most widely used informal assessment method in higher education. Each CAT is rated for (1) the amount of preparation work that instructors must do to use it, (2) the time students need to complete it, and (3) the time needed to analyze collected data. CATs are cross-referenced to the Teaching Goals Inventory (described in the context of developing learning outcomes; see p. 22), so instructors know which CATs are best suited to assess each goal on the inventory. Many CATs can be adapted to create formal assessments as well.

Three sample CATs are described below; all quotes are from Angelo and Cross (1993) This book's author uses all three of these techniques for informal assessment across the institution's entire laboratory program. The analytic memo additionally is used as a formal assessment in two different lab courses.

> **One-Minute Paper:** *At the end of class or lab, the instructor asks students to use a single sheet of paper to answer two questions. "What was the most important thing you learned during this class time? What important questions remain unanswered?" The great advantage ... is [that these papers] provide manageable amounts of timely and useful feedback for a minimal investment of time and energy. By asking students what they see as the most significant things they are learning, and what their major questions are, faculty can quickly check how well those students are learning what they are teaching. That feedback can help teachers decide whether any mid-course corrections or changes are needed...the instructor's feedback on their minute papers help students learn how experts in a given discipline distinguish the major points from the details....*
>
> *Despite its simplicity, the minute paper assesses more than mere recall. To select the most important or significant information, learners must evaluate what they recall. To come up with a question, students must self-assess—asking themselves how well they understand what they have just heard or studied. (p. 148)*

> **Analytic Memo:** *[This] is basically a simulation exercise. It requires students to write a one- or two-page analysis of a specific problem or issue. The person for whom the memo is being written is usually identified as an employer, a client, or a stakeholder who needs the students' analysis to inform decision-making.*
>
> *Analytic memos assess students' ability to analyze assigned problems by using the discipline-specific approaches, methods, and techniques they are learning. This CAT also assesses students' skill at communicating their analyses in a clear and concise manner.... This CAT works best when used early in the term, as a means to help students prepare for later graded assignments. (p. 177)*

> **Concept Maps:** *[These] are drawings or diagrams showing the mental connections that students make between a major concept the instructor focuses on, and other concepts they have learned....*
>
> *This technique ... allow[s] the teacher to discover the web of relationships that learners bring to the task at hand—the students' starting points. This CAT also*

> *helps the teacher assess the degree of 'fit' between the students' understanding of relevant conceptual relations and the teacher's concept map....*
>
> *By literally drawing the connections they make among concepts, students gain more control over their connection making. The concept map allows them to scrutinize their conceptual networks, compare their maps with those of peers and experts, and make explicit changes.... This technique can be used to assess and develop valuable metacognitive skills.* (Angelo, T., and P. Cross 1993, p. 197; © John Wiley & Sons. Reprinted with permission.)

Another source for informal assessment methods is the *Field-Tested Learning Assessment Guide* (*FLAG*), part of the College Level One program at the National Institute for Science Education (NISE) (*www.wcer.wisc.edu/archive/nise*). FLAG is both a primer on assessment and a collection of classroom assessment techniques that can be used for informal and formal assessments.

Is informal assessment worth the time? Absolutely. It provides students with valuable guidance that translates into quantifiable improvements in how much they learn and how long what they learn is retained. An extensive review by Black and Wiliam (1998) found that a well-designed assessment program that includes informal assessments may be the single most important factor for improving student learning. Routine informal assessment leads to significantly greater learning gains, with the greatest gains among low-achieving students. More recently, Ruiz-Primo and Furtak (2006) reported that informal assessment correlated with better performance on postcourse formal assessments and that students gained significantly greater insight into the thinking processes of scientific inquiry.

Formal Assessment Methods

Formal, or summative, assessment—that is, measuring learning gains at the end of an instructional cycle—is what instructors usually think of when they hear *assessment*. Typically, formal assessment is used to assign course grades. However, results from formal assessments are a vital part of course- and program-level assessment as well. They show what students can learn using a particular set of instructional materials and they provide benchmarks for future groups who will use the same materials.

The types of formal assessments used in a course largely determine how students approach that course (Handelsman, Miller, and Pfund 2007; McKeachie 2006). By selecting methods thoughtfully (i.e., with a view to the general learning outcomes), an instructor can steer students' learning efforts toward desired outcomes and goals. For this approach to work, though, students must know in advance what methods will be used.

For example, multiple-choice exams are widely used in the sciences. This method measures basic content knowledge and factual recall fairly well but not thinking-process skills or application skills. If multiple-choice exams are the only assessment used, students will deduce that factual recall is the priority for the course, and approach it accordingly. Conversely, when students know in advance that they will be expected to apply their content knowledge (e.g., by writing an essay or designing an experiment in which they must apply concepts and skills learned in lab), they approach studying very differently.

Like informal assessment, a robust formal assessment strategy measures gains in students' content knowledge, as well as thinking-process and application skills. There are innumerable methods available, yet science instructors tend to limit themselves to a few well-worn techniques. This is unfortunate because many of these commonly used methods measure only part of the range of skills that instructors intend to develop. The remainder of this section describes a variety of formal assessment methods, some more widely used than others, and compares their relative strengths and limitations. Those seeking more in-depth discussion should consult the references listed with each method. For in-depth reviews of general principles and practices consult Black and Wiliam (1998), Diamond (1998), Fox and Hackerman (2004), Handelsman, Miller, and Pfund (2007), McKeachie (2006), Palomba and Banta (1999), and Uno (1999).

Homework Assignments

Homework assignments appear to be informal assessments. They provide students with opportunities to practice skills they will need for a later exam or practical, give them feedback about their current knowledge and skills relative to the instructor's expectations, and encourage them to stay engaged with the course and keep up. Yet if the instructor grades homework assignments, they are by definition formal assessments. This distinction is important: If homework is intended for informal assessment, students' performance on it must not be counted as part of a grade. A compromise approach is to make half of the assignments count for class credit (formal assessment) and half not count (informal assessment); when using this approach, the ungraded assessments always should come first. Another strategy is to use homework assignments for both purposes simultaneously. Students are graded pass/fail based on whether or not they complete the assignment. This provides a formal assessment measure of their course participation and preparation. The content of the assignment is graded and returned to the students as an informal assessment, but the specific score is not part of their course grades.

Written Quizzes and Exams

Written quizzes and exams are the backbone of formal assessment. Many lab exams include both practical and written components. Developing effective exam questions is the subject of entire books (Downing and Haladyna 2006; Osterlind 1997). When designed carefully and written well, they can measure several different elements of learning simultaneously. Conversely, poorly or hastily written exams often focus entirely on content knowledge and factual recall and may actually hinder learning. Some advantages and disadvantages of various types of questions are outlined below.

Multiple choice. Multiple-choice questions are easily scored and can test content knowledge effectively. While useful in a proper context, many instructors rely too heavily on multiple choice. It is difficult to write questions that test higher-order synthesis or application skills. Good formal assessments will include other methods as well.

Two-tier multiple choice. A multiple-choice question is followed by a second question (usually free-response) that probes students' thinking processes in answering the prior question. Two-tier questions can be difficult to write but can show whether the exam is biased against students whose thinking patterns or interpretation differ from the instructor's.

True/false. In their traditional form, true/false questions are less informative than well-designed multiple-choice questions. With no knowledge of a topic, students can guess correctly 50% of the time. It can be difficult to develop unambiguous questions that test anything other than content knowledge. Many instructors compensate for the limitations of true/false questions by asking students to rewrite the statements they believe are false so that they are true. This technique shows how students are thinking about the questions and helps the instructor assess his or her own question-writing style.

Matching questions. Matching questions are most often used to assess content knowledge (i.e., students match terms to definitions or examples). Yet if the questions are well written, matching can assess other skills. Imagine a matching series in which students must match five sets of symptoms for diseases they have not studied to seven possible physiologic causes. This type of matching evaluates whether students can synthesize multiple concepts, then apply them to new situations. A follow-up, short-answer question might ask students to justify each of their pairings.

Concept mapping. Concept mapping is a good example of how an informal assessment can be converted to a formal one. The basic premise of concept maps was explained in the section on informal assessment (p. 46). Concept maps are ideal for assessing students' ability to compile, arrange, and synthesize content knowledge. One example might be, "Draw a concept map that visually summarizes the connections between respiration, circulation, and excretion." Concept maps are not particularly suited to assessing application skills. However, they are particularly useful for determining whether students have successfully made connections between ideas presented in several different lab meetings or in different units.

Word problems. Word problems are often used in physiology, ecology, genetics, and related courses to assess students' ability to apply mathematical or statistical models to real-life situations. If students are required to show their work, their thinking processes can be evaluated along with their application skills. Number-based word problems are not possible for all topics in biology. However, situational problems that require students to synthesize and apply their content knowledge and skills to real-life issues can be written for nearly any concept or topic. The primary drawback to word problems is the need to match the level of difficulty to student abilities. Instructors are prone to write problems that are too difficult. Remember, the goal in a constructivist learning environment is to provide challenges that students have a reasonable chance of accomplishing successfully.

Interpreting graphs or figures. Students will regularly need to interpret graphs and figures as literate adults. Most publishers offer a media library on disk that includes unlabeled versions of their textbook figures. These can be edited with a drawing or photo-editing program to fit the level of difficulty of the exam or practical. Not every question should ask students to identify features or structures in the figure. Questions should assess multiple cognitive skills. Say, for example, that a figure shows a schematic of a lipid bilayer, with an arrow pointing to an unlabeled phospholipid molecule. A low-order factual recall question is "What is this structure?" The question "How frequently does this structure flip to the opposite orientation?" requires the student to know and apply several pieces of factual knowledge ("It is a phospholipid" and "Phospholipids in bilayers do not flip over normally"). An even more complex question would be "Where is this molecule manufactured?" To answer, students must apply their knowledge of how cells manufacture and transport phospholipids. If the figure is a graph of experimental data, ask students to make a prediction based on it. A correct answer demonstrates the ability to interpret the figure and apply the general model it represents.

Short-answer and essay questions. Well-written free-response questions show whether students can synthesize ideas, apply their content knowledge to novel problems, or develop and support a coherent argument. Unfortunately, free response is easily misused, as happens when a short-answer question only asks students to write out definitions. A longer essay question that asks students simply to regurgitate the steps of a process is another common misuse. Multiple choice, matching, and diagrams are better methods for assessing factual content knowledge. Short-answer and essay questions should be reserved for assessing higher-order thinking skills. Another disadvantage to free-response questions is that a good answer requires time to think about and write; sufficient time may not be available if the majority of the lab exam format consists of practical questions or if many students must take the exam in a short period. In these situations, formal written assignments or take-home portions of an exam may be better choices.

Technical Writing

Lab reports and related forms of technical writing are ideal for assessing inquiry-based labs because they require students to demonstrate deep reflection on knowledge gained. Additionally they help students clarify their thinking; train them to seek out and evaluate evidence on which to base claims; and build analysis, interpretation, and prediction skills. The primary drawbacks are (1) the time required to prepare students to write effectively, (2) the difficulty of ensuring consistent grading between students and sections of a course, and (3) the demanding task of providing constructive comments. There are no easy solutions to these problems, but there are tools available that make all three processes less cumbersome.

Numerous books and web resources are available to assist students in learning to write scientific papers (e.g., Knisely 2002; Lipson 2004). Some instructors (this author included) opt to develop their own guides for teaching students the basic principles of technical writing. Simply providing students with writing guidelines is not sufficient, however. Instructors in the author's department have found that novice students (particularly nonmajors) will follow format rules and guidelines blindly, without ever developing the thinking processes that technical writing is meant to encourage. The author has experimented with a **directed writing format** (**DWF**) that helps students learn to analyze and interpret the results of experiments they conduct in inquiry labs. (See Keys et al.'s 1999 article on their "science writing heuristic," from which the DWF was derived.)

Briefly, a DWF assignment is two to three pages long and has eight sections: Beginning Ideas, Test Methods, Observations and Raw Results, Claims, Evidence, Related Reading, Reflection, and Information Sources. Each section has associated guide questions that students must answer. The sections of the DWF correspond with parts of a typical primary literature article, so students can use the DWF questions to read and understand articles more quickly. Students write their Beginning Ideas section at the start of an inquiry lab, before they conduct any experiments or collect any data. Students write down what they already know about their model system or organism, then what they do not know yet but would like to know. The last part is their general plan for finding an answer to at least one of their questions about the model.

Once they have performed an experiment and gathered new evidence, students write a short description of their actual test methods; then they write one or more claims that they are making based on their prior knowledge and experimental results. They outline their interpretation of the evidence they collected or obtained from external literature and conclude with a description of the relative strength of the evidence for their claim and a reflection on how their results fit with what other investigators have seen. The Information Sources section simply lists the primary literature and other sources.

Directed writing can be used alone or as a rough draft technique to prepare students to write traditionally arranged formal reports.

Grading Writing. The simplest way to normalize and speed grading is to develop rubrics for each assignment. A rubric is a scoring tool that lays out expectations for each of the elements within an assignment. Usually it is a table consisting of a short description of the overall task, dimensions (rows of characteristics to be evaluated), and columns listing three to five scale levels of performance. A blank template for composing a grading rubric is shown in Table 3 on page 52 (Stevens and Levi 2005).

Table 3

Format outline of a basic rubric. Dimension descriptors for each scale level go in each box.

	Description of the Task or Assignment		
	Scale level 1	**Scale level 2**	**Scale level 3**
Dimension 1			
Dimension 2			
Dimension 3			
Dimension 4			

The task description is just a short reminder of the assignment for the instructor or grader of what is to be evaluated (very useful after grading 100 papers!). The scale levels represent the range across which students can be rated. Ideally, one scale level is the highest level of performance possible and uses similar language as the course performance goals. The scale levels can take many forms—for example, "poor," "good," "excellent" or "not done," "done poorly," "done well." The dimensions are features or elements to be rated and can be mechanical ("Each section of paper adheres to length limits"), descriptive ("Discussion highlights relevant prior literature"), or some combination. The body of the rubric table contains dimension descriptors. The cells describe specific traits or features indicating the relative performance for each dimension. Ideally these descriptors incorporate the most common ways students fail to meet the highest level of performance. Table 4 is an example of a rubric for grading a short homework assignment. A rubric for grading a lab might be longer but would look essentially the same.

The rubric in Table 4 can be used in several ways to assign grades; one is to simply set point values for each scale level (say, 5 points for "Needs Work," 8 points for "Acceptable," and 10 points for "Excellent"), then add up the number of points earned for all descriptors and divide by the maximum number of possible points. Another strategy is to weight each dimension as a percentage of the total score (say, 25% for relevance to science and technology, 35% for factual accuracy, and 40% for connection to lab topics). Each scale level would have a corresponding percentage value as well (say, 50% for "Needs work," 75% for "Acceptable," and 100% for "Excellent").

Some instructors decry rubrics as being too "mechanical." Others simply have never learned to construct rubrics so have never thought to use them. Yet rubrics offer significant advantages to instructors. They reduce grading time while improving consistency because each student is held to the same objective standard. They help ensure consistency between sections of a lab, particularly when there are multiple instructors. Rubrics also provide students with constructive feedback about their writing. The rubric can be given to students as part of their writing guidelines; then rubrics completed by the instructor can be returned with students' papers so students know what specific portions of their assignments were not up to the instructor's expectations.

Table 4

Rubric for evaluating a written lab homework assignment

	Task: ***Read a newspaper article that relates science and technology to the topic of this lab unit and summarize the article.***		
	Needs Work	**Acceptable**	**Excellent**
Relevance of article	Science and technology not the main topic of article	Science and technology mentioned, but one of several themes of article	Science and technology is central theme of article
Factual accuracy of summary	Errors in facts learned during course	Errors in fact require outside reading to identify	No errors of fact
Relation of summary to a topic in this lab unit	No connections made to topics covered in this unit	Summary relates to one topic in this unit	Summary relates to two or more topics in this unit

Rubrics can be used for many types of assessment, not just writing assignments. Numerous resources explain how to design rubrics; Allen and Tanner (2006) and Stevens and Levi (2005) are good starting points.

Commenting on Student Writing. The most effective way for instructors to provide feedback is to make specific comments and suggestions directly on students' papers. Unfortunately, as the number of written assignments increases, the amount of time an instructor can afford to spend correcting each paper shrinks. The problem is magnified if students are still learning basic mechanics of technical writing. Some years ago a TA in the author's program devised an elegant way to make comments on student papers quickly and efficiently. She compiled and distributed to her students a numbered **comments list** describing the 20 to 30 most frequent mechanical, analytical, and logical mistakes that students make when writing lab reports. Rather than writing or typing out comments, the TA simply circled an error, then wrote over it the number of the relevant comment from the list (#13, #7, etc.). If a particular error was explained in more detail by the lab manual or writing guide, she added "See page so-and-so" and inserted the relevant page number(s). The only time the TA wrote out detailed comments was if an error was not covered by the writing guide or comments list. More serious problems were marked "See me," then discussed with the student one-on-one. Numerous instructors have since been trained to use this method; most say it allows them to make a greater number of substantive suggestions than do other grading methods, while spending significantly less time grading overall.

As group-based learning (see p. 28) becomes more common, many instructors are using peer review to give students useful feedback about their writing. One widely used method is Calibrated Peer Review (CPR), a web-based method developed at UCLA by Orville Chapman and Michael Fiore (2001) and funded by the National Science Foundation and the Howard Hughes Medical Institute (*http://cpr.molsci.ucla.edu*). According to the Chapman and Fiore's white paper on their website,

> *CPR is a method for enabling frequent writing assignments even in large classes with limited instructional resources. In a CPR assignment, students write short essays on a specific topic.... After electronic submission of essays, students read and assign a score to three "calibration" essays. When students demonstrate they are competent reviewers, they read and assign a score to three anonymous peer essays, and finally, to their own essay.... Instructors can use assignments from the CPR library or create their own. Regular use of CPR assignments teaches students to articulate ideas coherently and to critically evaluate both their peers' and their own work, while simultaneously reducing the time an instructor now spends reading and assessing student writing.*

Guilford (2001) has described a different method for employing peer review to teach thinking and writing skills and to conduct formal assessment. Students must complete the entire process of submitting a professional scientific manuscript for publication (with the instructor as editor-in-chief), from a letter of inquiry through peer review and revisions to final submission. Gratz (1990) and Koprowski (1997) also have described courses that use student peer review as an adjunct to the instructor's comments.

Practical Tests, Skills Tests, and Mastery Tests

These assessment methods are used almost exclusively in lab courses. Practical questions focus primarily on visual identification skills and basic content knowledge. Skills tests require students to demonstrate particular physical abilities that can be applied to other laboratory situations. Mastery tests are the most complex assessment of the three; students must demonstrate that they can successfully complete a predetermined activity that consists of multiple steps.

In practice, a single laboratory exam can use all three types of assessment. Imagine that students performed a lab unit in which they dissected an anesthetized earthworm, removed its midgut, then measured the levels of protease enzyme activity in the gut using a colorimetric assay. On the lab exam, students see the following three instructions:

1. The tray at Station #1 contains a dissected earthworm. Answer these questions:
   a. What is the structure under the T-pin marked "A"?
   b. What is the specific function of this structure?
   c. If this structure did not contain its normal enzymes, what do you predict would happen to the earthworm?

2. At Station #2, there is a spectrophotometer and a solution of dye that absorbs light at 421 nm. Determine the absorbance of the dye solution. All materials you need to answer this question are in front of you at the station.
3. At Station #3, there is a jar of earthworms that have been preserved in ethanol. Using the materials at the station, isolate the hindgut. Show the proctor once you have finished the dissection. There is no partial credit for this question.

The questions in item 1 are typical practical questions; questions a and b both assess basic content knowledge. Alone, practical questions are not very good for assessing thinking a skills. However, follow-up questions like c can measure students' ability to apply their basic content knowledge. An instructor's goal when writing practical questions should be to test students' knowledge of central concepts, not minutiae. Too often, instructors will use questions about minor details to separate students into A, B, and C grade groups. The better assessment strategy is to use application, skills, and mastery questions that test higher skills for this purpose instead.

Item 2 is a skills test; students must show they can operate a spectrophotometer properly. Like practical questions, a skills test can be extended by asking students to interpret possible outcomes or results. When setting up the station, include some distractors, that is, materials that are not actually needed to complete the task.

Item 3 is an example of a mastery test. Unlike the practical or skills questions, this test cannot be completed using prior knowledge alone. Instead, students must demonstrate an ability to apply their practical knowledge and skills. First, in this example, they must have mastered a complex set of dissection skills. Second, they must identify a structure that was not the focus of the original lab unit. Third, they must adapt the techniques used in the original lab to a new organ to isolate it properly. Finally, the question is graded pass/fail; students either accomplish the goal for the question successfully or they do not. Of the three types of assessments, mastery tests are the most difficult to write and administer. However, students who can answer them successfully have demonstrated unequivocally that they have learned more than students who can answer only 1 and 2.

Oral Presentations. Strong oral communication skills are more than just liberal arts ideals; they are critical job skills that employers expect of all graduates (NRC 1999, 2003). Yet instructors who try to develop these skills by using oral presentations for assessment often are disappointed at the outcome; a few students do well, while the majority of presentations are lackluster and poorly executed. This is not surprising; most undergraduates' experiences are limited to a lecture hall, where the goals and structure of the presentation are different. In the author's opinion, oral presentations (like technical writing) can be powerful tools for assessing students' thinking processes but only if students have been adequately trained. Teaching students to give good oral presentations requires significant time and is best done over the course of several weeks (Estrada et al. 2005; Houde 2000; Wiese, Varosy, and Tierney 2002) or even semesters. Therefore, instructors should consider very carefully the time investment required before choosing to use this assessment method.

In the author's department, nonmajors and majors give group oral presentations in every lab course about an experiment they designed as part of a guided inquiry. They are given written format guidelines that describe what is expected to be in each section of their presentations and a detailed grading rubric that spells out the most common mistakes. Copies of these materials are available online, along with a tutorial on developing visually effective PowerPoint presentations. Some institutions also post annotated video clips of good and poor student presentations as training guides (Wiese, Varosy, and Tierney 2002). For more information, see Smith's (2003) excellent topical bibliography on this subject.

Course Projects. The goal of a course project is for students to apply thinking skills and concept knowledge they have gained during the term to a question that interests them but is outside the scope of the course itself. While most instructors use projects in the context of lecture courses, the course project model can be adapted to multisection lab courses. Like lab experiments, course projects can provide valuable assessment data about students' thinking processes and application skills. Instructors frequently complain that students put off course projects until the last minute. A simple way to discourage procrastination is to set regular benchmarks throughout the term, then track students' progress toward them with informal assessments. Students know whether or not they are making appropriate progress overall but are not penalized if they miss a particular benchmark. Once again, rubrics are a useful grading tool.

Learning Portfolios. This is another formal strategy that was once restricted to lecture courses but is gaining popularity in labs. Learning portfolios are similar to the teaching portfolio that many faculty create for tenure review. A typical portfolio contains specific examples of a student's work that serve as evidence of learning, plus a narrative by the student that reflects on that evidence and ties it together. Evidence that a student might include in the portfolio is (but is not limited to) essays, completed problem sets, research papers, lab reports, recordings of oral presentations, concept maps, and excerpts from written or practical exams. Depending on the type of assessment being conducted, a student may include evidence from one or multiple courses.

Portfolios provide evidence for a suite of learning gains simultaneously: concept knowledge, thinking skills, and application skills. This method fits nicely with the outcomes-centered approach for course design (see Chapter 2). Moreover, the process of assessment is itself a significant learning experience for most students. In selecting examples of their work for portfolios, students begin to develop higher-order metacognitive and self-reflection skills that will be vital for lifelong learning. Portfolios are well-suited to evaluating students' progress over a course sequence. The chief drawbacks to portfolios are the time needed to train students and deciding how best to track and to score them.

There are extensive resources on the topic of learning portfolios. Bieshuisen et al. (2006); Wright, Knight, and Pomerleau (1999); and Zubizarreta (2004) provide good introductions to general principles and methods. Commander and Valeri-Gold (2001) provide more specific directions for implementing, monitoring, and evaluating portfolios.

Collaborative Formal Assessments. In the lab, students rarely work alone. Yet their grades typically are based on individual efforts. The increasing use and success of team-based learning, problem-based learning, and other collaborative and cooperative instructional methods (Cortright et al. 2003; Herreid 2007; McKeachie 2006; Michaelsen, Knight, and Fink 2002) have raised the question of whether students who work collaboratively in lab should be assessed as a group, too.

Many instructors resist giving all group members the same grade, believing that students should demonstrate individual thinking skills or that shared grades let less-motivated students hide behind the efforts of others. Yet there is overwhelming evidence to indicate that collaborative assessment leads to greater learning gains for individual students (Cortright et al. 2003). A reasonable solution is to compromise and use both. Michaelsen, Knight, and Fink (2002) describe several methods for mixing individual and collaborative assessments within a single course. Many of the methods for assessing individuals can be applied to collaborative groups, but the complexities of doing so are far beyond the scope of this book. Those who are (or are considering) using team-based, case-based, or problem-based learning methods should consult the above references for more information.

Methods for Assessing a Course or Curriculum

The goal of course and curriculum assessment is to determine whether current instructional practices and course structure are effective in helping students as a group meet specific performance goals and general learning outcomes. The mention of course assessment can raise academic hackles because it smacks of standardized testing and restriction of academic freedom. This view of course assessment is largely incorrect, for several reasons. While a well-designed course assessment plan may include standardized tests, it should always include multiple measures of effectiveness. Furthermore, a course's success is defined by how well it achieved the general learning outcomes that were established *by the instructor* at its beginning. From this perspective, course assessment does not restrict teaching freedom; rather, the resulting data inform the instructor's decisions about what topics to explore and what teaching methods and assessments to use.

Assessing Overall Course Structure

Final course evaluations are a widely used type of course assessment. They may consist of semi-quantitative Likert-scaled survey questions, multiple-choice or free response questions, or some combination thereof. Student surveys are useful for assessing course structure if they are written with the original course goals in mind. Often, however, an instructor is required to administer a generic course evaluation form that asks little about the effectiveness of particular instructional practices. The simplest solution is to hand out a second, course-specific survey.

When developing a course evaluation, use the general learning outcomes and specific performance goals as guides for developing questions. Avoid ambiguous questions like, What topic made the greatest impression on you? Ask for specific details; for example,

Which instructional method helped you most to understand amino acid structures?

a. Modeling them with sticks and balls in lab
b. Drawing derivatives
c. Working homework problems

After collecting the supplemental surveys, avoid reading the individual responses. Instead, collate the responses to each question. Compare the relative number of students who chose each option. Look for general trends, not isolated comments or complaints. Remember, the goal of course assessment is to make improvements in the course that will benefit as many students as possible, not to try and please each and every student.

Another useful course assessment instrument is the Teaching Goals Inventory (Angelo and Cross 1993), used earlier to identify general learning outcomes and informal assessment methods (see p. 22). Walter (2002) describes administering a very slightly modified version of their original Teaching Goals Inventory, called the Learning Goals Inventory, to students as a postcourse assessment.

> *The Learning Goals Inventory ... reveals the level at which the student ranks six learning goals: higher order thinking skills, basic academic success skills, discipline specific knowledge and skills, liberal arts and academic values, work and career preparation, and personal development. The students answer 52 questions with a numeric answer of 1 (not applicable), 2 (unimportant), 3 (important), 4 (very important), 5 (essential). The higher the value the more important is the goal to the student....* (p. 1)

By averaging student responses across a class for each of the questions on the modified inventory, the instructor can determine what students perceived were the actual general learning outcomes for the course. If students' perceptions do not match the intended goals of the instructor, the current course likely needs revision.

The National Institute for Science Education (*www.wcer.wisc.edu/archive/nise*) maintains the Student Assessment of Learning Gains (SALG), a validated online instrument that instructors can use to obtain feedback from students about how specific course elements are helping or hindering learning. This instrument, which provides a statistical analysis of students' responses, can be modified to fit a particular course design.

Assessing Group-Level Gains in Content Knowledge and Thinking-Process Skills

There are several methods for measuring aggregate learning gains after students complete a course or curricular sequence. Unfortunately, most focus on content knowledge, and provide less information about changes in thinking-process skills. Compounding this problem, students should be assessed both pre- and postinstruction to obtain an accurate picture of their gains in thinking skills. One strategy is to assess thinking skills of individual students, then look at aggregated outcomes from the entire group. As more instructors use course-level assessments, hopefully new methods for measuring group gains in thinking-process skills will become available.

The Major Field Tests (MFT) are standardized assessment exams administered by the Educational Testing Service. The MFT for biology consists of about 150 multiple-choice questions grouped into four topic areas: cell biology, molecular biology and genetics, organismal biology, and ecology and evolution. Most questions look at content knowledge, although some also test analytical skills and interpretation of data. The advantage of the MFT is that the results for local students can be compared to a normalized national dataset. The MFT can be used for evaluating learning gains of students who have completed a sequence of several courses, or for making pre/post comparisons of graduating students versus freshmen. At present, specialized topics are not covered by the MFT, and instructors have no control over what topics are assessed.

A similar test is the Collegiate Assessment of Academic Proficiency (CAAP), offered by the ACT. Separate modules are available for testing science reasoning, critical thinking, and writing skills. The chief limitation of the CAAP is that it does not focus specifically on biology content knowledge. However it provides similar data as the MFT regarding thinking skills.

Several colleges and universities have developed their own exams in lieu of tests like the MFT and CAAP. Homegrown exams can be precisely matched to the content and learning goals of the courses students have taken. Also, instructors in relevant courses can contribute individual questions to the exam. Some institutions incorporate questions from practice tests, retired questions, or prior editions of the Graduate Record Exam (GRE) for this purpose. The GRE is licensed and administered by the Educational Testing Service; permission to use these questions should be obtained in advance from ETS.

Another group-level tool is the Biology Concepts Inventory, developed by educators with Bioliteracy.Net. The aim of concept inventory-based assessment is to identify which common misconceptions about a topic students still hold, so that the instructor can address those misconceptions specifically. According to its developers:

> *Concept inventories have two complementary purposes. The first is to identify basic concepts…that instructors believe are well understood by students, but which are not. … [T]he second goal of concept inventories and related instruments is to*

identify student misconceptions, either innate or introduced in the course of instruction, in order to determine what approaches are most effective at leading students to conceptual mastery. (M. Klymkowsky, *http://bioliteracy.net*; reprinted with permission.)

For courses where the current inventory is not sufficiently comprehensive, it provides a model for instructors to develop similar instruments of their own for assessing students' functional comprehension of essential, topic-specific content knowledge.

The Views on Science-Technology-Society (VOSTS) (Aikenhead and Ryan 1992) was originally developed for use in high schools, but it is equally useful for undergraduate populations. VOSTS measures students' views on a range of topics relating to science in general, and is particularly informative when administered as a pre/post-course test. The test is constructed by the instructor, who selects around 20 multiple-choice questions from a pool of 114 choices. Each question consists of a lead statement, then a series of descriptive responses to that statement. Lead statements emphasize cognition (reasoned arguments) over affect (personal feelings). Each descriptive response represents a different point of view or attitude about the epistemology and process of science.

Students read the statement, then select the descriptive response that most closely describes their personal view or thinking. Questions range across the nature of science as a process, logical reasoning, scientific assumptions, science in the public sphere, science and ethics, the process of creating new knowledge, epistemology of science, and the relationship of science to larger society. Results from the VOSTS are calculated and expressed as the percentage of students choosing each of the descriptive responses. The number of students that select each descriptive response indicates the relative pervasiveness of the attitudes embodied by each response. VOSTS is particularly useful for assessing nonmajor students in lab courses that have been designed to promote general science literacy and understanding, rather than specific content.

WHERE TO LEARN MORE ABOUT ASSESSMENT

There has been increasing emphasis on assessment and accountability in postsecondary education in recent years. As a result, knowledge has grown rapidly in the last two decades, and will continue to do so for the foreseeable future. Those wanting to learn more can find good reviews in Fox and Hackerman (2004), McKeachie (2006), and Uno (1999). Details of specific strategies for developing course assessments have been published by Allen and Tanner (2004); Dancy and Beichner (2002); Dwyer, Millett, and Paynea (2006); Palomba and Banta (1999); and Walvoord (2004).

References

Aikenhead, G. S., and A. G. Ryan. 1992. The development of a new instrument: "Views on Science-Technology-Society" (VOSTS). *Science Education* 76: 477–491.

Allen, D., and K. Tanner. 2004. Approaches to biology teaching and learning: From assays to assessments-on collecting evidence in science teaching. *Cell Biology Education* 3: 69–74.

Allen, D., and K. Tanner. 2006. Rubrics: Tools for making learning goals and evaluation criteria explicit for both teachers and learners. *CBE-Life Sciences Education* 5: 197–203

Angelo, T. A., and K. P. Cross. 1993. *Classroom assessment techniques: A handbook for college teachers.* San Francisco: Jossey-Bass.

Beishuisen, J., P. Van Boxel, P. Banyard, A. Twinder, H. Vermeij, and J. Underwood. 2006. The introduction of portfolios in higher education: A comparative study in the UK and the Netherlands. *European Journal of Education* 41: 491–508.

Black, P., and D. Wiliam. 1998. Inside the black box: Raising standards through classroom assessment. *Phi Delta Kappan* 80: 139–148.

Chapman, O. L., and M. A. Fiore. 2001. Calibrated peer review: A writing and critical thinking instructional tool. Retrieved from *http://cpr.molsci.ucla.edu* on January 16, 2006.

Commander, N. E., and M. Valeri-Gold. 2001. The learning portfolio: A valuable tool for increasing metacognitive awareness. *Learning Assistance Review* 6: 5–18.

Cortright, R. N., H. L. Collins, D. W. Rodenbaugh, and S. E. DiCarlo. 2003. Student retention of course content is improved by collaborative group testing. *Advances in Physiology Education* 27: 102–108.

Dancy, M. H., and R. J. Beichner. 2002. But are they learning? Getting started in classroom evaluation. *Cell Biology Education* 1: 87–94.

Diamond, R. M. 1998. *Designing and assessing courses and curricula: A practical guide.* San Francisco: Jossey-Bass.

Downing, S. M., and T. M. Haladyna, eds. 2006. *Handbook of test development.* Mahwah, NJ: Lawrence Erlbaum.

Dwyer, C. A., C. M. Millett, and D. G. Paynea. 2006. *A culture of evidence: Postsecondary assessment and learning outcomes.* Princeton, NJ: Educational Testing Service.

Estrada, C. A., S. R. Patel, G. Talente, and S. Kraemer. 2005. The 10-minute oral presentation: What should I focus on? *American Journal of Medical Science* 329: 306–309.

Fox, M. A., and N. Hackerman, eds. 2004. *Evaluating and improving undergraduate teaching in science, technology, engineering, and mathematics.* Washington, DC: National Academies Press.

Gratz, R. K. 1990. Improving lab report quality by model analysis, peer review, and revision. *Journal of College Science Teaching* 19: 292–295.

Guilford, W. H. 2001. Teaching peer review and the process of scientific writing. *Advances in Physiology Education* 25: 167–175.

Handelsman, J., J. Miller, and C. Pfund. 2007. *Scientific teaching.* New York: W.H. Freeman.

Herreid, C. F. 2007. *Start with a story: The case study method of teaching college science.* Arlington, VA: NSTA Press.

Houde, A. 2000. Student symposia on primary research articles: A window into the world of scientific research. *Journal of College Science Teaching* 30: 184–187.

Keys, C. W., B. Hand, V. Prain, and S. Collins. 1999. Using the science writing heuristic as a tool for learning from laboratory investigations in secondary science. *Journal Of Research In Science Teaching* 36: 1065–1089.

Knisely, K. 2002. *A student handbook for writing in biology.* Sunderland, MA: Sinauer Associates.

Koprowski, J. L. 1997. Sharpening the craft of scientific writing. *Journal of College Science Teaching* 27: 133–135.

Lipson, C. 2004. *Doing honest work in college.* Chicago: University of Chicago Press.

McKeachie, W. J. 2006. *McKeachie's teaching tips.* 12th ed. Boston: Houghton Mifflin.

Michaelsen, L .K., A. B. Knight, and L. D. Fink. 2002. *Team-based learning: A transformative use of small groups.* Westport, CT: Praeger.

National Research Council. 1999. *Transforming undergraduate education in science, mathematics, engineering, and technology.* Washington, DC: National Academies Press.

National Research Council (NRC). 2003. *Bio2010: Transforming undergraduate education for future research biologists.* Washington, DC: National Academies Press.

Osterlind, S. J. 1997. *Constructing test items: Multiple-choice, constructed response, performance, and other formats,* 2nd ed. Berlin: Springer-Verlag.

Palomba, C. A., and T. W. Banta. 1999. *Assessment essentials.* San Francisco: Jossey-Bass.

Ruiz-Primo, M. A., and E. M. Furtak. 2006. Informal formative assessment and scientific inquiry: Exploring teachers' practices and student learning. *Educational Assessment* 11: 237–263.

Smith, C. B. 2003. Making presentations. ERIC topical bibliography and commentary. ERIC Clearinghouse on Reading, English, and Communication. Bloomington, IN. Retrieved from *http://ezproxy.wfu.edu:3000/login?url=http://search.ebscohost.com/*login. *aspx? direct=true &db=eric& AN=ED48089&site=ehost-live* on July 27, 2007.

Stevens, D. D., and A. J. Levi. 2005. I*ntroduction to rubrics.* Sterling, VA: Stylus Publishing.

Uno, G. E. 1999. *Handbook on teaching undergraduate science courses: A survival training manual.* Fort Worth, TX: Saunders College Publishing.

Walter, M. 2002. A window into students' learning goals: The learning goals inventory. Poster presented March 17, 2002 at American Association of Higher Education National Meeting, Chicago, IL.

Walvoord, B. E. 2004. *Assessment clear and simple.* San Francisco: Jossey-Bass.

Wiese, J., P. Varosy, and L. Tierney. 2002. Improving oral presentation skills with a clinical reasoning curriculum: A prospective controlled study. *American Journal of Medicine* 112: 212–218.

Wright, W. A., P. T. Knight, and N. Pomerleau. 1999. Portfolio people: Teaching and learning dossiers and innovation in higher education. *Innovative Higher Education* 24: 89–103.

Zubizarreta, J. 2004. *The learning portfolio: Reflective practice for improving student learning.* San Francisco: Jossey-Bass.

CHAPTER 4

Teaching Techniques for Inquiry Labs

Chapters 1–3 described general principles of inquiry and processes for developing and assessing an inquiry-centered course. At some point, instructors must actually go into a class and interact with students. This can be a scary prospect for those who are new to inquiry. They are likely to wonder, *How do I start off lab? When do I ask questions? Am I leading students too much? Telling them too little?*

Even when instructors are adept at leading inquiry-centered courses themselves, they may supervise teaching assistants (TAs) who are teaching for the first time, or they may mentor junior faculty. How can relative novices be trained to use inquiry effectively? Training new instructors is much like teaching undergraduates. There are general learning goals (i.e., behaviors and practices), specific methods of instruction for reaching these goals (training strategies), and assessments (teaching evaluation).

This chapter examines the process in greater detail. It outlines specific behaviors and practices that instructors can use to stimulate inquiry-based learning. It also includes strategies for training new faculty or TAs to use these methods and recommends tools for evaluating day-to-day instructional practices.

BEHAVIORS AND PRACTICES

The goal of an inquiry-based lab is for students to be active participants in exploring the topic at hand. They should be asking questions, devising ways to answer their questions, conducting investigations, and analyzing and communicating their results. These behaviors do not arise on their own; students must learn the process, and, until they do, the course instructor must provide concerted, ongoing direction. Instructors who teach this way for some time develop a distinctive suite of skills and behaviors they use to foster the inquiry mind-set in their students. Some will be unique to the individual, but many are shared by nearly all inquiry-oriented instructors. Widely shared behaviors and practices can be loosely divided into two categories: course process skills,and personal behaviors (Bohrer et al. 2008). Novice instructors can learn to lead inquiry-based courses very quickly by focusing on developing these shared features first. With experience, they will develop their own styles and favorite techniques.

Course Process Skills

Course process skills are skills that help the instructor guide students through the inquiry process itself. In an inquiry-oriented course, the instructor is not the ultimate source of content and process knowledge, but rather a learning facilitator. To use a common phrase, the instructor moves "from sage on the stage to guide on the side." Instead of providing pat answers to students' questions, the instructor uses a combination of active listening, contextual responses, and directed questioning to encourage students to think problems through and deduce answers for themselves.

Active listening and contextual responses are used routinely in counseling and mediation (among other situations). The listener (instructor) attends to the statements of a speaker (student), then recaps and summarizes key elements of those statements (Miller and Rollnick 2002). The listener follows up with a **contextual response,** that is, a statement or proposition that extends the speaker's original statements forward along a particular logical path. Active listening with contextual responses accomplishes two functions simultaneously. First, by hearing their own statements interpreted and reflected back, students learn that their words can be interpreted many ways. Students learn very quickly whether the language they are using is received and interpreted as intended and (ideally) will work to express themselves more precisely. Second, the listener (instructor) models a basic critical-thinking process for the speaker (student). In training new instructors, this book's author tells them to "ask students to go one step beyond their first thoughts." However, undergraduates often have not learned how to perform this cognitive action yet. When the lab instructor extends the initial ideas presented by the student, he or she is modeling this basic critical-thinking skill. With time, students will learn to take the next logical step themselves.

As students gain confidence, the instructor can offer a contextual response that does not follow logically or that contradicts some other piece of prior factual knowledge. Now the goal is for students to identify and correct the error or contradiction *themselves*. The instructor also can extend a student's idea in a way that is logically sound but is not applicable to the situation at hand. Again, the goal is for the students to recognize that the line of thinking is not applicable; they then self-correct, that is, back up and follow a different line of logic. These techniques should not be used with novice students but are very effective once students have some experience with the basic active listening process.

Directed questioning is closely tied to active listening. Rather than answering a question immediately, the instructor asks another question or makes a statement intended to elicit more questions, thereby guiding a student toward identifying answers for themselves. Directed questioning is often (incorrectly) referred to as Socratic teaching. The classical Socratic method is dialectic, follows a stricter format, and uses very specific questions. Socratic dialects begin with a central argument or thesis statement that the instructor poses to a particular student. The instructor uses subsequent questions to challenge the student's assumptions and probe the limits of his or her knowledge. The end point for that student (and other observers in the class) is insight into the limits of his or her current knowledge or the discovery of a new level of understanding. The true Socratic method is extremely powerful and is widely used by law schools. However, it is beyond the ability of most novice undergraduates, and can be so intimidating as to actually inhibit learning. McKeachie (2002) provides a concise example of the true Socratic method in action.

Unlike Socratic instruction, directed questioning is extremely useful with undergraduates. There are three types of questions: **divergent, convergent,** and **evaluative** (Lawson 2002). Divergent questions do not have a single correct answer and are open-ended; questions such as, What do you think happened? What are some ways you could test your hypothesis? and What would you do next? would be examples. These questions encourage a student to think creatively and to explore multiple possible solutions. Convergent questions have a predetermined or specific correct answer. An example would be, What volume of water did the plant lose by transpiration? Alone, convergent questions do not encourage much thinking, but when combined with divergent questions, they can point out key facts or variables that students did not take into account when making hypotheses or conclusions. Evaluative questions invite students to evaluate the facts presented, then defend their positions. An evaluative question would be, Which of these chemical reactions do you think is more likely to lead to cell death, and why? These types of questions are useful for stimulating discussion or debate.

Instructors should be aware that directed questioning frustrates undergraduate students immensely; most are accustomed to getting clear answers to their questions. In this author's experience, students frequently disparage the method on evaluations ("He never answered my questions outright, just asked more!"), even when their instructor uses the method extremely well.

In practice, it is difficult to separate active listening, contextual response, and directed questioning methods from one another. Fortunately it is not really necessary to do so. For the novice instructor, it is more important to understand how they work together to stimulate students' cognitive development. The following two scenarios show these techniques at work.

Scenario #1:

STUDENT: We just finished getting the weights of our plants that we gave saltwater last week. Now what do we do?

INSTRUCTOR: Put the data in the first column of the table in your lab manual, average your results, and transfer them to the table on the chalkboard.

STUDENT: We've done all that. What's next? Can we go home now?

INSTRUCTOR: Do you have all the data you need to complete the homework assignment?

STUDENT: Yeah.

INSTRUCTOR: Then you are done for the day.

The students have their immediate questions answered, but the instructor has no indication whether they have gained anything from the exercise. Contrast this with a different instructor's response to the same situation.

Scenario #2:

STUDENT: We just finished collecting the weights of our plants that we gave saltwater last week. Now what do we do?

INSTRUCTOR: Put the data in the first column of the table in your lab manual, average your results, and transfer them to the table on the chalkboard. *(Clarification questions like this can be answered outright, unless developing procedures was a planned part of the lab exercise.)*

STUDENT: We've done all that. What's next? Can we go home now?

INSTRUCTOR: Remind me again, what was the reason you collected those data? *(A good divergent question.)*

STUDENT: I dunno. *(Students often feign ignorance to avoid exposing gaps in their understanding or simply to avoid more work.)*

INSTRUCTOR: What question were you trying to answer to begin with? *(Convergent question seeks clarification of factual information.)*

STUDENT: Oh, that. We wanted to know if salt around the roots would change the rate of transpiration.

INSTRUCTOR: So what did you actually measure? *(Convergent question seeking additional informationn.)*

Student: Weights of each of the plants.

Instructor: Weights of the plants. How is that any different from what you did last week? *(Active listening; contextual response points out error in original statement.)*

Student: I meant we measured the weight of the plants after seven days of exposure to salt.

Instructor: So why measure them twice? *(Divergent question.)*

Student: If we compare the weights, the differences will tell us if salt affects the transpiration rate?

Instructor: How's that?

Student: If the weight change is less for plants given salt, there is an effect on transpiration.

Instructor: So less weight change means salt affects transpiration. That means the plants that weigh the least now have the lowest rate of transpiration. *(Active listening again, but the contextual response is a misdirection.)*

Student: Yeah. Wait, no. It's whichever plants had the smallest change in weight over the week that had the lowest transpiration rate. *(Student has self-corrected.)*

Instructor: So do you have all the data you need to answer that original question? It's part of the homework. *(A simple evaluative question. Students must decide if they have the necessary information.)*

Student: No, because we only did some of the plants. The rest of the class measured plants too. Our data are on the board, but we need their numbers too.

Instructor: Right. Everyone else should be done shortly. Wait a minute, and let me check on their progress. While I do, you and your partners think about this. Which do you think would affect transpiration more, salt, oil, or sugar in the soil, and why? *(Evaluative question, that could lead to testable hypotheses.)*

In Scenario #2, the students concluded on their own that they do not have all the necessary information. It is faster to answer students' question immediately, as occurs in Scenario #1, but remember that the goal of inquiry is to maximize student learning gains. From this perspective, the two-minute exchange in Scenario #2 has been a valuable learning opportunity. Students have seen a cognitive process modeled that they can copy in the future to answer questions without the instructor's input. Consider also what the instructor gained. He or she knows that this student understands the central purpose of the exercise and the relative level of thinking complexity that this student has currently. Assuming the instructor follows up with the student later in lab and asks how the student responded to the evaluative question, the instructor will be able to gauge that student's ability to apply new factual information to novel situations.

Other Useful Course Process Skills

The ability to ask questions effectively is just one of the characteristics that effective instructors share. They encourage students to explore a range of alternative ideas, consider multiple hypotheses, and look for more than one possible explanation for experimental results. They understand that uncertainty is a key part of the learning process, and they use it as a teaching tool. They expect students to discuss their ideas and uncertainties openly and to disagree with one another. They also help students understand and accept that some of their initial ideas will be naive, or simply wrong.

Effective inquiry-oriented instructors also have high but realistic expectations of students. They demand participation from all students; no one is allowed to hide behind others. They do not spend time rehashing basic factual knowledge that students should know already (or can reasonably learn on their own). Instead they spend class time challenging students to stretch beyond their current abilities. Whenever students meet these challenges successfully, inquiry-oriented instructors give genuine praise.

An inquiry-oriented instructor also is committed to ongoing assessment. As Scenario #2 demonstrated, asking students questions can provide valuable informal assessment data. Effective instructors use a variety of informal assessment techniques (discussed on pp. 45–47) to regularly check students' progress. They also turn formal assessments into another instructional opportunity.

Consider this situation: In a demonstration lab course the instructor asks students to graph transpiration rate versus soil salinity and turn the graph in for a homework grade. To grade the assignment, the instructor determines whether each student constructed the graph correctly, then checks off the assignment in the gradebook (essentially, a pass/no credit score). An inquiry-oriented lab instructor might make the same assignment but will add that the students must write a paragraph that interprets their graph. When grading the assignment, the second instructor reads the paragraphs to determine if each student knows what his or her results actually mean. Even if the second instructor grades the assignment as pass/no credit, the students will gain more from this formal assessment because they have an additional opportunity (the expository paragraph) to self-assess and self-correct before handing it in. Moreover, although the students are being graded, the grade penalty is quite small if they fail to meet the instructor's expectation. Low-stakes formal assessments like these serve as a bridge between strictly informal assessments and higher-stakes formal assessments such as exams and final papers. An instructor who is aware of the importance of ongoing assessment is more likely to make small changes like this, which can improve learning dramatically.

Personal Behaviors

While not every student responds to an instructor in the same way, certain behaviors tend to foster favorable responses toward inquiry in a majority of students. They are not part of the inquiry process but can still affect learning. A novice instructor would do well to cultivate them.

Undergraduates should be treated with basic courtesy. While there are instructors who are effective despite being inaccessible, rude curmudgeons, they are rare. Whenever possible, instructors should learn students' names and call on them by name. Instructors can also demonstrate respect for students and the teaching process by being organized and prepared. At the same time, they should remain flexible and adapt to the situation at hand. During class or lab, instructors should be actively engaged with students, asking questions and assisting them. Standing silently off to one side or, worse, leaving the lab entirely, sends the message that the instructor is not concerned about the learning process.

Effective instructors also stay composed in stressful situations. Even on the best days, an inquiry lab can seem chaotic at times, and invariably more than one student wants attention simultaneously. At other times, an instructor must handle a confrontational, injured, or ill student. Experienced colleagues who have taught a course previously can point out the most common day-to-day challenges and suggest positive ways to overcome them. Most institutions have formal policies that cover contentious issues such as sexual harassment or plagiarism. Similarly, many teaching lab programs already have set policies for dealing with students who are ill or want to argue a grade. When novice instructors know in advance what these policies are, they are less likely to be pushed off-balance when such issues arise. Many universities and community colleges offer professional development classes on practical conflict management, usually through the human resources office. Although these classes generally are designed for administrative supervisors, the general techniques for managing people are the same. If the course is not open to instructors, ask the trainer to share some tips or recommend good books on the subject. Another very useful training technique is role playing. Have a colleague imitate a disruptive student's behavior, for example, so that novice instructors can practice their responses.

Students respond well to an instructor who models their desired actions and behaviors. This applies both to physical and cognitive behaviors. For example, it is difficult for instructors to teach students to work safely in a teaching lab if they themselves do not wear safety glasses or if they mouth pipets or pour hazardous chemicals down the drain. Likewise, students are unlikely to seek alternative explanations for observations if instructors only want to hear the correct answer. Students are more likely to mimic instructors who admit they do not know the answer to a complex problem, then describe the process by which they would break the problem down into small, manageable questions. This principle extends to communication skills such as speaking and writing clearly. If the instructor cannot communicate clearly, can students reasonably be expected to do so?

Finally, effective inquiry-oriented instructors spend time reflecting on their own teaching and seek to improve it. They also try to see their teaching practices from the point of view of the undergraduate learner, then adjust them to meet students' instructional needs.

Potential Pitfalls

Just as certain skills and behaviors foster an inquiry mind-set in students, others can undermine it. A common mistake of novice instructors is to try to overlay inquiry methods and procedures on top of an already dense, content-centered course. Inquiry-based instruction is an alternative approach that needs to be kept in balance with didactic methods. Attempting to stack the two approaches atop one another exhausts both instructor and students and reduces overall learning gains. Those new to using inquiry should focus on making small changes at first, until they see how much students can accommodate realistically. The Instructional Methods Inventory in Appendix A can help instructors determine if they are trying to do too much at once.

Instructors who have not used inquiry before can be overwhelmed by the number of inquiry teaching techniques. It is important that new instructors start small and build up their repertoires over time. As is the case with assessment, adding one fundamental inquiry method per semester to personal teaching practices leads to enormous improvement over time, without sacrificing either student engagement or learning gains.

Another counterproductive behavior is devoting insufficient time to teaching preparation. Graduate student TAs are particularly prone to this if they feel their pedagogical duties take time away from their research. While it is true that inquiry-based teaching requires more time and concentrated effort than classical didactic methods, avoiding it is false economy, for two important reasons. First, more and more science departments are expecting new faculty hires to have a demonstrated ability to teach using inquiry and other nontraditional pedagogical methods. Second, the vast majority of graduate students who move into faculty positions will spend 50% or more of their professional time in the classroom, not the laboratory. Seen in this light, strong teaching skills are not a luxury; they are essential career skills.

Instructors often worry that inquiry-based instruction has no "right" answers—thus, no external metric with which to gauge their performances. Lacking that, they may begin to think they are not teaching successfully. Such anxiety can degenerate very quickly into overt resistance to inquiry as a teaching method. In courses with multiple sections and instructors, a weekly instructors' meeting goes a long way toward reducing anxiety. The leader of the meeting should encourage novice and experienced instructors alike to use the meeting as a safe space in which to voice concerns and questions and to discuss new teaching ideas with their peers. A summary of results from any postcourse assessments should be shared with all instructors so they can see exactly how well they are achieving the stated learning goals.

TRAINING NEW INSTRUCTORS

How can new instructors be trained to use inquiry methods? Most trainers do not rely on any one method, but rather use a combination of formal training sessions, directed practice opportunities, and individualized coaching and feedback (Bohrer et al. 2008). Often the greatest hurdle in training novices is developing a basic commitment to *teaching to the learner,* not to the course content. Once this commitment has been made, teaching students the mechanical elements is fairly straightforward. The process for designing a training program is beyond the scope of this book. For more information or assistance, readers should consult with their local teaching resource centers. Alternatively, the Sheridan Center for Teaching at Brown University has an extensive outline of its instructor training program (*www.brown.edu/Administration/Sheridan_Center*). A similar program is sponsored by the Center for Teaching and Learning at the University of Georgia (*www.isd. uga.edu*). Links to other programs can be located via the University of Kansas's Center for Teaching Excellence (*www.cte.ku.edu/cteInfo/resources/ websites.shtml*), which maintains a master list of active teaching centers worldwide.

Nationally there is the Preparing Future Faculty Program (PFF), which according to its website aims to "transform the way aspiring faculty members are prepared for their careers. PFF programs provide ... students ... with opportunities to observe and experience faculty responsibilities at a variety of academic institutions with varying missions, diverse student bodies, and different expectations for faculty" (*www.preparing-faculty.org*; quote used with permission).

The Wisconsin Program for Scientific Teaching also has recently published guidelines for developing research-based college science teacher training programs (Handelsman, Miller, and Pfund 2007).

EVALUATING TEACHING

The purpose of undergraduate assessment is to guide students and measure their learning gains. Instructor evaluation serves a similar purpose, in that it provides individualized feedback that teachers need to develop and maintain a reflective teaching practice. Well-designed evaluation procedures also help ensure consistency among instructors who are teaching different sections of the same course. Evaluation can help instructors determine if they are using their class time effectively—that is, in ways that promote their general learning outcomes and specific performance goals. For curriculum coordinators, data from properly designed teaching assessments can be used to ensure consistency among instructors in multisection lab courses.

Like learning, teaching should be evaluated both while in progress and again upon completion, using several methods in tandem. Several evaluation methods, with their advantages and disadvantages, are summarized on the following pages.

In-Progress Evaluation

Grade Comparisons

Grade comparions are a simple method to compare instructors across multiple lab or lecture sections. To be useful, all instructors should be using the same assignments and grading scale, and they should be scoring assignments using similar rubrics. Grades cannot be the only method for evaluation because they do not distinguish between differences that are the result of student effort versus instructors' teaching methods. However, differences in grades between sections can be used as an early warning sign of problems.

Some course coordinators continuously compare grades by using shared electronic gradebooks. A shared gradebook lets all instructors see how their scores compare to peers and, ideally, self-correct. However, instructors see grades for students who are not in their sections. Some gradebook software allows the program administrator to hide individual students' names from everyone except their primary instructor; only average grades for assignments are visible.

Classroom Assessment Techniques (CATs)

Many CATs (see p. 45) can be modified to evaluate the teaching process. Ten of them—including chain notes, assignment assessments, exam evaluations, and reading rating sheets—specifically assess learner reactions to instructional methods. Consult Angelo and Cross (1993) for more information.

Classroom Observations

Directly observing an instructor while he or she teaches is the gold standard for evaluation. The course coordinator or other trained observer arranges in advance to attend the instructor's class. Using a standardized protocol, observers document specific teaching methods used, types of questions students are asked, and other teaching behaviors. They also make note of how much time students spend in various activities during the class. Many observers simultaneously videotape the observed class for additional documentation. Afterward, the instructor and observer watch the tape (if one was made) and review the observer's findings. If the observer is an instructional specialist, he or she may suggest ways for improving the instructor's teaching or course management.

The chief disadvantage of classroom observations is the time required. Fortunately, many institutional teaching centers operate formal consultation programs that send trained, independent observers to instructors' classes. Usually, observers are not familiar with the factual content of the course, freeing them to focus on the teaching process. The quality of evaluation also depends on the instrument used and the training and experience of the observer. One observation can be informative, but two or more (such as the first and third class of a semester or in succeeding semesters) are better. If records are available from a prior evaluation, the observer can look for specific suggestions and document whether they were implemented successfully. Ideally, the same observer will conduct the evaluation each time, in order to minimize observational bias. Alternatively, two observers may collaborate to perform each evaluation.

Peer observation is another option for conducting classroom observations. Observers are trained to use a standardized observation protocol, then watch other instructors teaching the same course. Instructors can be paired so they observe one another; this approach has the advantage of fostering greater dialogue. The observer additionally benefits from seeing how another instructor handles the same classroom situations he or she does. Alternatively, experienced instructors can observe novice instructors in the same course. Here the chief advantage is that the more experienced observer can offer more detailed suggestions. However, the observer may focus too much on content issues and not enough on teaching methods.

Another variation is peer observation within the context of a training workshop. One instructor is videotaped, then a group of instructors reviews the tape using a standardized observation protocol. The group discusses their observations, then makes recommendations for the individual who was videotaped. Archived videotapes can be used if the goal is to train observers or novice instructors rather than provide feedback to a particular individual.

Several previously validated observation protocols are available for those who would prefer to conduct their own evaluations (Benford and Lawson 2001; Bohrer et al. 2008; Lawson et al. 2002; MacIsaac, Sawada, and Falconer 2001; Sampson 2004; Sawada et al. 2000; Sawada et al. 2002; Wainwright et al. 2004; Young et al. 1995). Most institutional teaching centers will share their protocol and observer training materials with curriculum coordinators as well.

When planning a classroom observation program, one source of reliable data that is frequently overlooked is undergraduates' behavior during class. Observers should look for behaviors that indicate students are actively engaged in the inquiry process. For example, when an instructor uses inquiry routinely, students should be asking each other the types of questions described on pp. 66-67. Conversely, if students are not engaged in inquiry, their questions are likely to center on procedural details or issues unrelated to the topic at hand. Other behaviors to watch for are described in Chapter 1, A Brief Introduction to Inquiry.

Postcourse Evaluation

A postcourse instructor evaluation is much like formal assessment of students. The goal is to determine how well the instructor has achieved the predefined general learning outcomes and specific performance goals of the course. Both the evaluator and instructor need to rein in their egos and approach the process with a positive, collaborative attitude.

Student Surveys and Questionnaires

End of course student surveys or evaluations are ubiquitous in higher education (Algozzine et al. 2004). Faculty often claim that these surveys are of little value, and a surprising number of faculty are openly hostile, claiming that undergraduates do not even know what good teaching is. Unfortunately many surveys reinforce this opinion because they consist of generic questions that focus on the mechanics, not the substance, of teaching. Numerous studies have established that *properly designed* student surveys can be very informative and provide a very accurate picture of the learning processes with the course (Algozzine, et al. 2004; Felton, Mitchell, and Stimson 2004; Fox and Hackerman 2004; Hellman 1998; Ouimet et al. 2004). Students tend not to volunteer the kind of information about teaching that is being sought unless they are asked specifically about it. Thus, the quality of data obtained is directly related to the quality of questions asked.

Consider the excerpts from two course evaluations (Table 5 and Table 6). Both are intended to measure the instructor's use of inquiry in the course, and both use a combination of Likert-scaled and open-response questions. In light of the guidelines above, one version clearly will provide more useful data about teaching effectiveness.

Table 5

Example 1 of a course evaluation

Question #	Question	Score
1	Responds well to students' questions.	1 2 3 4 5
2	Encourages active participation.	1 2 3 4 5
3	Right amount of interaction for this size and type of class.	1 2 3 4 5
What other comments, either positive or negative, would you like to make about your lab instructor for this semester?		

Both sets of questions seek the same basic information. Example 1 exemplifies the typical survey designed to serve multiple courses, with generic questions that can be interpreted in many ways. In contrast, the questions in Example 2 focus on specific behaviors that are central to the inquiry teaching process. Question 3 in Example 2 asks for details about specific instructor behaviors during the course. That question is intended to evaluate an instructor teaching a laboratory course, so it would not even be appropriate for a lecture course evaluation. Finally, the open-response questions ask for specific evidence of teaching effectiveness. Overall, questions like those in Example 2 provide a more useful picture of the instructor's teaching methods for this course.

Ideally, the framework of the teaching evaluation is created at the same time as the rest of the course assessment program. Each course should have its own evaluation instrument. When the instrument is first drawn up, approximately half of the questions should focus on whether the instructor used general practices that are common to all inquiry-oriented instruction. The remaining questions should focus on how well the instructor achieved the general learning outcomes and specific performance goals of that particular course. Once the teaching evaluation has been administered to students a few times, the types and balance between general and specific questions can be refined to provide the maximum amount of useful data.

Table 6

Example 2 of a course evaluation

Question #	Question	Score
1	Used students' questions as opportunities to probe their current understanding and improve upon it.	1 2 3 4 5
2	Ensured that all students participated fully and remained engaged in the inquiry process.	1 2 3 4 5
3	Used lecture sparingly; spent the majority of class time interacting with students and discussing problems.	1 2 3 4 5
What things did your lab instructor do this semester that improved your ability to reason and think through problems?		
Did your instructor do anything that hampered your ability to learn or to think critically about novel problems?		

One concern many curriculum coordinators and instructors have when designing a teaching evaluation is the amount of time required to collate the data. Fortunately, many institutions now have the infrastructure to administer teaching surveys and questionnaires online, using courseware such as Blackboard. Alternatively, instructors might ask their institutions' instructional technology groups or computer science programs to design and host a dedicated online survey program. Several vendors also offer web-based survey hosting and data collation services for free or a very modest monthly subscription fee. As with other elements in teaching, it is a good idea to start at the end: Design a robust teaching evaluation first, then search online with the phrase *survey hosting* to find a vendor that offers the proper combination of services needed to implement it.

Fox and Hackerman (2004) discuss this topic extensively and describe many examples of evaluation instruments. Sudweeks (1998) describes a more detailed strategy for designing evaluation surveys. The Office of Educational Assessment at the University of Washington has validated instructional assessment forms for a variety of course formats posted online (*www.washington.edu/oea*). These forms can be a good starting point for developing new, course-specific instruments.

Instructor Self-Assessment

Self-reflection is an important habit for inquiry-oriented instructors. However, novices may not know what questions they should be asking themselves. The purpose of formalized self-assessment is to help new instructors develop this habit concurrently with their other skills. The general principles for designing a self-assessment are much the same as for creating a student evaluation. Questions should focus on how well instructors think they are helping students meet the general learning outcomes and performance goals.

Self-assessments can be informal or highly structured; it is more important that instructors self-assess at regular intervals. For example, a colleague of the author's at another institution administers a short questionnaire to all new TAs in an introductory biology lab at the start of their first semester of teaching, then again at the end of their first and second semesters. The questionnaire asks about their prior knowledge or experiences with inquiry, overall confidence in using inquiry while teaching, and other parameters that affect teaching. After completing the questionnaire for their third time, instructors are given back their two prior sets of responses. Comparing their responses shows them how their teaching has changed and (ideally) improved.

A more structured instrument for self-assessment is the *Instructional Methods Inventory (IMI)* (Bohrer et al. 2008). The IMI consists of a list of 31 teaching practices that do or do not promote inquiry. For each practice, instructors indicate how often they experienced it in the past, use it currently, and would use it in an ideal teaching situation. The summary results of the inventory indicate the relative amount of class time an instructor actually spends using instructional methods that promote inquiry and his or her overall knowledge of the process. An abridged and modified version of the original IMI, with instructions for administering and scoring it, are in Appendix A.

Focus Groups With Feedback

For a focus group, the evaluator recruits a subset of students from each instructor's sections of the course or lab, then meets with them privately to discuss their classroom experiences. The evaluator writes a summary of their comments and passes it along to the instructor. Ideally the evaluator will do more than just collect observations and comments from students. They will meet with the instructors for a course (individually or as a group) for the express purpose of discussing teaching practices. The evaluator will make specific suggestions for improvements and get feedback from the instructor(s) on the feasibility of implementing those suggestions.

Focus groups are particularly informative when conducted as a follow-up to a written student survey or instructor self-assessment. The evaluator can use written comments as the starting point for exploring particular teaching behaviors in greater detail. As with classroom observations, the chief disadvantage of focus groups is the time required. Furthermore, a disgruntled or nonmotivated student can skew opinions in the group. At the same time, students who were not engaged by their instructors sometimes provide the most insight into how the instructors' teaching practices might be improved.

Standardized Tests

It cannot be stressed enough: A standardized test cannot be the only metric for evaluation. Students, instructors, and courses alike must be evaluated on the basis of multiple sources of evidence. No one instrument or assessment can ever provide all of the necessary data. That said, standardized exams that are used properly provide valuable information that cannot be obtained any other way.

The SALG, BCI, VOSTS, and GRE are nationally standardized instruments that are used for formal student assessment. The main benefit of using a nationally standardized exam is the large pool of data from peer institutions for comparison. If students' scores are grouped by instructor, these same data can be used to evaluate relative teaching performances. However, not all of these exams can provide a breakdown of student scores by instructor.

Although a few programs have developed their own standardized exams for assessing teaching effectiveness, this is not generally recommended. Developing an exam that can accurately evaluate multiple instructors is a lengthy process that must be done in consultation with an educational testing specialist. In addition, the data set for making comparisons between instructors is much smaller than with a national exam. There also is a significant risk of a few outlier scores dramatically affecting the results for an individual instructor.

For individual instructors, using standardized questions to evaluate teaching effectiveness between semesters is less problematic. It is vital, however, to establish procedures for making comparisons and for pruning outliers. For example, the author teaches two large-enrollment biology survey courses annually. The final exam in each course includes standardized questions that make up ~50% of the total points on the exam. These questions are designed to assess specific higher-order thinking skills, not factual recall. To evaluate teaching success, students' scores on these standardized questions are summarized in the following three ways, then compared to the same values from preceding semesters.

Unadjusted mean score. The average score of all students on all standardized questions. The target goal is less than 3% change in the mean from year to year. A 3% limit was selected because it is the difference between letter grade designations (i.e., B+ versus A- or C versus C+) in the standard grading scale for the author's department.

Mid-quintile mean score. Students' grades for the standardized questions are sorted from highest to lowest, then divided into five bins (quintiles) of equal size. Scores in the highest and lowest quintiles are eliminated, then the mean score for standardized questions is recalculated using the individual scores of the remaining three quintiles. This method removes both high and low outliers and provides a center-weighted estimate of the mean for the semester. The target goal is less than 3% difference in mid-quintile mean score between the current and prior semesters.

Mean score for each quintile. The average score is calculated for each quintile. The target goal is less than a 3% change in quintile scores between semesters.

In addition to overall scores, the author tallies the number of students that miss each of the standardized questions. If 3% more students miss a particular question than in prior semesters, it suggests that the author did not encourage development of that particular thinking skill sufficiently during the current semester.

TO LEARN MORE ABOUT EVALUATION AND TRAINING

Those wanting to learn more about general principles and methods for training new instructors should consult Brookfield (1995), Marincovich, Prostko, and Stout (1998), and Mintzes and Leonard (2006). Discussions and examples of additional methods for instructor evaluation and feedback can be found in Fox and Hackerman (2004), Hampton and Reiser (2002), and McBeath (1992). Specific examples of successful professional training programs for college science teachers have been described by Baumgartner (2007), Erklenz-Watts, Westbay, and Lynd-Balta (2006), and Winternitz and Davis (2000). These programs can be useful models for educators who aredeveloping new programs of their own.

References

Algozzine, B., J. Beattie, M. Bray, C. Flowers, J. Gretes, L. Howley, G. Mohanty, and F. Spooner. 2004. College teaching: A practice in search of principles. *College Teaching* 52: 134–141.

Angelo, T. A., and K. P. Cross. 1993. *Classroom assessment techniques: A handbook for college teachers.* San Francisco: Jossey-Bass.

Baumgartner, E. 2007. A professional development teaching course for science graduate students. *Journal of College Science Teaching* 36: 16–21.

Benford, R., and A. E. Lawson. 2001. *Relationships between effective inquiry use and the development of scientific reasoning skills in college biology labs.* Arlington, VA: National Science Foundation.

Bohrer, K., A. Ferrier, A. D. Johnson, and K. Miller. 2008. Teaching the teacher: Introducing and training TAs in inquiry-based learning methods. *Proceedings of the 29th Conference of the Association for Biology Laboratory Education*.

Brookfield, S. D. 1995. *Becoming a critically reflective teacher.* San Francisco: Jossey-Bass.

Erklenz-Watts, M., T. Westbay, and E. Lynd-Balta. 2006. An alternative professional development program: Lessons learned. *College Teaching* 54: 275–279.

Felton, J., J. Mitchell, and M. Stinson. 2004. Web-based student evaluations of professors: The relations between perceived quality, easiness, and sexiness. *Assessment & Evaluation in Higher Education* 29: 91–108.

Fox, M. A., and N. Hackerman, eds. 2004. *Evaluating and improving undergraduate teaching in science, technology, engineering, and mathematics.* Washington, DC: National Academies Press.

Hampton, S. E., and R. A. Reiser. 2002. From theory to practice: Using an instructional theory to provide feedback and consultation to improve college teaching, learning, and motivation. Presented at the annual meeting of the American Educational Research Association, New Orleans, LA (April 3).

Handelsman, J., J. Miller, and C. Pfund. 2007. *Scientific teaching.* New York: W.H. Freeman

Hellman, C. M. 1998. Faculty evaluation by students: A comparison between full-time and adjunct faculty. *Journal of Applied Research in the Community College* 6: 45–50.

Lawson, A. E. 2002. *Science teaching and the development of thinking.* Belmont, CA: Wadsworth/Thompson Learning.

Lawson, A., R. Benford, I. Bloom, M. Carlson, K. Falconer, D. Hestenes, E. Judson, M. Piburn, D. Sawada, J. Turley, and S. Wyckoff. 2002. Evaluating college science and mathematics instruction: A reform effort that improves teaching skills. *Journal of College Science Teaching* 31: 388–393.

MacIsaac, D., D. Sawada, and K. Falconer. 2001. Using the reformed teaching observation protocol (RTOP) as a catalyst for self-reflective change in secondary science teaching. Presented at the annual meeting of the American Education Research Association, Seattle, WA (April 10).

Marincovich, M., J., Prostko, and F. Stout, eds. 1998. *The professional development of graduate teaching assistants.* Bolton, MA: Anker Publishing.

McBeath, R. J., ed. 1992. *Instructing and evaluating in higher education: A guidebook for planning learning outcomes.* Englewood Cliffs, NJ: Educational Technology Publications.

McKeachie, W. J. 2002. Facilitating discussion. In *W.J. McKeachie's Teaching Tips.* 11th ed., ed. W. J. McKeachie, 38–40. Boston: Houghton Mifflin.

Miller, W. R., and S. Rollnick. 2002. *Motivational interviewing.* 2nd ed. New York: The Guilford Press.

Mintzes, J. J., and W. H. Leonard. 2006. *Handbook of college science teaching.* Arlington, VA: NSTA Press.

Ouimet, J., J. C. Bunnage, R. M. Carinia, G. D. Kuh, and J. Kennedy. 2004. Using focus groups, expert advice, and cognitive interviews to establish the validity of a college student survey. *Research in Higher Education* 45: 233–250.

Sampson, V. 2004. The science management observation protocol: using structured observations to improve teachers: Management of inquiry-based classrooms. *Science Teacher* 71: 30–33.

Sawada, D., M. D. Pibum, K. Falconer, J. Turley, R. Benford, and I. Bloom. 2000. Reformed teaching observation protocol (RTOP). *ACEPT Technical Report No. IN00-1.* Tempe, AZ: Arizona Collaborative for Excellence in the Preparation of Teachers.

Sawada, D., M. D. Piburn, E. Judson, J. Turley, K. Falconer, R. Benford, and I. Bloom. 2002. Measuring reform practices in science and mathematics classrooms: The reformed teaching observation protocol. *School Science and Mathematics* 102: 245–253.

Sudweeks, R. A. 1998. Questions for evaluating a college course. In *Designing and assessing courses and curricula: A practical guide*, ed. R. M. Diamond, 241–246. San Francisco, CA: Jossey-Bass.

Wainwright, C., L. Flick, P. D. Morrell, and A. Schepige. 2004. Observation of reform teaching in undergraduate level mathematics and science courses. *School Science and Mathematics* 104: 322–35.

Winternitz, T., and W. E. Davis. 2000. Lessons learned during five years of the UC Davis Program in College Teaching. *Journal of Graduate Teaching Assistant Development* 7: 69–75.

Young, M. J., B. Brett, S. Squires, and N. Lemire. 1995. Symposium: A new observation tool for looking at inquiry-based teaching and learning. Presented at the annual meeting of the American Educational Research Association, San Francisco, CA (April).

Part II

Introduction: Tested, Inquiry-Based Laboratory Exercises

Part II contains 40 exercises, grouped into 16 separate lab units. Each unit has been tested, revised, then retested over several semesters in multisection laboratory courses by multiple instructors at the author's home institution. Many units also have been tested by faculty at other institutions. Each unit was developed and tested for students at a particular level. The units range in difficulty and incorporate inquiry methods of varying degrees, depending on whether they were designed for nonmajors, freshman majors, or sophomore majors. The units are presented at the same level at which they are used most frequently. However many units have been tested with more than one audience; in those cases, additional information is provided in the Instructors' Notes that appear in each unit. **A summary of each unit appears on pages 86–91.**

Some exercises and model systems have not been published previously; others are well-known model systems that have been revised or updated for an inquiry-oriented lab instructional format. Those wanting to revise existing labs to make them more inquiry-based likely will find several useful examples for doing so in this latter group. For example, in many lab courses students must demonstrate the principles of Mendelian inheritance by setting up crosses of wild-type and mutant Drosophila. In Unit 2, Mendelian Genetics, students again set up and analyze crosses of wild-type and mutant Drosophila. The general methods used are no different from what most instructors are accustomed to; what is different is that the outcome is unknown. Even when students think they know the answer already, novel mutations appear in the F2 generation that force them to wrestle with genetic principles to a much greater degree than they do in a typical mono-hybrid cross lab. Other examples of units that use traditional lab methods or models in novel ways are Unit 5, Applied Enzymology; Unit 6, Energetics and Photosynthesis; and Unit 13, Animal Behavior.

In many institutions (including the author's), biology majors continue taking large, multisection lab courses through their sophomore year. That said, most sophomore-level units can be simplified fairly easily to suit freshman majors. Similarly, many units targeted to freshmen majors can be modified to fit the instructional goals for nonmajors courses or made more challenging and used with sophomores.

Readers should know that the units incorporate inquiry methods to varying degrees. In an inquiry-based—or learner-centered—course, thinking and analytical skills are built over time; thus units originally designed for the earlier stages of students' academic careers will be correspondingly less demanding or complex. Also, many instructors have not used inquiry previously. To ease their transition, most of the units retain more didactic features than are actually necessary or ideal. As instructors gain confidence and learn more about their student population's responses, these didactic elements should be replaced with more inquiry-oriented activities. More specific suggestions are provided in the Instructors' Notes for each unit.

Instructors who want to provide an even richer inquiry-based experience can use these units as a starting point for further experimentation and development. For example, consider eliminating the Procedures (step-by-step instructions) from each exercise and have students develop the basic procedures as prelab activities in a learning center or through interactive online tutorials. Turn the lab period into dedicated time for students to conduct their own experiments. There are innumerable ways to use the models described; the key is to focus on the learning goals first (as discussed at length in Chapter 2), *then* choose the exercises in each unit that will help students successfully achieve those goals.

GENERAL FORMAT

Each of the units begins with background material and a set of exercises for students. Following the student pages are the teacher pages, made up of Instructors' Notes and Preparatory Notes.

Structure of the Student Pages

The background information for students has been written for a particular audience, as was noted above. Instructors and coordinators should revise this information to match their particular audience's prior knowledge and current course work. Some student audiences may require considerably more information than is provided, while others with more experience with inquiry might reasonably be expected to locate much of this information on their own.

Instructors should note that the Goals for This Unit section does not appear at the beginning of the student pages of each unit. Putting the goals *after* the Background section was done intentionally. Some instructors using this book may have preferred that the lab goals appeared at the beginning of each unit, to help students focus on important elements. However, the first step in an inquiry-based class is to create engagement, not establish goals. Putting the goals first sends an implicit message that lab should be goal-oriented, not exploratory. (Those wanting to know more about the theoretical underpinnings of this point of view are urged to consult Bybee and Van Scotter [2006] or other references that describe the 5E Instructional Model developed for the BSCS program.) Also, little explanation or rationale is provided for the goals. This is intended to encourage students to think more deeply about *why* these goals are important. Instructors can revise the text to provide supplemental information to their students if the level of a unit as written is too high for their students, use the listed goals as starting points for prelab warm-up discussion, or eliminate them entirely.

Interspersed in the units are italicized questions that are designed to guide students' thinking as the unit unfolds. These should be answered during lab as students reach that step. *Students should not skip these questions or wait until later to answer them.* Similarly there are Unit Summary Questions at the end of the student pages. These questions ask students to identify critical factual information or to think about the implications of the exercises they have completed up to that point. Ideally, these summary questions will be answered in a laboratory notebook (for later evaluation) or turned in as part of a homework assignment. For more advanced students, the questions can be deleted and the students asked to summarize their learning gains on their own.

Following the exercises in most units is a section describing options for student experiments, called Your Own Experiment. The experimental option can be deleted without loss of continuity, but the author *strongly* urges instructors to use at least some of them. The list of available materials and suggestions for experiments are guidelines based on past experiences at the author's institution. Students often have other exceptional ideas for experiments and should be allowed to perform them if at all possible. If students will not be conducting their own experiments, the references to experiments in the Background section should be deleted to avoid confusion.

Instructors' Notes

The Instructors' Notes provide background information on the model system and interesting examples that demonstrate the key concepts underlying the exercises. The section General Teaching Strategy and Common Problems outlines the pedagogical goals and typical outcomes in detail and suggests how the instructor can guide students through the common trouble spots in the unit. There also are suggestions for modifying units to fit a variety of course meeting times, sizes, and audiences.

The section Recommended Prelab Skills lists the basic skills students will need to complete the unit successfully. This information is intended to help instructors sequence labs. For example, if a particular unit recommends basic microscopy skills, but students do not have those skills, the instructor should plan to include a general training exercise on microscopy prior to conducting that unit.

For instructors using pre- and postlab quizzes as part of their routine formal assessment strategies, the Assessment section describes reasonable pre- and postlab expectations of students. However, instructors should always use additional assessment methods besides quizzes, exams, and homework questions. The specific methods that will work best depend on the instructor and audience; readers should refer back to Chapter 3 for more specific suggestions.

The Other Tips section has miscellaneous suggestions and observations from other instructors that will help the unit run smoothly. The Safety and Housekeeping section describes biological or chemical safety hazards associated with the lab, general safety precautions, and recommendations for proper disposal or destruction of hazardous materials. These guidelines reflect general prudent practices and federal safety standards; those wanting to know more should consult references by Furr (2000), the National Research Council (2000), and Summers (2006). Specific rules on safety training for undergraduates and proper handling and disposing of chemical waste and biologically hazardous materials vary by state and institution. Readers should contact their institution's office of environmental health and safety for details.

The Supplemental References section lists selected articles relating to the unit. Some of the references date back farther than might be expected, but they were deliberately included for several reasons. Some are classic articles in which a particular phenomenon or model was described for the first time. Others are easier for students to understand than more recently published articles, or they provide a concise review that is not available in a more recent article. Finally, focusing only on recent literature can bias students against articles and books that are more than a few years old. Students can gain valuable insights into a model system by delving more deeply. The listed references are just a starting point; instructors should always search their local resources for other reference material that is available to their students.

The Preparatory Notes section lists organisms, reagents, equipment, glassware, and consumables needed to conduct the unit as written. Separate lists are provided for each week of multiweek units. Quantities listed are for 20 students working in groups of 3 to 4. The lists include extra material to accommodate inevitable spills and other accidents. All quantities can be scaled up for larger or multiple lab sections. Shared materials can be kept on a central bench or divided between two to three workstations. Materials for each workstation should be available on or close to each group of students' desks or work areas.

If the instructor is not keeping photocopies of students' experimental proposals (see Appendix B for details), he or she should keep some kind of log or database describing prior students' experiments and a brief summary of their results. Over time, this log can help the instructor predict what supplemental materials might be required. Moreover, students can build on past observations or repeat experiments having unusual or intriguing outcomes.

The Preparatory Notes close with instructions for how to obtain, grow, or prepare unusual materials, especially those that cannot be purchased directly from education suppliers.

WHAT IS *NOT* INCLUDED

A concerted effort has been made to provide extensive documentation for each of the laboratory units. However, instructors and coordinators must provide students with guidelines and procedures for their particular instruments and equipment. For example, it has been left to the instructor or lab coordinator to show students how to use their institution's particular brands of microscopes and spectrophotometers. Check the manufacturers' websites; most post student guides for their instruments as free PDFs. Similarly science education supply houses are the best source of information on how to house and care for study organisms.

There is no general primer on statistics in this book. Instructors often have strong opinions on which methods they want students to use. Some favor a rigorous formal approach, while others prefer that students develop an intuitive understanding of statistical analysis first. Regardless of philosophy, instructors need to plan for how they will implement statistical training into their courses. At the minimum, instructors should provide students with a printed or online guide to the statistical methods and tests they will use most often in the lab course. An undergraduate statistics textbook can provide a reasonable introduction to formal statistical reasoning. Those interested in trying out a less formal approach should consult Motulsky (1995).

References

Bybee, R. W., and P. Van Scotter. 2006. Reinventing the science curriculum. *Educational Leadership* 64: 43–47.

Furr, A. K. 2000. *CRC handbook of laboratory safety,* 5th ed. Boca Raton, FL: CRC Press.

Motulsky, H. 1995. *Intuitive biostatistics.* New York: Oxford University Press.

National Research Council (NRC). 2000. *Prudent practices in the laboratory: Handling and disposal of chemicals.* Washington, DC: National Academy Press.

Summers, J., J. Texley, and T. Kwan. 2006. *Science safety in the community college.* Arlington, VA: NSTA Press.

Summary of Units

Unit 1: Designing Scientific Experiments
Audience: nonmajors
Model and questions: Do termites follow trails? If so, why?
Major concepts: cyclic nature of science; differences between observation, hypothesis and prediction; negative controls; replicates; summarizing data; basic statistics
Prior skill, knowledge: none. A good introduction to inquiry for a nonmajors course.
Degree of difficulty: basic Novice students are led stepwise through the process of developing a hypothesis and making predictions. To create a more advanced experience, most of the background information can be deleted.
Unit 2: Mendelian Genetics
Audience: sophomore majors
Model and question: What type(s) of mutation causes the white-eyed phenotype in Drosophila?
Major concepts: mono- and dihybrid crosses, dominance and recessiveness of alleles, Drosophila care and breeding.
Prior skill, knowledge: none. Prior experience analyzing monohybrid crosses with a Punnett square is helpful.
Degree of difficulty: intermediate Four different mutant strains have the same white-eyed phenotype, but when outcrossed, produce different inheritance patterns. Students must compare their observations to known inheritance patterns to determine the type of mutation their fly line carries. To make the unit more challenging, remove the guidelines for setting up test crosses. Advanced students can be given strains with more complex mutations or inheritance patterns.
Unit 3: DNA Isolation and Analysis
Audience: sophomore majors
Model and question: How do mutations in DNA coding for bacterial beta-galactosidase affect expression and activity of the enzyme?
Major concepts: DNA isolation, DNA sequencing, mutations, restriction mapping, operons, DNA library, bioinformatics, electrophoresis
Prior skill, knowledge: Students must use serological and micropipets. Basic computer skills (word processing, copying and pasting data) are required.
Degree of difficulty: intermediate to advanced Students must use multiple pieces of evidence to explain how a DNA mutation changes gene expression in *E. coli.* Parts of this unit can be converted into separate one- to two-week inquiry exercises. Deleting most of the technical background information will produce a more advanced exercise.

Summary of Units (cont.)

Unit 4: Properties of Enzymes
Audience: nonmajors
Model and question: How do changes in environment, enzyme concentration, and substrate concentration affect activity of the enzyme beta-galactosidase?
Major concepts: reporter molecules, spectrophotometry, indirect measurements, enzyme activity
Prior skill, knowledge: Students must use serological pipets. Prior experience performing calculations and graphing data with a spreadsheet program is helpful but not essential.
Degree of difficulty: basic The unit emphasizes conceptual understanding instead of specific terms. Students can be challenged further by requiring them to devise the two concentration series on their own or to determine enzyme activities in unknowns. Advanced majors can be asked to create a Michaelis-Menton plot and calculate Km, Vmax for bGAL without specific guidelines.
Unit 5: Applied Enzymology
Audience: freshman or sophomore majors
Model and questions: How is phosphatase activity affected by enzyme or substrate concentration, pH, and inhibitors? How are those characteristics used to diagnose clinical conditions?
Major concepts: same as for Unit 4; also enzyme inhibitor, isoforms, standard curve
Prior skill, knowledge: Students must use serological and micropipets. Prior experience graphing data with a spreadsheet program is helpful but not essential.
Degree of difficulty: intermediate This unit examines enzyme properties in a way that quickly engages students interested in clinical science. Advanced students can be given a list of possible conditions but be required to locate—on their own—supplemental information about phosphatases in general, the diagnostic differences between the enzyme in each disease, or both.
Unit 6: Energetics and Photosynthesis
Audience: freshman or sophomore majors
Model and questions: How do isolated chloroplasts capture solar radiation, then use it to power chemical reactions? How can those reactions be disrupted?
Major concepts: chloroplasts, differential centrifugation, redox reactions, electron transfer, Hill reaction, environmental effects on photosynthesis
Prior skill, knowledge: Students must use serological and micropipets. Prior experience with spectrophotometers is helpful but not essential.
Degree of difficulty: intermediate For a richer inquiry activity, students can be asked to devise their own chloroplast extraction and quantitation protocols. The main assay can be used in conjunction with other units if students want to measure the effects of abiotic factors on photosynthesis.

Summary of Units (cont.)

Unit 7: Signal Transduction
Audience: sophomore majors
Model and questions: How do signals from the external environment trigger cellular changes in Chlamydomonas? What signaling pathways are required for these responses?
Major concepts: signaling pathways, agonists, antagonists, cell movement, phototaxis
Prior skill, knowledge: Students must use a compound microscope. Prior experience using a hemacytometer to count cells or particles is helpful but not essential.
Degree of difficulty: intermediate to advanced This unit is a less daunting introduction to a difficult topic than appears in some texts, but it should be reserved for majors with some experience. Advanced students should be able to devise their own data collection procedures and phototaxis assays. Several of the agonists and antagonists are hazardous chemicals; students should not be allowed to use them unsupervised.
Unit 8: Animal Hormones
Audience: freshman nonmajors and majors
Model and questions: How do hormones control growth and development of fruit flies? What happens if the hormonal balance is disturbed?
Major concepts: juvenile hormone, ecdysone, insect development, hypothesis testing, dose response curve, normalization.
Prior skill, knowledge: none. A spreadsheet program can speed up data analysis, but it is not essential.
Degree of difficulty: basic to intermediate This unit is a good introduction for majors to the scientific method; supplemental information is available in Unit 1. Advanced students could reasonably be expected to decide which hormones to test on their own, to devise protocols similar to those outlined, and to deduce that their data require normalization. If the hormones are replaced with insecticides or other compounds, the unit can be used as a toxicology assay with endless variations.
Unit 9: Neuromuscular Control
Audience: freshman nonmajors and majors
Model and question: How can one muscle, controlled by one nerve, produce different amounts of force, for different lengths of time?
Major concepts: action potential, myogram, twitch, motor unit, summation, ion channels.
Prior skill, knowledge: Basic computer skills (opening and closing files, copying data) are needed. Prior dissection experience is helpful.
Degree of difficulty: basic to intermediate Nonmajors tend to be more engaged if they are not restricted to mechanistic, hypothesis-driven experiments, and can ask, What does this chemical do? Students engage more if they know specific names of compounds they are testing (malathion insecticide, for example) rather than the general class (acetylcholinesterase inhibitors). Advanced students should be able to develop their own questions and find background information, but they still will need to be trained to use the data collection system.

Summary of Units (cont.)

Unit 10: Homeostasis and Heart Rate
Audience: nonmajors or freshman majors
Model and question: How do neurotransmitters and drugs affect the heart rate of blackworms?
Major concepts: physiological response, homeostasis, set point, normalization, baseline, neurotransmitter
Prior skill, knowledge: Students must use a compound microscope.
Degree of difficulty: basic Since many foods and medications are described as "bad for your heart," most students are readily engaged by this unit, and their experiments often are very creative. For a more challenging exercise, delete the explanation of normalization and encourage students to derive it. Other neuroactive compounds can be tested, and it is a robust model for students wanting to conduct eco-toxicology experiments.
Unit 11: Metabolism and Oxygen Consumption
Audience: freshman majors
Model and questions: Do changes in temperature or other environmental conditions affect oxygen consumption by crayfish? How is this different from endotherms?
Major concepts: oxygen consumption, specific metabolic rate, indirect measurements, ectothermy.
Prior skill, knowledge: none. Students can complete calculations more quickly with a spreadsheet program, but it is not required.
Degree of difficulty: basic. Advanced students should be able to perform the calculations without guide tables, but probably cannot design a full protocol without assistance. Another option is to teach students the procedures with crayfish, then explore further with a more charismatic aquatic organism such as tropical fish. The core procedure can be combined with Unit 13 to look at oxygen consumption by bettas during different behavioral displays.
Unit 12: Transpirational Control
Audience: nonmajors or freshman majors
Model and questions: Does salinization of soil slow down transpiration rate in mung bean seedlings? If so, at what salinity will transpiration stop?
Major concepts: transpiration, salinization, soil water potential, plant physiology, point counting
Prior skill, knowledge: none
Degree of difficulty: basic to intermediate. This unit is a good choice for those who are new to inquiry or who want to attempt a pure guided inquiry for the first time. Beans are inexpensive, and thousands of plants can be grown in a small space. There are few technical limitations; students can measure leaf size in many ways besides point counting and still will get interpretable results. For more advanced students, change the focus from salinity to look instead at the effects of various nutrient overloads or deficiencies on transpiration.

Summary of Units (cont.)

Unit 13: Animal Behavior
Audience: nonmajors or freshman majors
Model and questions: How do male and female bettas respond to each other's behaviors? What elements in a betta's behavior are important to other members of its species?
Major concepts: ethology, agonistic behavior, intraspecific behavior, interspecific behavior, stimulus filtering, habituation, minimum stimulus, rules of observation.
Prior skill, knowledge: none
Degree of difficulty: basic to intermediate This is another good "first" unit for instructors who are new to inquiry. Betta behavior is fascinating to watch and extremely robust. Most students have minimal observation skills (and little patience); this unit is designed specifically to develop those skills. Instructors can give students considerable latitude with experimental design in the second week. For a more advanced version, delete the description of basic behavioral elements and require students to discover them through direct observation.
Unit 14: Resource Allocation in Plants
Audience: freshman majors
Model and questions: Do different species of crop plants allocate resources in different patterns? Do environmental conditions affect that allocation pattern?
Major concepts: allocation, resource limitation, phenotypic plasticity, root:shoot ratio
Prior skill, knowledge: Students gain more from the unit if they have discussed energy and nutrient capture beforehand in lecture. The ability to use a spreadsheet program is useful.
Degree of difficulty: basic to intermediate This unit can be used with students at all levels. Novices will need the introductory material to guide them. However, more advanced students can find or deduce most of the introductory information, including use of the root:shoot ratio. Removing the data tables forces students to decide for themselves how to report and evaluate their results.
Unit 15: Population Ecology
Audience: nonmajors or freshman majors
Model and question: Do molds grow differently when plated in isolation versus when they are plated together? How does their environment affect the interactions?
Major concepts: population, abiotic factors, biotic factors, interspecific interactions, intraspecific interactions, carrying capacity, point counting.
Prior skill, knowledge: Some experience with basic sterile or microbiological technique is helpful, but not required. Students can ask more complex questions if they have been introduced to the logistic growth model in advance of the unit.
Degree of difficulty: basic Molds are part of students' daily experience, so most will quickly become engaged by this unit. Unless they have had a prior course in microbiology, students will need a detailed description of the plating procedure. For a more advanced challenge, delete the introductory material and data summary tables, require students to develop their own quantitation procedures, or ask students to fit their results to a logistic growth curve.

Summary of Units (cont.)

Unit 16: Measuring Biological Diversity
Audience: nonmajors or freshman majors
Model and questions: What is the difference in biological diversity between two different bodies of water? What can those differences tell us?
Major concepts: high- and low-quality habitat, stable and unstable habitat, diversity indices, richness, dominance, sampling bias
Prior skill, knowledge: Students must use both compound and dissecting microscopes.
Degree of difficulty: basic This is another extremely flexible unit. The basic procedure can be adapted to any habitat for which there is an established biotic sampling method. To provide more challenge, delete the introductory information, and require students to find their own resources for keying out the invertebrates. Most students will need some assistance when devising methods for estimating diversity; species richness is fairly intuitive, but they are unlikely to be able to derive the dominance model themselves.

UNIT 1

Designing Scientific Experiments

Student Pages*

BACKGROUND

In this unit, you will learn about designing sound, testable hypotheses and experiments. No doubt you memorized the steps of the scientific method long ago, and you probably think they represent how scientific research is done. Published scientific papers reinforce this view. Their questions are well thought out and stated precisely. The methods are clear, and unambiguous results flow in a logical progression leading to clearly stated conclusions.

We're going to let you in on a secret. A scientific paper is a good way to communicate the *results* of experiments, but it is a poor model of how scientific research *actually* gets done. No scientist is talented and intuitive enough to ask a perfect question, choose exactly the right methods, then get enough data from just one set of experiments to reach an unambiguous, publishable conclusion. In reality, scientists continuously refine the questions they ask. They try different methods for testing their hypothesis, and they perform multiple trials in order to get sufficient data for analysis. Once they obtain their data, they may find that some results conflict with others and require that the experiment be redesigned. At other times, the data may point in an entirely different direction than the original question that spawned the study. Technical problems may cause an experiment to fail miserably and produce no useful data at all. Even if everything goes right, the results of one set of experiments usually raise two questions for every one that they answer. In short, like everything else in life, scientific researchers have their share of false starts, surprising outcomes, and, occasionally, just plain bad luck. Only at the very end do the researchers have sufficient hindsight and understanding to package their data into the tidy summary we see in a scientific paper.

This does not mean it is impossible to get answers to scientific questions. Rather, our understanding of the world grows by an ongoing process. As our knowledge increases, old ideas get displaced, and the way we view the world changes. The biggest challenge scientists face is to look at *all* the data they collect, compare it with published observations by fellow scientists, then think about the data critically and creatively, and come up with the simplest explanation possible that accounts for all of the observations.

*Teacher Pages begin on page 103.

The Scientific Method Is Cyclic, Not Linear

Another misconception about science is that it has a clearly defined end point. In reality, each experimental result becomes a new observation that leads to new or revised hypotheses about how the world around us works. The process is summarized in Figure 1.1.

Figure 1.1

Cycle of scientific method

The scientific method is not a linear process, but cyclic. Each result becomes a new observation that leads to more experiments, and so on.

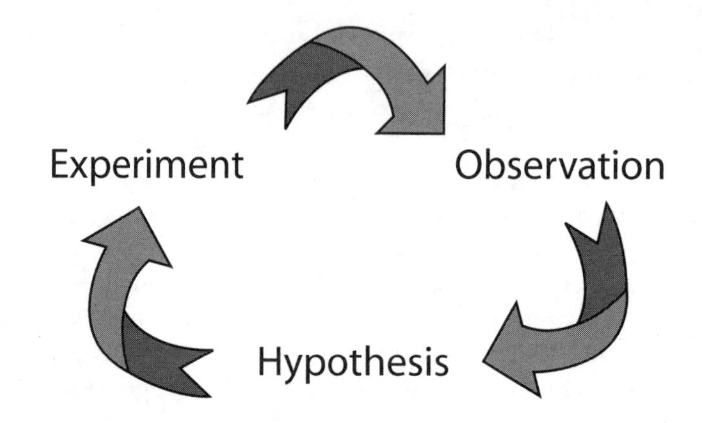

The typical steps are as follows:

1. Observe the natural world and ask questions about what you see.
2. Based on your observations (and the observations of others), formulate a testable hypothesis.
3. Design and conduct controlled experiments that test your hypothesis. Do not just design experiments to support your hypothesis. Design experiments that could prove your hypothesis is wrong. This reduces the chances that someone will come along and disprove your hypothesis for you (talk about embarrassing!).
4. Analyze the results of your experiments to determine whether your original hypothesis was supported. Look for alternative explanations for your observations (and do more experiments if necessary) until you are sure that your hypothesis is the *only* rational explanation for your initial observations and for the results of your experiments.
5. If your original hypothesis is supported, use that knowledge as the basis for a set of new observations that will generate additional hypotheses.

6. If your original hypothesis was not supported, combine your new observations with your original observations, and develop a new or modified hypothesis. Think of a new set of experiments that will test your modified hypothesis and produce new results that become yet another set of observations, and so on.

The following sections describe the first three of these steps in more detail.

Making Observations

There are many ways to hone observational skills; biologists do so by looking for patterns in the natural world. Everyone has casually observed animals and plants. However, you may not have looked at them in a systematic way before, or looked specifically for repeating or unusual patterns of behavior or physiological responses.

Most scientific hypotheses come from trying to understand these patterns. Other very important scientific discoveries have been made when investigators already knew the typical pattern of responses of their model systems but observed an unusual event that they did *not* expect to see.

The ability to make systematic observations is a valuable skill to cultivate, even if you do not plan to become a scientist. Financial auditors often must identify a single irregularity within thousands of numbers. A lawyer or forensic investigator must meticulously comb through evidence for repeating patterns or anomalies. Artists must pay close attention to the interplay of lighting, colors, and textures. All of these professionals rely on strong observational skills.

Formulating a Testable Hypothesis

A hypothesis is any potential explanation for observed phenomena. A **testable hypothesis** is one that is stated in a way that makes predictions that can be tested. Usually, a testable hypothesis can be worded as an *if/then* statement. The hypothesis is the "if" part, while the "then" part provides one or more predictions that are based on that hypothesis.

For example, it may seem obvious to us that nerves control muscle movements, but someone had to discover this fact. Egyptian embalmers knew the brain existed 3000 years ago, but prior to the 1760s, few believed it controlled muscle movements. Then in 1766, Albrecht von Haller made observations that led him to hypothesize that the brain controls muscle movements through nerves. His testable hypothesis might have been, "If the brain controls muscles via nerves (hypothesis), then cutting a nerve should prevent movement of a muscle (prediction)." Von Haller performed several experiments in which he severed nerves in living animals (and in several friends!) then observed the outcome. Over time he collected considerable evidence to support his hypothesis. He also observed little or nothing to indicate that his hypothesis was incorrect. Thus he concluded that his original hypothesis (nerves control muscle) was probably correct.

When hypotheses are tested, they can be shown to be incorrect (disproven). Yet the reverse is not true. Observations and data can support a hypothesis, but they can never *prove* that a hypothesis is true. That is the fundamental nature of science. Everything we have learned about the world around us is based on our observations. Scientists can state hypotheses that are based on those observations, then make predictions about what they would expect to see given their hypothesis. If enough observations support their hypothesis, eventually that hypothesis will become accepted as the most logical explanation for the data available to us. Subsequent replications by other scientists may further strengthen the support for a particular hypothesis; for example, von Haller's original hypothesis has been supported by more than two centuries of observations. So it is very likely that his explanation for how muscles are controlled is how it really happens. Yet even today an experiment could be reported that shows nerves cannot in fact control muscles. If that ever happened, scientists would have to conclude that von Haller's hypothesis was incorrect, and start looking for another explanation.

Designing Experiments to Test a Hypothesis

If an experiment is to provide meaningful data, it must be designed properly. Typically it includes a negative **control group** as well as one or more **experimental groups**. The negative control group is your baseline for comparison with the experimental group. Ideally the experimental group differs from the negative control group by a single factor. To give you an example, go back to the earlier hypothesis: "If the brain controls muscles via nerves (hypothesis), then cutting a nerve should prevent movement of a muscle (prediction)." You might set up two groups of frogs to test this hypothesis. The negative control group would contain three frogs that are anesthetized, have the sciatic nerves leading to their leg muscles exposed surgically, and then are sewn back up without any treatment. The experimental group would also contain three frogs that are anesthetized and have their sciatic nerve exposed. In this group though, the sciatic nerve would be cut before the leg incisions are sewn back up. When both groups have recovered, you could then test to see how far the three frogs in each group could jump when startled by a loud sound. If the three frogs in the control group jumped an average of four times as far as the three frogs with severed sciatic nerves, you might conclude that the leg muscles are probably controlled by the sciatic nerve.

In most experiments, you will manipulate one variable, and measure the effect on another. The manipulated variable is called the **independent variable**, while the response of the experimental group that you actually observe is the **dependent variable**. Going back to our frog experiment, the independent variable we control is whether or not the sciatic nerve is cut. The ability to jump in response to a loud noise is the dependent variable that we measure.

Replication: How Many Samples Do I Collect?
When designing an experiment, how many times do you need to repeat the experiment to be sure of your results? If you take too many samples you may needlessly waste time, resources, and money. On the other hand, if you take too few samples, it may be difficult to draw meaningful conclusions from your results. Anytime you must compare groups, try to conduct at least three replicate trials of your experiment. Then report the means for any dependent variables you measure.

Summarizing Raw Data
Raw data usually are not very meaningful to someone who is unfamiliar with the project. Data must be organized in a way that presents your significant findings clearly and concisely. The most common ways to summarize results are figures, graphs, and tables. You can create graphs and tables by hand, but a spreadsheet program allows you to create and revise figures more quickly. There are also statistical methods for summarizing and presenting data that are widely used and understood. Your instructor may ask you to complete a statistics tutorial as part of this unit.

Goals for This Unit
Like ants, bees, and other social (colony-forming) insects, termites exhibit some very complex behaviors. A few years ago, an entomologist spilled a few termites onto her desk, and noticed that some followed lines drawn in her lab notebook. In some subsequent trials, though, the termites seemed to ignore drawn trails, and just wandered aimlessly. This week your goal is to try to unravel this simple biological observation. As you work, focus on developing a reasonable hypothesis, testing it, and learning to interpret and communicate your results. You will use these skills extensively in later lab sessions.

General Safety Precautions
The termites (Class Insecta; Order Isoptera) you will be studying today are harmless. They do not bite or carry human disease. Be careful not to get wood splinters in your fingers or eyes when prying apart decayed wood to search for termites.

UNIT EXERCISES

Exercise 1: Do Termites Follow Trails?
In this first exercise, you will try to replicate the original observation that worker termites follow trails.

1. Plain white paper, pencils, and pens of various kinds should be at your lab bench. Take a single sheet of paper, and draw a "test trail" in the middle of the sheet with a pencil or pen.
2. Obtain a single termite from the colony in the container at the front of the lab. Termites are soft-bodied, so handle them gently. The easiest way to catch them is to let one crawl onto a dissecting needle, then carry it over your hand back to your lab station.
3. Place the termite near the trail. Watch it for five minutes, and see whether or not it tries to follow the trail. Record your overall impressions and observations in your lab notebook. If the termite crawls off or to the other side of the paper, use the dissecting needle or a brush to gently move it back to the center of the paper.
4. Repeat the above experiment, using the other side of the paper and a different pen or pencil. Observe the termite for five minutes and record your observations.

What did your termite do when offered trails to follow? Did its behavior the second time differ from the first time?

Based on your observations, do you think that termites will in fact follow a trail?

Ask your classmates about their termites' behavior. Based on their data and your data combined, do you think that termites will follow a trail?

Exercise 2: Generating a Hypothesis

Exercise 1 illustrated an essential step in the scientific method that is often forgotten. Before basing a hypothesis on someone else's observations, you should always try to replicate that person's results. At the least, you will have some baseline observations of your own from which to construct a more reasonable hypothesis. You might also discover that, in your hands, something entirely different occurs. For example, you might have observed that termites seemed to *avoid* drawn trails. If so, you might generate a hypothesis about *why* they avoid a trail, and your hypothesis would probably be very different than if you simply assumed the previous observer was right.

To generate your own hypothesis, look back at your preliminary observations from Exercise 1, and try to answer the following questions as well (record your answers to be turned in later).

Again ask others in your lab: Did every termite follow every trail offered?

What specific behaviors did you and your classmates use to conclude that a termite was or was not following a trail?

If every termite followed every trail, why do you think termites follow trails? If they only followed certain trails, what factor(s) do you think they use to select one trail to follow, but not another?

Based on your observations so far, think of a testable hypothesis for why termites follow trails. Remember, a testable hypothesis consists of a tentative explanation that accounts for your observations, followed by a prediction that can be tested experimentally.

What is your testable hypothesis?(Record your response before going on.)

Exercise 3: Designing Your Experiment

Background

Once you have developed a testable hypothesis, you must design an experiment to test it. What are the major characteristics of a well-designed experiment? The following example illustrates some basic principles to follow when designing experiments.

Bella and Lenny observed that their termite ignored a red pencil trail, but seemed to follow a purple pencil trail for a short distance (2–3 cm). The only difference they came up with was that the second trail was a mix of the colors red and blue. Since the termite did not follow red, they decided:

"Termites choose to follow blue trails. If termites prefer blue trails, then termites will follow a blue pencil line further and more often than trails drawn in other colors."

This is a reasonable, testable hypothesis, but how can Bella and Lenny test it? Furthermore, how can they design their experiment to produce quantifiable data, rather than just random observations?

Experiments Should Produce Quantifiable Data

Bella and Lenny came up with two different experiments. For the first experiment, they drew three lines, each 15 cm long. The first trail was drawn with red pencil, to serve as a negative control. The second test trail was drawn with purple ink. The third trail was drawn with blue highligher. For the experiment, the group placed a single termite at the beginning of the red trail five times and made a pencil mark at the point where the termite left the trail. They then measured the distance from the starting point to each mark and averaged the five replicates. Next, they repeated the experiment twice more with the same termite, except they used the purple and blue test trails. See Table 1.1, page 100, for their final results.

Table 1.1

Effect of trail color on termite trail following

	Average Distance Followed (cm; mean±s.d.)		
	Red	**Purple**	**Blue**
Termite #1, 3 replicates	0.8±0.2	1.4±0.1	5.5±1.2

For their second experiment, Bella and Lenny focused on how often their termite selected one trail versus another. They drew a diagram like the one in Figure 1.2.

Figure 1.2

Setup for trail color choice experiment

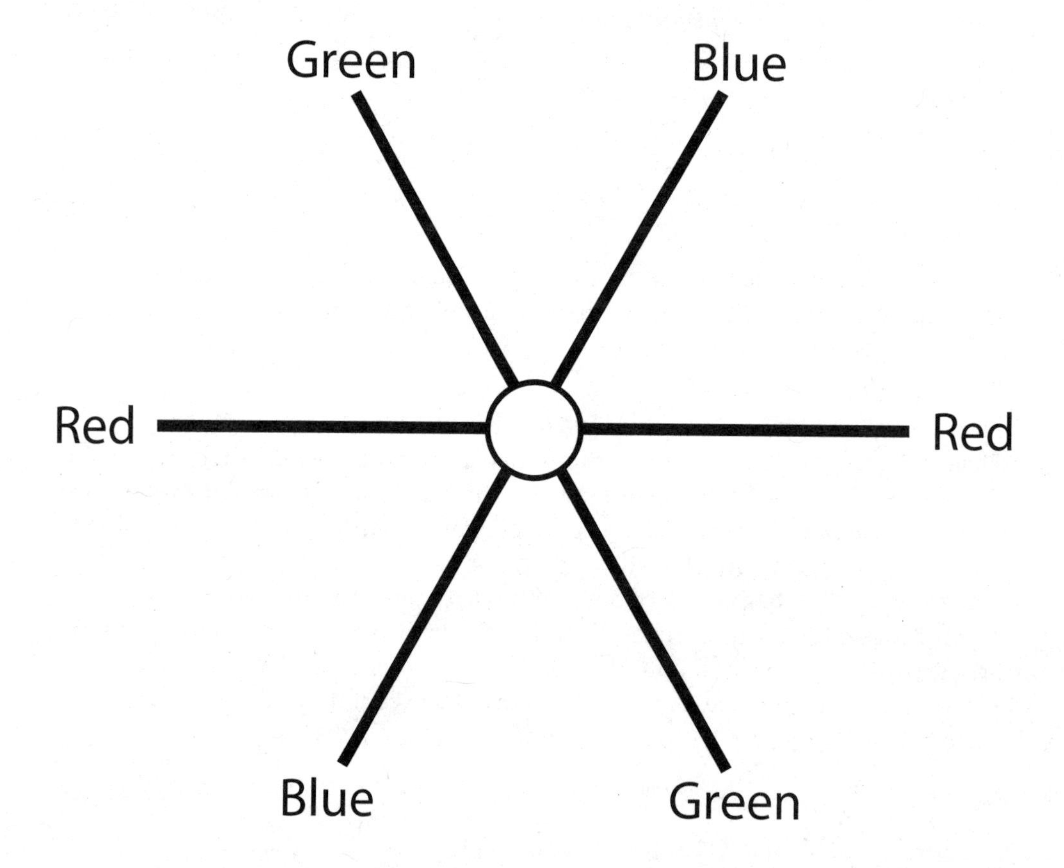

They placed the same termite into the central circle 10 times. If it crawled at least halfway out a trail, they put a check mark beside the end of that trail. After 10 trials, they tallied the scores for each color. Their summarized data are in Table 1.2.

Table 1.2

Effect of trail color on termite trail following

	Number of Times a Color Was Selected		
	Red	**Purple**	**Blue**
Termite #1	1	3	6

Based on their two experiments, Bella and Lenny concluded that their termite preferred blue trails and that their original hypothesis was correct. Or was it?

Replicate Measurements Improve Most Experiments

The experiments outlined above used a single termite. So how do Bella and Lenny know that all termites prefer blue trails? The best way to determine this is simply repeat the experiment with additional animals. Usually, data from more **replicates** give you greater confidence in your results. So Bella and Lenny repeated their first experiment using five termites instead of one, and obtained the results shown in Table 1.3.

Table 1.3

Effect of trail color on termite trail following

	Average Distance Followed (cm)		
	Red	**Purple**	**Blue**
Termite #1	0.8	1.4	5.5
Termite #2	0.9	1.8	6.0
Termite #3	0.2	1.1	6.2
Termite #4	2.3	1.3	6.5
Termite #5	0.8	0.8	5.6
Mean, s.d.	1.0±0.7	1.3±0.3	6.0±0.4

Bella and Lenny also repeated their second experiment (the one summarized in Table 1.2) using five termites instead of one. Each termite was tested 10 times. The data are summarized in Table 1.4, page 102.

Table 1.4

Effect of trail color on trail chosen by termite

	Number of Times a Color Was Selected		
	Red	**Purple**	**Blue**
Termite #1	1	3	6
Termite #2	2	0	8
Termite #3	1	0	9
Termite #4	0	3	7
Termite #5	1	1	8
Mean, s.d.	1.0 ± 0.6	1.4 ± 1.4	7.6 ± 1.0

Based on the summary data in Tables 1.3 and 1.4, Bella and Lenny were much more confident in concluding that termites follow blue trails.

Procedure

1. Working with your lab partners, design an experiment to test your hypothesis. Make sure that your experiment includes
   - quantifiable measurements,
   - replicate animals, and
   - a negative control of some kind.
2. Write a summary of your experimental design.
3. Have your instructor check your design. If he or she approves it, use the materials available in the lab to perform the experiment.
4. Record your results and observations in your lab notebook.
5. Once you have recorded your results, discuss them with your lab partner. What are reasonable conclusions you can make about the data? What are future experiments you might do? What did other lab groups discover? In the future, how could you extend or improve the experiment you just completed?
6. Again, write down your conclusions in your lab notebook.

Unit Summary Questions

Ask at least two other groups in your class what their hypotheses were and whether their observations supported or disproved the hypotheses. Did their observations support your general conclusions or were they contrary to your results? How do their results change your interpretation of your own data?

UNIT 1

Designing Scientific Experiments

Teacher Pages

INSTRUCTORS' NOTES

Background and Key Concepts

Termites use olfactory cues to identify nest mates. At least one compound used widely in ballpoint inks, phenoxyethanol, triggers their natural tracking behavior (Chen, Henderson, and Laine 1998). This compound is not in the formulations of most gel, rollerball, and felt-tip inks. As a result, termites will follow trails drawn with most ballpoint pens, but ignore trails drawn with pencil or other types of pens. The color of the ink is not important, but termites will not follow ink trails more than a few hours old. Once a termite has walked along an ink trail a few times, both it and other termites of the same species will follow that trail faithfully because subsequent termites mark and reinforce the trails.

At the start of the unit, groups are given two or three ink pens and colored pencils, but only one should be a "followed" ballpoint; ideally, the colors of the ballpoint-pen ink differ among work groups in a lab section. Each work group must formulate a testable hypothesis based on its observations of termite behavior. Students usually hypothesize that color or darkness/opacity of the line is the cue, despite the fact that termites live in darkness. When groups compare hypotheses, they often conflict with each other's observations. This lack of agreement forces them to rethink their initial hypotheses and, as time permits, to perform additional experiments.

Other major points of this unit:

- The process of scientific discovery is not linear, but cyclic. Data may have many possible interpretations, and results of a single experiment often spawn new questions and additional experiments.

- Replicates and proper controls are essential parts of any experiment.
- Numerical data may be either discrete or continuous. There are methods for summarizing and comparing both types of numerical data.

General Teaching Strategy and Common Problems

This unit is particularly useful for introducing nonmajors to inquiry. The exercises encourage students to ask new questions, build their critical-thinking skills, and raise their awareness of the natural world. Students should be able to complete the exercise in two hours.

The lab instructor *must* resist the urge to tell students the solution. Instead, encourage students to hypothesize and experiment further. Ask them to develop as many alternative hypotheses as possible; do not let them stop with just one. Let them share observations and hypotheses between groups.

Some students are intimidated by the discussion of quantification and replication. Let students collect some general observations first, then discuss how their random observations could be refined later in the session.

Students may complain, "This exercise is too simple/silly to teach me anything useful." Students' lab reports clearly show they do not know how to develop, state, and test a robust hypothesis. The instructor should point out that this exercise is a model for how students should work through more complex questions in later units.

Assessment

There is no factual content suitable for prelab quizzes in this unit. Once this unit has been completed, students should be able to critique a basic experimental design. On a postlab quiz the author often gives students a short scenario, then asks them to develop a reasonable hypothesis and explain how they could test the hypothesis.

Safety and Housekeeping

Termites pose no safety risks to students. If there is no queen, the termites cannot escape and colonize nearby structures.

Other Tips

- There is another variable that students might uncover as part of this exercise. Soldier termites ignore ink trails and focus solely on defense. Soldiers are less common than workers. They have larger heads and jaws, which often are dark reddish brown.
- If termites are not available, substitute small ants. They also follow ink trails, but are more active and easily startled, so they are more difficult to work with.
- This is a good unit for teaching students how to use a lab notebook properly.

Reference

Chen, J., G. Henderson, and R. A. Laine. 1998. Isolation and identification of 2-phenoxyethanol from a ballpoint ink pen as a trail-following substance of *Coptotermes formosanus* Shiraki and *Reticulotermes* sp. *Journal of Entomological Science* 33: 97–105.

PREPARATORY NOTES

Quantities listed are for a lab section of 20 students working in pairs.

Shared Materials

- White paper
- Termites, ~100/lab section of 20 students
- Extra ink pens of different colors

Materials at Each Work Station

- 1 Bic brand ballpoint ink pen
- 1–2 gel, rollerball, or felt-tip pens or colored pencils
- Dissecting needle or blunt probe
- Ruler

Sources of Materials

Termites may be purchased from a biological supply company or collected locally. Most North American termites maintain a nest in soil where eggs and reproductive adults are found. Soldiers guard the nest, while workers forage in adjacent wood and bring food back to the colony.

Look for fallen pines or hardwoods in direct contact with the soil, preferably in a shady or damp area. Use a pick or pry bar to turn and split logs. Pale to white workers congregate at the boundary between softer, semi-rotten wood and drier, harder wood. Transfer workers with pieces of rotting wood to a plastic storage box. One brick-sized piece of wood may contain 200 or more workers. Add a paper towel moistened with spring or distilled water to the box to maintain humidity.

In the classroom, keep the towel moist, and store the container covered and in the dark. Under proper conditions, workers survive for one to two weeks. If students cannot find workers on the surface of the wood, split larger pieces to expose more tunnels. At the end of the lab, place the wood remnants in the freezer overnight to kill any remaining termites.

UNIT 2

Mendelian Genetics

Student Pages*

BACKGROUND

In 1866, Gregor Mendel published a model that explained how seven characteristics are inherited in pea plants. His model established three laws: (1) the law of discrete inheritance, (2) the law of segregation, and (3) the law of independent assortment. According to Mendel, the seven physical features of pea plants are each controlled by two units of inheritance, which are passed from parental plants to offspring. One unit comes from each parent, and these units can be either dominant or recessive to each other. His law of segregation states that the two units of inheritance separate from one another when a pea plant produces offspring (seed). Each offspring inherits one of each parental plant's two possible units of inheritance. Finally, Mendel stated that the units of inheritance for the traits he studied assort independently. For example, a trait like seed shape passes to offspring independently of another trait like flower color. Using his simple rules, he could predict both what types of offspring would develop when he crossed any two pea plants and the relative numbers of each type. What made Mendel's rules so exciting to the scientific community was that they predicted inheritance patterns for many other eukaryotic organisms as well.

*Teacher Pages begin on page 123.

More than 30 years later, Walter Sutton finally described a physical mechanism that explained Mendel's observations. Sutton demonstrated that most cells are **diploid**, that is, they have two **homologous** chromosomes. Homologous chromosomes have the same overall shape and size, and genetic information is usually arranged in about the same order along the DNA strands. Sutton found that prior to mitosis (normal cell division), cells duplicate every chromosome. As part of mitosis, a copy of every chromosome passes to each of the two diploid daughter cells. The daughter cells still have one copy of each chromosome in the homologous pair. In preparation for reproduction, plant or animal cells undergo a distinctive form of cell division called **meiosis**. In the first stage of meiosis, the two homologous chromosomes separate, and only one goes into each of two daughter cells. The maternally and paternally derived member of a pair of homologous chromosomes can go randomly to either of the two daughter cells during meiosis. Sutton had uncovered a physical process that explained Mendel's law of segregation and independent assortment.

Chromosome Structure and Inheritance

Eukaryotic organisms have two types of chromosomes: (1) autosomes, which are inherited independently of the sex of the parent or offspring, and (2) sex chromosomes, which are inherited in a specific pattern from each parent and determine the sex of the offspring.

Genetic information is not scattered randomly along the strands of DNA that make up chromosomes. The DNA code for a specific enzyme, structural protein, or RNA is found at a discrete physical location on the strand, called a **locus** (plural is **loci**). Figure 2.1 shows the double-ended arrow pointing to a specific locus.

Figure 2.1

Basic genetic terminology

The left side of each panel shows the chromosomes before DNA replication in S-phase, the right side, after replication

Panel A shows two different, non-homologous chromosomes. They have different physical structure and banding patterns, and do not encode the same information.

Panel B shows two homologous chromosomes. They have the same general structure, and encode the same general information. The specific version of information (i.e., allele) at each position (locus) may be different, but the loci will be in the same positions on the two chromosomes.

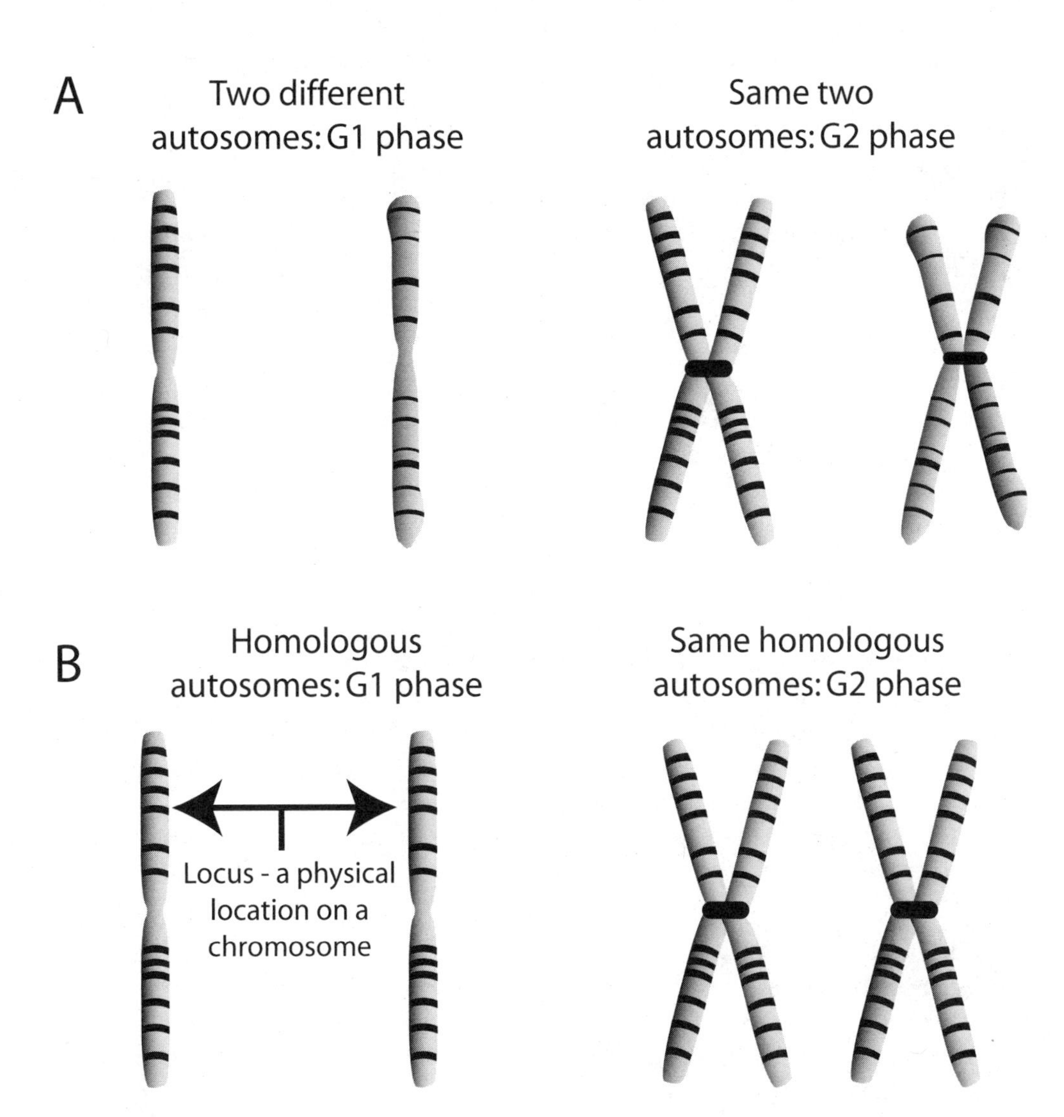

In a normal homologous pair of autosomes, a particular locus will be at the same physical location on both of the chromosomes, regardless of which parent contributed the autosome. Yet the autosomes in a homologous pair may not be exactly alike. At any particular locus, there can be slightly different versions of the DNA code, called **alleles**. Each allele encodes the same general protein. However, the protein produced when each allele is translated may be slightly different.

Going back to the earlier example, imagine the locus shown in Figure 2.1 codes for the enzyme "greenase," which catalyzes production of a green eye pigment. The most common allele (version) of that locus codes for a very active form of greenase. An organism that is homozygous for this most common allele makes lots of pigment and so has dark green eyes. A second, less common allele for that locus codes for a version of greenase that makes half as much pigment. As a result, an organism that is homozygous for the second allele probably will have light green instead of dark green eyes. A third allele may contain an early stop codon, so that no greenase is produced and no green eye pigment is made. The result: Organisms that are homozygous for this third allele have pale eyes that are not green at all.

Any given locus can have two, three, or dozens of alleles. However, an individual organism can only have two alleles for each locus: the one on the chromosome inherited from the female parent and the one inherited from the male parent. Which two alleles were inherited from the parents at that locus is what determines the eye color of the offspring.

Alleles Behave Differently on Sex Chromosomes

In diploid organisms, the sex chromosomes do not always form a homologous pair. If an organism has two copies of the X chromosome (which is the larger sex chromosome), it is usually a female. However, if it has one X chromosome and one smaller, Y chromosome, the organism is usually male. In some cases (like bees), there is no Y chromosome, and the male organisms carry just one X chromosome.

Like autosomes, X and Y sex chromosomes have loci that control certain functions, and there may be two or more alleles at each locus. However, the rules for expression of dominant/recessive phenotypes are different for sex chromosomes. Imagine there is a recessive allele for a locus on an X chromosome. A female organism (which has two X chromosomes) must inherit two copies of that recessive allele before it will express the trait. A male organism does not have a second X chromosome to compensate for any recessive alleles. Therefore, male organisms express nearly all recessive alleles they inherit that are on the X chromosome.

Goals for This Unit

Mendel, Sutton, and many others relied on experimental crosses between organisms to uncover the principles of inheritance. Today, experimental crosses still are an essential tool for genetics research. In the clinic, physicians and genetic counselors must understand the principles of inheritance before they can accurately advise patients who have a family history of genetic disease.

To begin this unit, you will complete two exercises that teach you how to handle and sort fruit flies (Drosophila) and how to recognize wild-type and mutant strains. Once you master these core skills you will be given two sets of vials containing separate stocks of wild-type and mutant strains of flies. The mutant strains are homozygous for one or more alleles and are true-breeding. However, not everyone in the lab will be assigned flies of the same genotype.

Over the next several weeks, you must accumulate sufficient data to answer these questions about your assigned mutant:

What is the mutant phenotype of my assigned strain of flies?

Is this mutant phenotype caused by a dominant or by a recessive gene? An autosomal or a sex-linked gene?

Is this mutant phenotype caused by one gene or by interactions among two or more genes?

If more than one gene is involved, does the inheritance pattern suggest they are on the same chromosome or on separate chromosomes?

Answering these questions will require you to perform several test crosses and tabulate the phenotypes of offspring through at least the F2 generation. It is your responsibility to decide on crosses, set them up, and monitor their progress. To assist you, the General Procedures section on page 116 describes all the methods you will need to complete this project.

General Safety Precautions

Drosophila, their media, and the carbon dioxide tablets used for anesthesia are harmless. Escaped flies can become a nuisance if they are not kept under control. Make sure all vials and containers are plugged well to prevent escapes, especially before disposal. Destroy any vials of flies that you will not use again. To eliminate food sources for escapees, clean your workbench and the shared areas thoroughly each time you finish working with flies.

UNIT EXERCISES

Exercise 1: Sorting Flies by Sex

Background

The major differences between male and female flies are shown in Figure 2.2.

Figure 2.2

Sexually dimorphic features used to distinguish male and female Drosophila

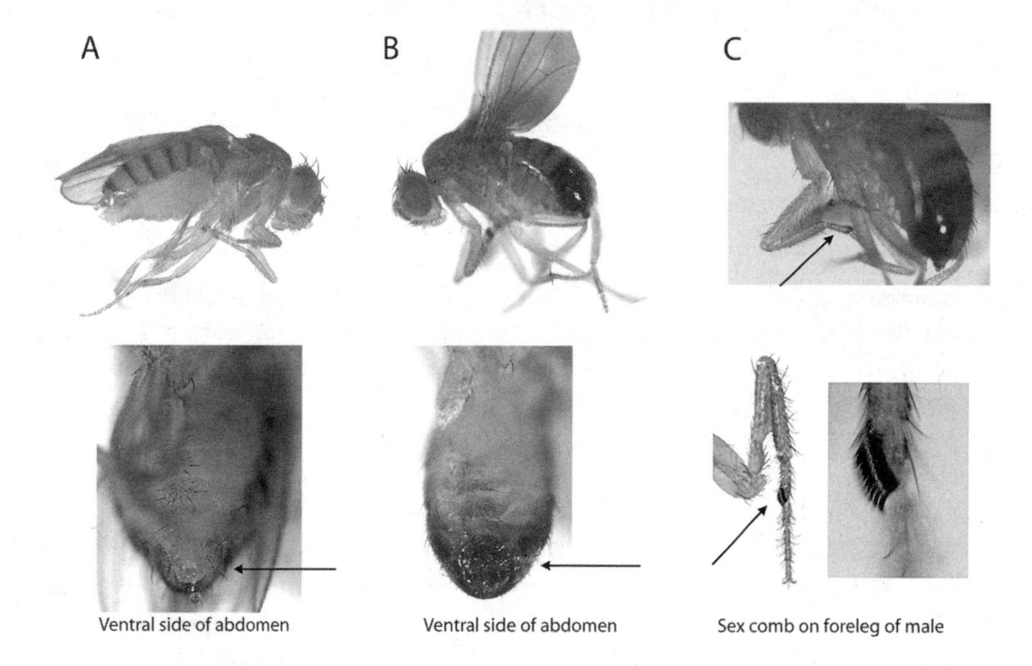

- Males have a dark sex comb on each of their front legs. This is the most reliable marker to use.
- On the posterior, ventral side of the abdomen the male has two claspers; females lack claspers.
- The male's abdomen tends to be more rounded and blunt with darker markings at the tip, especially on the ventral surface.

If you are not certain if you see sex combs, the other features should allow you to consistently separate males from females. If you are ever in doubt though, you should kill a fly of indeterminate sex and place it into the fly morgue, rather than risking adding a male to a vial of virgins you plan to use for test crosses.

Procedure

1. Take a vial of flies labeled "Practice: Wild Type" back to your desk. These flies are still awake, so you will need to anesthetize them. Instructions are given in General Procedures (p. 116) and your instructor will demonstrate this procedure for you.
2. Empty the vial of flies onto a white index card attached to a cold block, then place the block on the stage of a dissecting microscope equipped with an overhead light. Using a soft brush, gently push and turn flies on the card.
3. Look at several flies until you are certain that you can identify each of the three dimorphic features previously listed. Sort out six flies that you are certain are male and six more that are female. Then ask your instructor to see whether you have sorted them accurately. If you made a mistake, go back and double-check the features on each one.
4. After you have successfully sexed your first group of 12 flies, sort the remainder of the flies in the practice vial into three groups: males, females, and undetermined. Record your counts in Table 2.1.

Table 2.1

Sex ratios of sorted flies

	# Males	# Females	# Undetermined
Your count			
Partner's count			

5. Have your partner double-check your sorted flies. If your partner doesn't agree with you the way you have sorted the flies, ask your instructor to check your work.
6. When you are finished, return all of the wild-type flies to the practice vial so others may use them.

Exercise 2: Identifying Mutant Phenotypes

Procedure

1. Take one of the vials labeled "Practice: Mutants" back to your bench. This vial contains flies with several different mutant phenotypes. Anesthetize them as before.
2. Using the microscope, look for differences between the mutants and wild-type flies in any of the following features: eye shape, eye color, body color, wing size or shape, or body bristles.

3. Each time you think you have identified a distinct mutant phenotype, give it a descriptive name and write it down in Table 2.2.
4. As you work, you will find other flies with the same mutant phenotype. Sort your flies into phenotypes, then tally up the number and sex for each phenotype you find, and write them down in Table 2.2.
5. When you are finished, have your instructor double-check that you have sorted the flies correctly.

Table 2.2

Summary of the fly mutations in your mixed stock

Description of Mutant Phenotype	# Males	# Females	# Undetermined

Exercise 3: Determining the Inheritance Patterns of a Drosophila Mutant

Procedure

1. Your instructor will give you four vials of flies.

    - Two vials are marked "+" and contain normal, wild-type flies. You can assume that their genotype is "+/+" at every locus.
    - Two vials will be marked "Mutant" and have some other designation letter or number. All the flies in your mutant vials should have the same phenotype; flies with different phenotypes were not mixed.

2. Identify the mutant phenotype of your assigned flies. Choose a descriptive term for any mutation you see (e.g., eyeless or purple-striped).

3. Record which mutant you were assigned and your description of it in your lab notebook. Check with the instructor to confirm that you have correctly identified its mutant phenotype.

From here on, you must care for your own flies. Written procedures for setting up cultures, anesthetizing flies, collecting virgins, and other procedures are provided. However, more specific instructions are not possible. Your instructor can guide you, but it is your responsibility to decide which test crosses you need to identify the inheritance pattern of the mutation(s) in your flies.

Hints for Planning Out Test Crosses

Your instructor will show you how to use a Punnet square and the known genotypes of two parents to predict the genotypes (and phenotype ratios) of their offspring. If you need additional assistance, there is a tutorial at the end of the General Procedures section.

In a test cross, virgin flies of an unknown genotype are mated with flies of a known genotype. The first and second generation of offspring are counted and sorted, and the phenotype ratios calculated. Given the known genotype of one parent and phenotype ratios observed in the offspring, you can determine the genotype of the unknown parent.

1. Start by drawing a Punnet square for one possible cross and inheritance pattern (e.g., flies that are homozygous for a single autosomal recessive mutant crossed to wild-type flies). Calculate and record the expected normal and mutant phenotype ratios for the F1 and F2 generations.
2. Do the same for other possible patterns of inheritance.
3. Based on your Punnet squares, determine which crosses you can use to differentiate between various patterns of inheritance. Record them in your notebook, along with your rationale. If you will need to perform more than one test cross, explain why.
4. When you have determined which test crosses allow you to distinguish the various inheritance patterns, you and your partner should set up those test crosses using your assigned mutant flies. Set them up as soon as possible.

    - Determine the number of males and virgin females you will need of your wild-type and assigned mutant flies.
    - Look for pupae in your vials, and try to estimate when you can begin to collect virgins (it should be within five to seven days).
    - Once flies begin to emerge, you will need to come to the lab every six to eight hours. Make up a schedule of the times outside of the normal lab meetings when you or your partner will come in to collect virgins.

Keep Up With Your Notebook Entries

Your notebook is a running record of everything you do for lab. It should contain every bit of the data you collect. When your instructor looks through your notebook, there should be entries for every time you came to the lab to work. Your notebook should contain all of your daily observations, a copy of any data your partner(s) collected, detailed descriptions and rationale for all experimental crosses you do, the Punnet squares you used to calculate genotype and phenotype ratios, the raw numbers from when you count flies, and other data you collect. In short, based on what you write, the instructor should be able to reconstruct every last thing you did, why you did it, and what the outcome was.

General Procedures

Routine Care

Once you receive your vials of flies, you must maintain your own working stocks. Working stocks are essential; if your first test crosses fail, you can start again using your working stocks.

1. Every week, transfer 10–15 flies of each type to new vials with fresh food. You do not even need to anesthetize flies to transfer them to new vials.
2. Keep older working stocks until adults start emerging from newer working stock vials, then dispose of the older stocks.

Anesthetizing Flies

1. Transfer flies to be anesthetized into an empty plastic vial. It requires some skill to transfer flies between vials, then plug both vials. Your instructor will demonstrate the proper technique. Some flies may escape initially, but you will become better at doing this with practice.
2. Place the vial of flies to be anesthetized into a vial holder or tube rack.
3. Get a cold pack from the freezer. Secure a white index card onto the cold pack with rubber bands.
4. Take an anesthetizer, and add 10 mL of water to the vial. Place the vial next to your vial of flies in the holder.
5. Carefully slide the end of the anesthetizer tube between the foam plug and the wall of the vial containing your flies. Do not remove the foam plug!
6. When you are ready to anesthetize flies, add half of one CO_2 tablet to the water in the anesthetizer vial. Immediately plug it with the stopper, so the gas produced is directed into the vial of flies. Do not allow liquid to bubble up and push through the tubing into the vial of flies; they will drown.
7. As the CO_2 is being administered, gently tap the vial of flies on the counter so they drop to the bottom of the vial. As soon as flies are immobile, pour them onto the white index card attached to the cold pack.

8. Place a paper towel on the viewing stage of a stereomicroscope to catch condensation. Place the cold pack on top, then use a benchtop lamp to illuminate the flies from above.
9. Once chilled, you can work with the flies for up to one hour. Use a soft brush to move them around.
10. If you need to keep flies after observing or sorting them, carefully brush them from the card into a fresh vial containing food. Immediately plug the vial, and lay it on its side to keep the flies from getting stuck to the food. You can stand the vials upright when the flies revive after one to two minutes. Dump unneeded flies into the "fly morgue" on your bench.

Preparing Food Vials

1. Determine the number of vials you need for the day. Do not make up a large number of extra vials of media. Without flies present, they become overrun with bacteria.
2. Label each plastic vial with your initials, lab section, date, and contents.
3. Working at the sink, put 10 mL of dehydrated fly media in the bottom of each tube. Add 10 mL of spring water or distilled water to each vial, and immediately swirl it gently to mix the media and water.
4. Once the media solidifies (less than five minutes), add one grain of dry yeast to each vial. WARNING: If you add too much yeast to a vial, your flies can die of carbon dioxide poisoning.
5. Place a plug in the top of each tube, with about half of the plug sticking out of the vial.
6. Clean up all media or yeast you spilled and sanitize the counter with ethanol.

Disposing of Used Fly Vials

Flies are grown in inexpensive plastic shell vials. The vials cannot be adequately cleaned for reuse, so do not try to recycle them.

1. To destroy a culture, push the foam plug down so that it is flush with the top of the vial. DO NOT push plugs all the way down into the food!
2. Place vials in the box marked "Discarded Vials." They will be frozen to kill the flies prior to disposal.

Controlling Bacteria

If there are not enough larvae present, bacteria can overrun the media. Heavy infections usually kill the flies. To control bacterial growth:

1. Do NOT open any fly vials you find containing a large amount of bacteria, which appear as a pale to tan slimy film on the surface of the food. Push the stopper down in the vial and place the vial into the waste box for disposal.

2. Do not keep old stocks around. Once you are finished with a cross or a stock, put it in the waste box.
3. When you are making up vials of fly food, make sure to use clean vials and stoppers. Do not reuse vials and stoppers.
4. Do not put more than one or two grains of yeast into a vial of media. Excess yeast provides nutrients that encourage bacterial growth.
5. If a vial has only a small amount of bacteria, it can be treated with antibiotics. Using a disposable transfer pipet, add three drops of 100x antibiotic solution to the surface of the culture media. After three days, add another three drops. It is not necessary to unplug a vial to add antibiotics. Simply slide the transfer pipet between the vial wall and foam plug.
6. If the bacteria continue to spread, do not try to rescue the flies. Instead, destroy the vial and reset the cross or stock.

Collecting Female Virgin Flies

The general life cycle of fruit flies is shown in Figure 2.3.

Figure 2.3

Life cycle of Drosophila

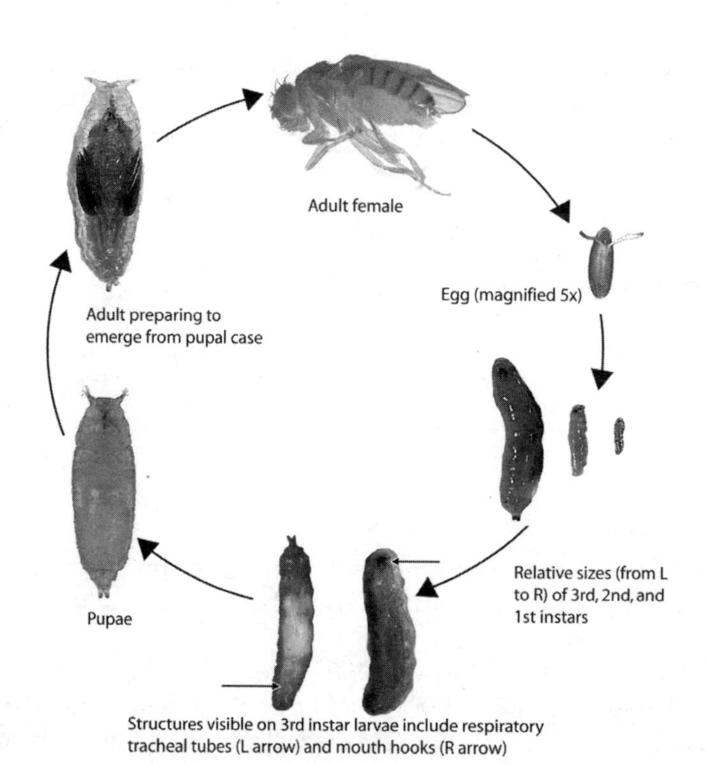

Female flies mate for the first time 10–12 hours after emerging from their pupal cases. They mate with more than one male, and store sperm for fertilization later. The only way you can ensure that you know which male they mate with is to use virgin females for test crosses.

You obtain virgin female flies by removing all of the adult flies from a culture, waiting six to eight hours, then collecting and separating the newly emerged females from the males. To collect virgins for crosses, you will need one or more vials of flies that are 9–11 days old. Look for dark pupae on the sides of the vial; these are flies getting ready to emerge.

It can take several days to collect enough virgins for crosses, and you will need to return to the lab every six hours or so. Divide the workload equally between you and your partner, decide in advance who is collecting when, and exchange e-mail addresses and telephone numbers, in case you must get in touch quickly.

1. Remove all adults from the vials from which you plan to collect virgins. If live flies are stuck to the food, push them down into the food with a probe or pencil.
2. Return the vials to the incubator for six to eight hours.
3. After six to eight hours (*no longer*), collect all the newly emerged adult flies. Anesthetize them with carbon dioxide, and separate the sexes into two different vials. The females will be virgins that can be used for crosses. The males can be used in an appropriate cross or discarded in the fly morgue.
4. If you did not obtain enough virgins during the first collection, put the stock vials back in the incubator. Return every six to eight hours; each time, collect and sort the adults by sex.

Setting Up a Test Cross Vial

Usually you will need two to three vials for each test cross to produce sufficient offspring for counting. If you do not have enough offspring, your phenotype ratios may be skewed.

1. Collect five to seven female flies (use virgins if necessary), and three to five males of the appropriate phenotype. These will be the parental, or P1, generation. Label the vial with the genotypes and sexes of the parents, your initials, and the date.
2. Add the flies, and plug the vial. When the flies wake up, place the vial in your rack in the incubator.
3. After four to five days, place the entire vial on its side under a stereomicroscope. Focus on the food near the wall of the vial. You should see small larvae crawling around within the food.
4. Once larvae are present, remove the P1 parents from the vial. Transfer them to a fresh vial of food (you do not need to anesthetize them), and label it carefully. This is a backup vial for the P1 mating. After four to five

days in the second vial, larvae should be present in the backup vial too. At this point, discard the adult flies in the morgue.

5. Twelve to fourteen days after you first added adults, the first generation of offspring (the F1 generation) will begin emerging from pupal cases along the side of the vial. Anesthetize them and separate them by phenotypes under the stereomicroscope. Record the phenotypes and sexes of all the flies. Keeping counting flies in the F1 generation until you have scored 100–200 flies.

   Frequently you will need to cross siblings from the F1 generation to produce an F2 generation. This is known as an **inter-se cross** (brother-sister mating). For inter-se crosses, the females need not be virgins.
6. After sorting, and while they are still anesthetized, transfer five to seven female and three to five male F1 flies to a fresh, labeled vial of food. Some of the females will already have mated and will begin laying eggs shortly after being transferred to the new vial.
7. After four to five days, transfer the F1 flies to a fresh vial of food, to make a backup of the F1 cross. After another four to five days, remove the adults and discard them.
8. About 12–14 days after you added the F1 adults, F2 progeny will begin to emerge from pupal cases. Anesthetize them, separate the flies by sex and phenotypes, and count them. Record the phenotypes and numbers in your lab notebook.
9. Continue to sort and count emerging F2 flies for several more days. You should count as many flies as you can (ideally, 200 or more). Do not count vials beyond six days, because early F3 flies may begin to emerge that will confound your results. If necessary, count the adults that emerge in your backup vials.
10. Once you have counted your F2 flies, dispose of them in the fly morgue. Do not put flies that have already been counted back into their vial.
11. Use a chi-square test to determine whether your observed results are significantly different from the expected phenotype ratios for a particular pattern of inheritance.

Using Punnet Squares to Determine Expected Genotype and Phenotype Ratios

To determine whether the mutation(s) in your unknown flies follow a particular inheritance pattern, you first must know what the potential patterns could be. You should already know that when wild-type flies and flies with a recessive mutation are crossed, the mutation may disappear from the F1 generation. If F1 siblings are mated though, the mutation usually reappears in some of their offspring. The precise ratio of normal and mutant phenotypes can tell you if the mutation is dominant or recessive, sex-linked or autosomal, single or double, or some other type of inheritance pattern.

For example, wild-type Drosophila have brick red eyes and flat wings that extend beyond the tip of the abdomen. There is a recessive mutation in a locus on Chromosome #2 called *apterous*; homozygous recessive flies never develop wings. Another recessive mutation called *eyeless* is found on Chromosome #4; homozygous mutants never develop compound eyes.

Suppose an apterous fly was crossed with an eyeless fly and, after two weeks, their F1 offspring were crossed. What would be the predicted outcome? Using a Punnet square (see Table 2.3 and Table 2.4), the crosses break down as follows:

First cross: EEaa (normal eye, apterous) × eeAA (eyeless, normal wing)

Table 2.3

Punnet square for the first cross

		Potential Alleles From *Apterous*	
		Ea	**Ea**
Potential alleles from *Eyeless*	eA	EeAa	EeAa
	eA	EeAa	EeAa

F1 inter-se cross: EaAa (normal eye and wing) × EeAa

Table 2.4

Punnet square for the F1 inter-se cross

		Potential Alleles From First Parent			
		EA	**Ea**	**eA**	**ea**
Potential alleles from second parent	EA	EEAA	EEAa	EeAA	EeAa
	Ea	EEAa	EEaa	EeAa	Eeaa
	eA	EeAA	EeAa	eeAA	eeAa
	ea	EeAa	Eeaa	eeAa	eeaa

Now assuming that the alleles E and A are fully dominant to the recessive e and a alleles, there are 16 flies with nine possible genotypes, that lead to four possible phenotypes (Table 2.5, p. 122).

Table 2.5

Genotypes and phenotypes of the F2 generation

Genotype	# Flies w/ Genotype	Description of Phenotype	# Flies w/ Phenotype
EEAA	1	Normal eyes, normal wings	9
EEAa	2		
EeAA	2		
EeAa	4		
EEaa	1	Normal eyes, wings absent	3
Eeaa	2		
eeAA	1	Eyes absent, wings normal	3
eeAa	2		
eeaa	1	Both eyes and wings absent	1

The expected **phenotypic ratio** would be 9:3:3:1. This means that, on average, for every 16 flies in the F2 generation there should be 9 normal flies, 3 flies without eyes, 3 flies without wings, and 1 fly with both eyes and wings missing.

You can use phenotype ratios to work backwards as well. For example, if you count flies in a test cross and find there are four phenotypes that occur in a 9:3:3:1 ratio, it is very likely that the phenotypes are the result of two mutations on two different autosomes. If you generate Punnet squares for each of the other possible mutations (such as autosomal, sex-linked, and so on) that can occur, you can calculate phenotype ratios for each of them as well. If you subsequently see a particular ratio of phenotypes in a test cross, you will know the most likely mutation(s) to have caused that phenotype ratio.

Frequently the observed phenotype ratios in a test cross do not exactly match the expected phenotype ratios. For a detailed explanation of how to perform a chi-square analysis that compares observed and expected phenotype ratios, consult a statistics textbook or online source.

UNIT 2 Mendelian Genetics

Teacher Pages

INSTRUCTORS' NOTES

Background and Key Concepts

In a typical Mendelian genetics lab, students cross flies or other organisms of known genotypes, score phenotypes of the offspring, and determine if their results are significantly different from expected phenotype ratios. For this unit, the traditional dihybrid cross lab has been reformatted into an inquiry-based exercise. Students receive two vials of wild-type (Canton S or Oregon R) flies and two vials of mutant flies, but they are *not* told the genotype of assigned mutants. During the first week they must identify the mutation(s) present by comparing the adults in the two vials. Subsequently students must cross mutant flies to wild-type flies and, based on phenotype ratios in the F1 and F2 generations, determine the pattern of inheritance and most likely genotype of the original mutants. Students decide for themselves what crosses must be done to uncover the pattern of inheritance.

Other skills that students learn during this unit are

- Drosophila culture and care,
- how to sort flies by sex and identify mutant phenotypes, and
- how to use Punnet squares to predict genotype and phenotype ratios.

For unknowns, there are four different strains of white-eyed flies. Eye color in Drosophila depends on two pigments—one bright orange and the other dark brown. In wild-type flies, these two pigments are present in about equal concentrations, making the eyes brick red. There also is a central point where ommatidia are much darker than the surrounding ones. A mutation that inactivates any enzyme in the brown pigment pathway causes flies to have bright orange or scarlet eyes. Usually they also lack the dark central spot, a feature that helps in distinguishing older orange-eyed flies from young wild-type flies. Conversely, a mutation that inactivates any enzyme in the orange pigment pathway causes the eyes to be dark brown.

A white-eyed phenotype can occur for several reasons. There is an X-linked, single allele mutation that inactivates the ABC transporter that carries both pigments to their final destination in the ommatidia; as a result, the eyes are white. White-eyed flies also may be homozygous recessive for inactivating autosomal mutations in *both* pigment paths. Other fly strains have an autosomal mutation in one pigment pathway and the X-linked mutation in the ABC transporter. When students cross their particular strain of white-eyed mutants to wild-type flies, different ratios of orange-, brown-, white-, and brick red-eyed progeny will emerge in the F1 and F2 generations. Based on the phenotypic ratios, students can deduce the pattern of inheritance and, from this, the most likely genotype of their original parental strain of white-eyed mutant flies.

General Teaching Strategy and Common Problems

The author's program uses this unit as a self-paced, half-semester lab project in a genetics course for sophomore majors. The open format works best because collecting virgins, transferring adults, setting crosses, and backing up stocks invariably needs to be done at times other than when lab meetings are normally scheduled. At the first lab regular meeting students complete Exercises 1 and 2, then are given their stock vials of white-eyed mutant and wild-type flies. Subsequently students work mostly on their own. They are responsible for completing Exercise 3 and maintaining their stocks. Students are required to come to lab weekly for 30 minutes so the instructor can check their progress and notebooks. Students who are on track to solve the problem may leave or work independently, while those who are not making progress spend additional time with the instructor to get back on track. Alternatively, the students and instructor can use part of the lab meeting time to solve and discuss genetics word problems.

All of the eye color mutants can be identified using the same two crosses. In the parental generation, students should mate wild-type males to virgin mutant females, and mutant males to wild-type virgin females. If students score the phenotypes of the F1 progeny, then cross siblings to each other, they should have sufficient information to determine the inheritance patterns for any mutant strain. However, the instructor should NOT tell students what crosses to do. The ultimate success of this unit depends on the students selecting the required crosses themselves.

Students respond in two very different ways to this unit. One group will claim that "we did this in high school," decide the phenotype is the result of a single, sex-linked mutation, and plan their crosses accordingly. When these students count offspring, there will be two entirely new phenotypes, which is inconsistent with their starting idea that the phenotype was the result of a single mutant locus. Use their confusion as an opportunity to explain how a single phenotype can result from many alternative mechanisms. Their task is to determine which of all these options is actually operational in *their* mutant line.

The unknowns overwhelm other students because they think there are endless possibilities. Tell students to assume for a moment that eye color is due to a single autosomal mutation. Ask them, What phenotypic ratios would you expect in the F1 generation if you crossed wild-type and homozygous mutant white-eyed flies (i.e., a monohybrid cross)? Now what about in the F2 generation? Once they can answer these basic questions, ask them to repeat the thought process for dihybrid autosomal crosses, then sex-linked monohybrid crosses. Most students soon realize there are a limited number of discrete possibilities.

Recommended Prelab Skills

Students should be able to generate a Punnet square for a monohybrid cross. They will learn to generate Punnet squares for dihybrid and X-linked crosses as the unit progresses. It is useful if students have a basic working knowledge of a dissecting microscope, but the skill can be learned quickly if not done previously.

Assessment

If pre- and postlab quizzes are used as part of formal assessment, both the prelab and postlab quiz should focus on problem-solving skills. Since students work primarily on their own, class participation is difficult to evaluate directly in this unit. Fortunately, students should be keeping notebooks, which provide an indirect measure of the relative effort put in by the students in each pair. If both students put forth the same amount of effort, their data will be evenly distributed or the same between notebooks; if one student is not participating, most of the data analyses will be found in only one student's notebook.

On postlab quizzes, students should be able to use basic genetic terminology correctly. To test this, they could be asked to differentiate between an allele and a locus or genotype and phenotype. Students should be able to calculate phenotypic ratios from raw counts, and know which ratios to expect from monohybrid, dihybrid, and other crosses. If the instructor includes chi-square analyses as part of the unit, students should be able to calculate and interpret the statistic.

Students' lab reports (if assigned) should state which type of inheritance pattern (monohybrid, dihybrid, sex-linked) their particular eye color mutation follows; the lab reports should also contain both summary data and a detailed explanation of how the data led the student to that conclusion.

Safety and Housekeeping

Old vials of flies should be plugged tightly, collected, and frozen overnight to kill the flies before disposal in general trash.

Other Tips

- The same mutant strains may be used more than once within a single lab section, as long as they are given different name or number designations.
- This unit describes how to anesthetize Drosophila using carbon dioxide plus cold. An alternative anesthesia method is FlyNap (triethylamine), used as described in Unit 8: Animal Hormones. Both work well, but each has disadvantages. Flies wake up more quickly from carbon dioxide/cold, and may escape, while FlyNap is effective for longer periods of time; however, stale material can sterilize or kill flies. Ultimately the choice comes down to instructor preference.
- Students may try to use other groups' phenotypic ratios to explain their own results. Remind them that each group may be working with a different strain or mutation. Students who worked with the X-linked white-eyed mutant in high school may assume their flies have the same genotype again. Remind them that different genotypes can lead to the same phenotype.
- This project requires considerable out-of-class work by students. Do not schedule a second inquiry unit so it runs concurrently with this one.

Supplemental References

These references explain the biochemical pathways underlying eye color in Drosophila.

Ferre, J., F. J. Silva, M. D. Real, and J. L. Mensua. 1986. Pigment pattern in mutants affecting the biosynthesis of pteridines and xanthommatin in *Drosophila melanogaster*. *Biochemical Genetics* 24: 545–567.

Lloyd, V., M. Ramaswami, and H. Kramer. 1998. Not just pretty eyes: Drosophila eye-colour mutations and lysosomal delivery. *Trends in Cell Biology* 8: 257–259.

Mackenzie, S. M., M. R. Brooker, T. R. Gill, G. B. Cox, A. J. Howells, and G. D. Ewart. 1999. Mutations in the white gene of *Drosophila melanogaster* affecting ABC transporters that determine eye colouration. *Biochimica et Biophysica Acta* 1419: 173–185.

PREPARATORY NOTES

Quantities listed are for a lab section of 20 students working in pairs.

Week 1: Shared Materials

- 12 vials of live, wild-type flies, labeled "Practice: Wild Type"
- 12 vials of live mutant flies (equal mix of white, brown, orange, and wild-type eye colors), labeled "Practice: Mutants"
- 22 stock vials of flies labeled "Wild Type," seven to eight days old
- 22 stock vials of flies labeled "Mutant #N," seven to eight days old
- White or clear labeling tape
- Seltzer tablets (1 box of 36 tablets)
- 100x penicillin-streptomycin solution (store 10 mL aliquots in refrigerator)
- Plastic-coated ice blocks (used for shipping; store in refrigerator freezer)

Week 1: Materials at Each Work Station

- Resealable box or bag of dry food
- 2 measuring scoops (10 mL)
- 25 plastic shell vials (2–3 cm diameter)
- 25 foam plugs for shell vials
- Microtube with 0.2 mL of granulated dry bread yeast
- Soft paintbrush
- Permanent marker
- Dissecting microscope
- White index cards
- Anesthesia apparatus
- Fly morgue (jar one-quarter full of mineral oil, with a funnel taped in the top)

Weeks 2 to 7: Shared Materials (keep stocked)

- Shell vials
- Foam plugs
- Fly culture media (powder)
- White or clear labeling tape
- Seltzer tablets
- Penicillin-streptomycin solution (stored in 10 mL aliquots in refrigerator)
- Ice blocks (stored in refrigerator freezer)

Sources of Materials

Flies can be purchased from the Bloomington Stock Center or an education supplier. It is more economical to order stock vials to arrive eight weeks in advance of the lab and amplify them locally than to purchase all the required vials. Although not essential, it is best to maintain stocks locally between semesters, as some strains are only intermittently available. The mutant strains, eye color, and chromosomal locations are summarized in Tables 2.6 and 2.7.

Table 2.6

Features of single mutant fly lines

Strain	Location of Mutation on Chromosome	Eye Color
Wild type	NA	Brick red
Vermilion	Chr. 1 (X)	Orange
White	Chr. 1 (X)	White
Brown	Chr. 2	Brown
Cinnabar	Chr. 2	Orange
Scarlet	Chr. 3	Orange
Sepia	Chr. 3	Brown to black

Table 2.7

Features of double mutant fly lines

Strain	Mutation	Chromosomes	Eye Color	Comments
Mutant #1	Vermilion X Brown	Chr. 1, Chr. 2	White to pale apricot	Can be confused with orange single mutant.
Mutant #2	White X Sepia	Chr. 1, Chr. 3	White	
Mutant #3	Brown X Scarlet	Chr. 2, Chr. 3	White	
Mutant #4	Cinnabar X Brown	Chr. 2, Chr. 2	White	Loci are far enough apart to assort independently.
Mutant #5 (optional)	Vermilion X Sepia	Chr. 1, Chr. 3	Light at eclosion; darken w/ age	Can be hard to score if students do not check flies when young.

Instructors are encouraged to experiment with other strains carrying mutations that affect body color or eye, wing, or bristle morphology.

Solutions, Reagents, Equipment

100x Penicillin-Streptomycin Solution

Purchase premixed antibiotic solution designed for tissue culture; do *not* use formulations that contain glutamine or anti-fungal agents (amphotericin or ketoconazole) Break the solution into 10 mL aliquots and store refrigerated.

Students should regularly inspect stock and cross vials using a dissecting microscope. They should see numerous larvae feeding near the surface of the media. The fly larvae churn the surface of the medium enough to limit bacterial growth. When there are not enough larvae, a slimy tan or pale yellow scum forms on the surface of the medium. Severe infections will kill the entire vial of flies.

The best strategy is prevention. If only one or two adults are available to set up a cross or new vial, students should add more of the same types as soon as possible to increase the number of eggs and larvae. In the early stages of an infection, cultures can be salvaged by adding antibiotics. As soon as bacterial slime appears, prepare a new vial of media, then add three drops (~150µL) of 100x antibiotic solution directly to the media. Transfer adults from the contaminated vial to the new vial. Treat the old vial with another three drops of antibiotic solution to try and save existing larvae. Treat both tubes again two days later. Note that the antibiotic treatment can delay eclosion up to two days.

Anesthesia Apparatus

Obtain a one-hole black rubber stopper that fits the brand of shell vials used. Wet the narrow end of a 1 or 2 mL disposable polystyrene serological pipet and insert it in the hole of the stopper so that the narrow end of the pipet projects 1 in. beyond the larger, outer side of the stopper. Cut off the pipet flush with the inner face of the stopper. Attach an 8 in. piece of plastic aquarium air tubing to the narrow end of the pipet. Add 10 mL of water to the shell vial, and mark point of the meniscus with a permanent marker.

Wine Traps

Fill empty wine or beer bottles one-quarter full of red wine. Add a pinch of dry yeast. Place a funnel into the top of the bottle, and tape the funnel into place. Flies that escape are attracted by the smell of the wine and yeast. They will fly or crawl down the funnel, and drown. Three to four traps are sufficient for one large lab for an entire semester. At the end of the semester, flush dead flies and wine down the sink with copious water.

DNA Isolation and Analysis

Student Pages*

BACKGROUND

Geneticists have identified numerous mutations that cause or contribute to disease. We now know the precise genetic mutations responsible for several forms of muscular dystrophy, anemias, cystic fibrosis, Huntington's chorea, and other inherited genetic disorders. For some diseases, we even know how the mutation affects the structure and normal function of the encoded proteins. But how were these mutations identified?

Investigators use multiple methods to identify general regions in the genome that may differ between normal and disease-prone individuals. Once a candidate region has been identified, it is isolated and sequenced using high-speed automated instruments. Next the nucleotide sequence of the DNA region in disease-prone individuals is compared to an equivalent region in normal individuals. The sequences for normal individuals are downloaded from a massive computer database containing complete genome sequences for dozens of organisms, including humans. Then the two sequences are aligned using bioinformatics software, allowing the investigator to identify changes in the DNA that may be involved in the disease process.

As you will learn in this unit, the computational methods for analyzing DNA sequences are available to anyone with an internet connection and a web browser. However, a computer does not take the place of careful experimentation and observation. It can take years of intellectual effort and lab work for an investigator to demonstrate that a specific genetic variation contributes to a particular disease.

*Teacher Pages begin on page 179.

In this unit you will isolate DNA suitable for sequencing and other analyses. You will use bioinformatics software to find differences between normal and mutant DNA sequences, then use DNA restriction mapping to confirm the differences you find. So that the lab procedures can be performed in a reasonable amount of time, you will work with bacterial rather than human DNA. However, the methods and thinking process are much the same as for characterizing genes linked to human disease.

The *Lac* Operon Is the Model System

Free-living bacteria like *Escherichia coli* use many carbon sources for energy and raw materials. Their preferred sources are glucose, lactose, maltose, or other sugars. To catabolize sugars, bacteria must transcribe genomic DNA into mRNA, then translate it to produce the necessary enzymes. *E. coli* does not continuously synthesize every enzyme and transporter protein needed to metabolize every possible sugar. Instead, it transcribes mRNAs for the required enzymes only when that particular sugar is present in the local environment. This process can be demonstrated in lab using the ***lac* operon**.

Operons are regulatory units in bacterial genomes. Each operon has a single promoter region that controls transcription of one large mRNA. As ribosomes translate the large mRNA, they make two or more separate proteins. Most of the time all of the proteins encoded by a single mRNA are used in a single specific metabolic path. The *lac* operon controls production of *β-galactosidase* (bGAL for short), the enzyme that hydrolyzes the disaccharide lactose to produce the monosaccharides glucose and galactose. The mRNA also encodes two accessory proteins (LacY and LacA) needed to take up and metabolize lactose. Figure 3.1 shows the annotated DNA sequence for part of the *lac* operon.

Figure 3.1

Annotated sequence for the promoter region of the *lac* operon

This sequence matches the original submitted to Genbank (Accession #146575), but some functional site boundaries have been revised based on later studies.

This sequence begins with the Sfo I restriction site (GGC^GCC) upstream of the 5′ end of the operon. Marked sites in the promoter region include CAP and LacI protein binding sites (both underlined), the -10 and -35 binding sites for sigma factor and RNA polymerase (all capitals), and the ATG start codon (italicized, bold capitals).

tGGC^GCCcaa tacgcaaacc gcctctcccc gcgcgttggc cgattcatta

atgcagctgg cacgacaggt ttcccgactg gaaagcgggc agtgagcgca

acgcaattaa tgtgagttag ctcactcatt aggcacccca ggcTTTACAc
(CAP binding site) (-35 site)

tttatgcttc cggctcgTATGTT gtgtggaattg tgagcggata acaatt
(-10 site) (Operator/LacI binding site)

tcacacagga aacagct–***ATG***–ACC–ATG–ATT–ACG–GAT–...
^ ^
(Start site for RNA transcription)
(Translational start site; codon is for the first amino acid "Met" that starts βGal protein)

Most bacterial promoters contain -35 and -10 sites where general transcription factors bind then recruit RNA polymerase for transcription. Upstream of the -35 site in the lac operon is a binding site for an activator protein called cAMP activator protein (CAP). Immediately downstream of the -10 site is an operator region, where a repressor protein called LacI can bind and block transcription. Downstream of the operator is where RNA transcription starts. Just downstream of that is the "ATG" codon for the first methionine in the bGAL protein.

How Is the *Lac* Operon Regulated Normally by Bacteria?

The entire process is summarized in Figure 3.2, page 134. When *E. coli* is growing in an environment without lactose, LacI protein binds tightly to the operator region of DNA and prevents RNA polymerase from binding to the DNA and transcribing it. As a result, the *lac* operon is "off," and very little mRNA coding for bGAL (and the other two proteins) is produced. When lactose becomes available in the environment, it enters the cell, binds to the LacI protein, and causes LacI to leave the operator region. This lets RNA polymerase bind and synthesize a few molecules of mRNA.

Figure 3.2

Summary of how production of mRNA coding for bGAL is controlled by the *lac* operon

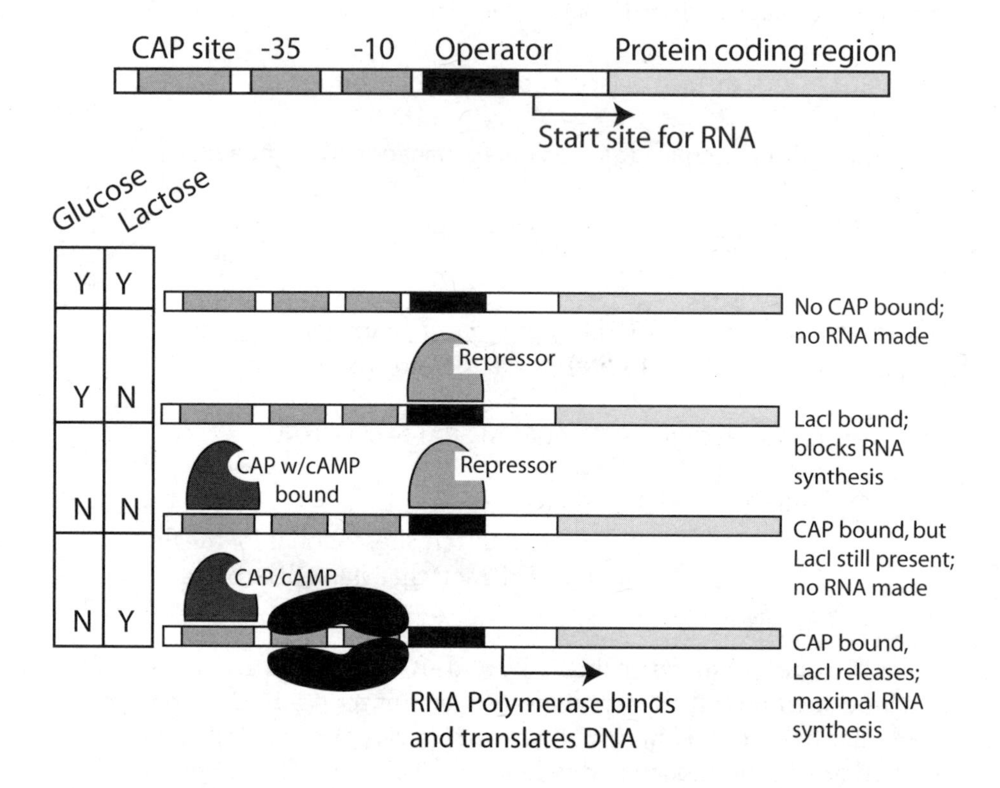

Lactose alone is not sufficient to induce maximal transcription of mRNAs coding for bGAL. Even if lactose is present, *E. coli* still prefer glucose as a carbon source, and as long as glucose is available, the bacteria will not synthesize large amounts of bGAL. This response is controlled by the operon as well, via CAP protein. CAP is an activating transcription factor that only binds to the CAP site on the promoter when cyclic AMP (cAMP) is present. As long as glucose is available in their environment, *E. coli* synthesize ample ATP, and cAMP levels stay low in the cells. Anytime glucose is unavailable, bacteria deplete their ATP and begin to synthesize cAMP. cAMP binds to the CAP protein, which in turn binds to its site on DNA. CAP strongly stimulates RNA polymerase to transcribe new mRNA coding for bGAL.

Mutations Can Affect the *Lac* Operon in Many Ways

You should already know there are several types of mutations. Deletions or insertions of large regions, or even one to two bases, can shift the spacing or positions of protein binding sites relative to each other, change the reading frame, or create early stop codons. Transposition can move entire segments of DNA from their normal positions to new ones. Each type of mutation has very different effects on bGAL expression and activity, depending on its location. Think what would happen if each of the preceding mutations occurred in these sites in the operon:

- CAP binding site
- Promoter (-35 or -10 sites)
- Operator (binding site for the LacI protein)
- Transcriptional start site
- Coding region for the bGAL protein itself

You Will Analyze a Part of the *Lac* Operon That Was Inserted in a Plasmid

Isolating and manipulating genomic DNA directly from any organism (even bacteria) is very time-consuming. For that reason, molecular biologists routinely take DNAs of interest and insert them into small carriers called **plasmids**. In nature, plasmids are small DNA molecules found in many bacteria. Like bacterial genomic DNA, plasmids are double-stranded and circular and have an origin of replication. However, plasmids are much smaller; an average plasmid is 3–12 kilobase pairs (Kb) in size, while the genome of E. coli is 4.7 × 106 megabase pairs (4.7 Mb). Most plasmids contain one or two genes coding for an enzyme that makes the bacteria carrying that plasmid resistant to a particular antibiotic (like ampicillin, tetracycline, or chloramphenicol). While each bacterial cell has only one copy of its genomic DNA, it may contain 300+ identical copies of a small plasmid. Bacteria often exchange plasmids with each other or take them up from their environment.

Plasmids are essential laboratory tools as well. They let us copy, purify, manipulate, and analyze DNA without having to cultivate, feed, grow, or otherwise handle the whole organism. Think about them this way. If you need to study one diagram and two pages of text from a 1000-page textbook, which would you rather carry around—the whole book or a photocopy of just the three pages you need?

Figure 3.3 (p. 136) shows the process used to create the plasmids you will analyze in this unit. First, genomic DNA was extracted from *E. coli* cells that have a normal (wild-type) *lac* operon. The DNA was cut with the restriction enzymes (more about them later) to release a DNA fragment containing just the promoter and coding region for bGAL (LacZ).

Figure 3.3

Schematic diagram summarizing how a fragment of the *lac* operon was inserted into a plasmid to create pLac/WT, the model DNA for this unit

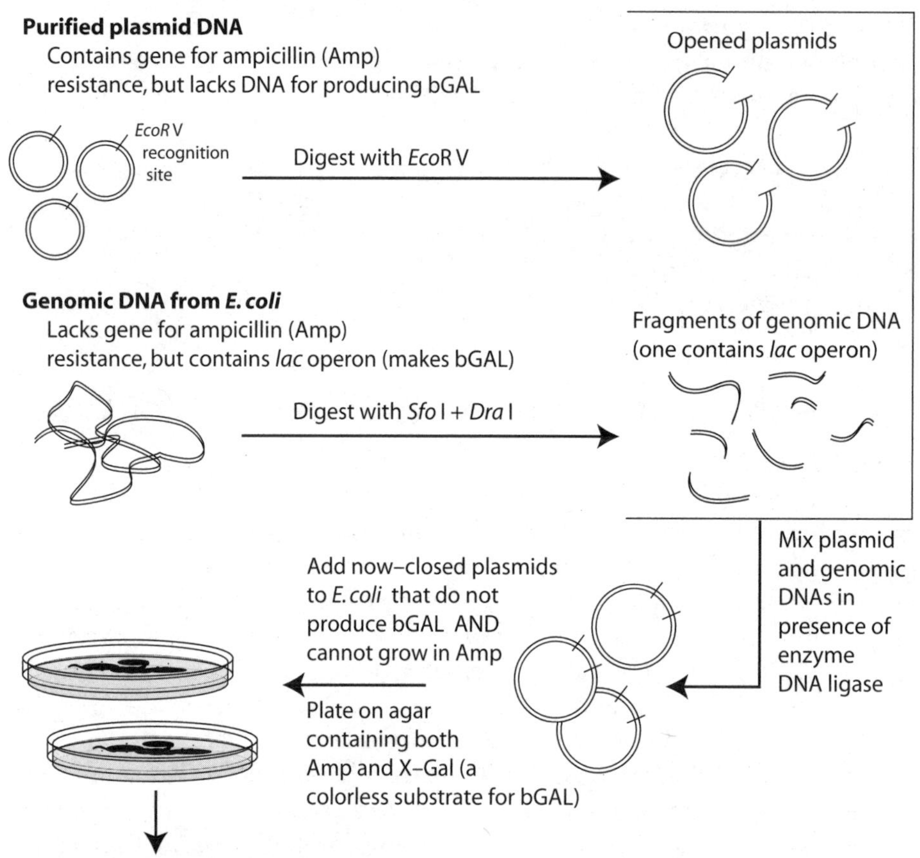

The DNA fragment was mixed with a backbone plasmid that codes for ampicillin resistance. The two pieces were covalently bonded together using the enzyme DNA ligase, to create a new plasmid called **pLac/WT**. This new plasmid was added to 71-18, a mutant strain of E. coli that has no *lac* operon of its own and cannot grow in the presence of ampicillin. When 71-18 bacterial cells were plated on agar containing ampicillin, most died. Any cells that took up the pLac/WT plasmid became ampicillin resistant and gained the ability to produce bGAL enzyme in response to lactose.

After constructing plasmid pLac/WT, several mutations were made in the *lac* promoter and LacZ coding region. Different mutant versions of pLac/WT were named pLac/m# (where the "#" is a specific number). Mutations could be as small as a single base or a larger insertion or deletion of hundreds of bases. The mutation could be in the CAP binding site, the promoter (-35 or -10 sites), operator, transcriptional start site, or coding region for the bGAL protein itself.

Goals for This Unit

Your overall goal in this unit is to characterize the pLac/m# mutant plasmid that you have been assigned and to compare its sequence to pLac/WT. At the start of this unit you will be given two plates of bacteria. The colonies on one plate carry the original plasmid pLac/WT. The other plate has bacteria carrying a mutant pLac/m# plasmid. Over the next several weeks you will do the following:

- Amplify bacteria from the two plates, extract their plasmid DNA, and estimate the quantity.
- Use sequence analysis to identify mutations in the promoter, operator, or coding regions of your pLac/m# plasmid DNA.
- Use restriction enzymes to confirm the presence of these mutations.

UNIT EXERCISES

Exercise 1: Isolating Plasmid DNAs

Background

Alkaline lysis is the most common method for extracting plasmids from bacteria. *E. coli* that contain plasmids are grown to maximum cell density, then lysed with alkaline detergent solution (sodium hydroxide plus sodium dodecyl sulfate, or SDS). This mixture ruptures cell walls and membranes and causes strands of chromosomal and plasmid DNAs to denature (separate). Next the lysate is neutralized by adding mild acid plus potassium ions, which causes the detergent, cell proteins, and other debris to fall out of solution. Now that the pH is neutral again, DNA molecules try to re-form hydrogen bonds with their partner strands. The numerous small circular DNAs from plasmids quickly pair with a complementary strand, but the larger chromosomal DNA invariably gets tangled in the protein debris, which is removed by centrifugation. Soluble plasmid DNA is then forced to precipitate by mixing the solution with ethanol or isopropanol, which makes the solution more hydrophobic, which causes the plasmid DNA to be less soluble. Precipitated plasmid DNAs can then be collected by centrifugation.

Safety Reminders

The strain of *E. coli* used in this lab is not pathogenic, but you should still make sure all tubes are tightly capped when mixing the solutions with bacteria. Wash your hands thoroughly before you leave the laboratory. Solution 2 contains sodium hydroxide, which can burn your skin. If you spill Solution 2 on your hands, wash them immediately. You will be using a microcentrifuge to separate solids from liquid supernatants. If a rotor becomes unbalanced, it can ruin the centrifuge or hurl dangerous shrapnel at those nearby. When you load the instrument, make *sure* that tubes containing equal volumes are positioned symmetrically around the rotor. If there is a rotor lid, be *sure* the lid is screwed on. Once you start a centrifuge, stay close by until the rotor reaches its correct speed.

Prelab Procedure: Inoculating Overnight Cultures

You must start overnight cultures of bacteria carrying pLac/WT and your pLac/m# mutant on the afternoon or in the evening before your regularly scheduled lab time. Your instructor will demonstrate the basic procedure the week beforehand.

1. Remove your two bacterial stock plates and four culture tubes containing 3 mL of LB media plus ampicillin from the refrigerator.
2. Place the four culture tubes in the incubator or a water bath until they come to room temperature.
3. Label all four tubes with your name and the date. Label two as "Plasmid pLac/WT" and the other two as "Plasmid pLac/m#" (put the number of your assigned mutant on the tube).
4. Light a Bunsen burner or alcohol lamp. Flame the tips of your forceps, then remove a sterile toothpick from the tube provided. Close the toothpick container.
5. Open your plate of pLac/WT cells. Press the TIP of the toothpick into a single, well-isolated colony on the plate. Close the cover.
6. Open the first culture tube labeled "pLac/WT," and drop in the entire toothpick. Close the tube loosely.
7. Flame the forceps again.
8. Repeat Steps 4–6, and inoculate the second tube of pLac/WT.
9. Take out a third sterile toothpick and press it into a colony on your pLac/m# stock plate.
10. Place the third toothpick in one of your tubes labeled "pLac/m#." Close the tube loosely.
11. Flame the forceps again.
12. Repeat Steps 9–11 to inoculate the second tube of pLac/m#.
13. Flame the forceps one last time, then extinguish the lamp or burner.
14. Check that all four tubes are loosely capped, and place them in the tube rack inside the shaking incubator. If it is not already on, turn on the switch

on the side of the incubator. Do not change the temperature or shaker speed settings!

15. Your cultures will be ready for DNA extraction 24 hours after inoculation. If your cultures incubate more than 36 hours, destroy them with bleach and start over by picking fresh colonies from your stock plates.

Extracting Plasmid DNA From Bacteria

You will prepare plasmid pLac/WT from one tube of bacteria and your assigned pLac/m# plasmid from a second tube. One 3 mL culture of bacteria is sufficient for each isolation. The second tube was a backup, in case one of your overnight cultures did not grow.

1. Retrieve your four tubes of overnight cultures. Split one 3 mL culture of pLac/WT between two labeled 1.5 mL microcentrifuge tubes. Split one 3 mL culture of pLac/m# between two additional labeled microtubes.
2. Close all four tubes, and place them in a microfuge. Spin the tubes for one minute at top speed (~13,000 rpm).
3. Remove the supernatant using a P-1000 pipet. Keep the pelleted cells, and do not disturb them. Discard the liquid media in the marked waste container.
4. Thoroughly resuspend the cells in each of the four microtubes in 100 μL of Solution 1. This causes an osmotic shock, helping to burst the cell wall and cell membrane. If necessary pipet the solution up and down several times to break up the pellet.
5. Add 200 μL of Solution 2 to each tube. The SDS breaks open the plasma membrane, while NaOH denatures the DNA. Gently invert the tubes two to three times to mix.
6. Add 150 μL of Solution 3 to each tube. Gently invert the tubes two to three times to mix. Small, circular plasmid DNA molecules will re-anneal and stay in solution. Large chromosomal DNA strands get tangled in the precipitating proteins and cannot renature properly. As a consequence, the bacteria's chromosomal DNA precipitates out of solution as part of the opaque white mass.
7. Spin all four tubes for 10 minutes at top speed in the microfuge. Be sure to arrange the tubes in the rotor so that they are balanced across the main axis.
8. The supernatant (liquid on top) contains plasmid DNA and soluble RNAs, so this is the part you need. Using a P-1000, transfer all of the supernatant from each tube to a clean, labeled microtube. Be very careful; remove as little of the white precipitate at the bottom of the tube as possible.
9. Put the tubes of supernatant back into the microcentrifuge, and spin again for five minutes. This will pellet any remaining debris that you transferred accidentally.

10. This time remove just 450 µL of supernatant from the first tube, and transfer it to another clean microtube. If you have less than 450 µL of supernatant, just remove as much of it as you can get without disturbing any pelleted debris. Transfer 450 µL of liquid from each of the other three tubes to clean, labeled microtubes.
11. Add 1 mL of cold 100% isopropanol to each of the four tubes (this makes the final concentration of alcohol ~66%). Incubate the tubes at -20°C (refrigerator freezer) for at least 15 minutes (longer will not hurt the samples) to precipitate the plasmid DNA. If you must store the DNA, this mixture can be kept for several weeks in the freezer.
12. To pellet the DNA, spin the tubes for 10 minutes at top speed in a microfuge.
13. Use a P-1000 to gently remove the supernatant. If necessary, use a P-200 to remove the last remaining supernatant.
14. Gently add ~1 mL of 70% ethanol to each tube. Spin tubes for one minute at top speed in the microcentrifuge. Gently remove the ethanol with a pipet. Be very careful not to disturb the glassy-to-white pellet. It contains your DNA. You must remove all the ethanol from the tube this time. It is easiest to remove most of it with a P-1000 pipetor, then spin the emptied tubes again for two to three seconds, to collect any remaining drops of ethanol. Remove the last traces with a P-200 pipetor.
15. Leave the cap open and let the pellets of DNA dry for five minutes. The DNA pellets should remain stuck to the bottom of each tube. If you still see droplets of alcohol on the walls of a tube, let the pellet continue to dry until they are gone.
16. To resuspend the DNA pellet in each tube, add 50 µL of sterile distilled water. Gently flick the tubes with your finger to encourage the DNA pellet to dislodge and go into solution.
17. Combine your two tubes of pLac/WT DNA into one tube. Combine your two tubes of pLac/m# DNA into a second tube. Your yield of DNA will be around 1-5 µg for 3 mL of bacterial starting culture.
18. Make sure that your tubes of plasmid DNA are labeled with your name, the date, and the names of the plasmids. Your instructor will store them frozen until you are ready to run check gels and perform restriction analysis.

Checking the Concentration of Your DNAs

Later you will learn in detail how agarose gel electrophoresis works. For now, just know that it can be used to determine the quality and quantity of plasmid DNA you just purified. If time is short, freeze your DNAs and run the check gel later.

1. Transfer 15 µL of your pLac/WT DNA to a 0.5 mL microtube, and add 3 µL of 6x DNA sample buffer. Repeat this step with pLac/m#, in a second tube. Mix the tubes by gently flicking them with your finger.

2. Your instructor will make several check gels that consist of 1% agarose melted in TAE buffer. The gels will have individual wells in which you will load your DNA samples and a DNA mixture of known concentration. The instructor also will make a dummy agarose gel and mock samples that you can use to practice loading DNA samples. If you have never loaded small sample volumes onto an agarose gel before, try loading several wells on the practice gel. Do not try to load your DNA samples on your check gel until you can consistently and accurately load mock samples on the practice gel.
3. To load a gel, use a P-20 micropipetor to remove 18 μL of precut, lambda-phage DNA standard solution from the stock tube provided. Put the DNA standard solution into a well on your 1% agarose check gel.
4. Transfer the 18 μL of premixed pLac/WT DNA plus dye into the well beside the DNA standard. Load the 18 μL of pLac/m# DNA (plus dye) in the third well. (*Note:* To save materials, more than one group can load samples onto the same check gel. Be sure to note whose samples are in which lanes.)
5. Once the gel is loaded, attach the lid to the gel box, and plug the electrical leads to the power supply. Remember that DNA is negatively charged, and thus it will run toward a positive electrode. Make sure the electrode from the end of the box opposite the wells is connected to the (+) electrode.
6. Set the power supply to 125 V and run the gel until the first dye front is near the lower edge of the gel (~60 minutes).
7. Turn off the power supply, unplug the electrodes, then remove the lid.
8. Your instructor will show you how to transfer the gel to a UV transilluminator or other gel documentation apparatus.

Warning: A UV light source can cause sunburn or eye damage. Always wear safety glasses and a face shield if you must work near an uncovered UV light source.

Estimating DNA Concentration From Lambda-Phage DNA Markers

The check gel contains ethidium bromide, a DNA intercalating agent that inserts between bases and makes DNA fluoresce when excited with UV light. The brighter the fluorescence of a DNA band, the more DNA is present. Each band in the precut lambda-phage DNA marker lane contains a different amount of DNA. You can estimate the amount of DNA in your pLac/WT and pLac/m# solutions by comparing them to the brightness of bands in the standard.

1. In the gel shown in Figure 3.4 (p. 142), purified linear and uncut plasmid DNAs (left and middle lanes, respectively) were separated along with precut lambda-phage DNA as a standard.

Figure 3.4

A typical check gel

Left lane: a single fragment of purified DNA (circled). Middle lane: plasmid DNA purified from bacteria. Right lane: phage lambda DNA digested with BstE II. The DNA concentrations in the standard lane are listed to the right of the gel.

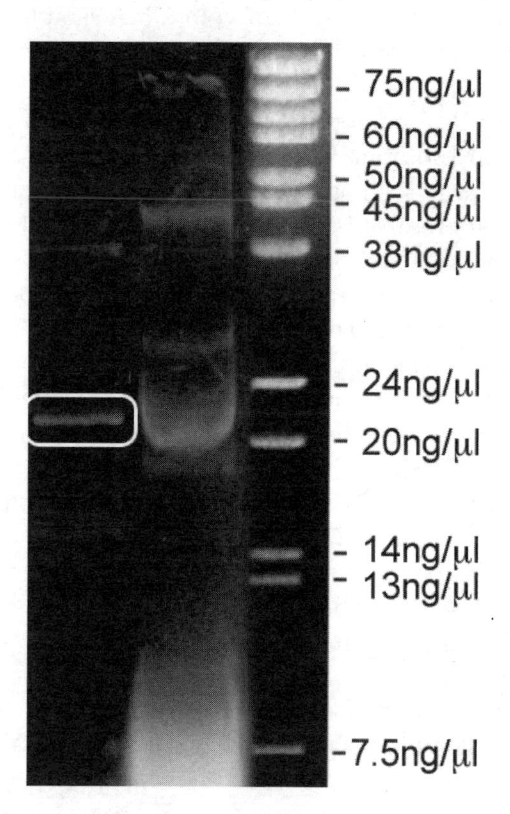

2. Focus on the circled band. Compare the brightness of this band to the various bands on the right side of the gel. For now, do not worry about how far the unknown band migrated; just look at its relative brightness.

   Which band in the standards lane has about the same brightness as the circled band?

3. Suppose you find that the brightness of the circled DNA band is about the same as the 7.5 ng/mL standard band. You can reasonably assume that your original sample contained about 7.5 ng/ mL Of DNA.

   What is the concentration of DNA in the circled band in Figure 3.4?

4. Look at your own check gel, and compare the brightness of the DNA bands in your samples versus the standard lane. Record your estimated DNA concentrations in Table 3.1.

Table 3.1

Estimated concentration of purified plasmid DNAs

Sample	Estimated DNA Conc. (ng/µL)
pLac/WT	
pLac/m#	

5. To set up restriction digests, your solutions of pLac/WT and pLac/m# plasmid DNAs must be at least 20 ng/mL and preferably 50 ng/mL or more.

If you did not obtain enough DNA, arrange a time with your partner to come back and extract additional DNA. You will repeat the same procedure you performed today.

Storing Your Bacterial Plates and Purified Plasmid DNAs

1. Label your two stock tubes of DNA with your name, the date, and the concentration of DNA in each tube. Place them in the rack provided for your lab section in the refrigerator. Your DNAs will be stored frozen until you need them again.
2. If not already wrapped, seal the edges of your pLac/WT and pLac/m# stock bacterial plates with a strip of Parafilm. Store the plates in the box in the refrigerator for your lab section.

SEQUENCE ANALYSIS

Background

How Is a DNA Strand Sequenced?

Typically DNA is sequenced using some modification of the chain termination method. The general procedure is outlined in Figure 3.5 (p. 144). First, a short oligonucleotide (oligo for short) is synthesized that is complementary to a region ~50 bases upstream (that is, on the 5′ side) of the region to be sequenced. The DNA to be sequenced is heated to separate the two strands, then mixed with a combination of oligo, the four normal deoxynucleotides (dNTPs), a smaller quantity of dideoxynucleotides (dideoxyNTPs), and purified DNA polymerase. The oligo base pairs with its complementary region on the DNA strand and acts as a primer for DNA polymerase. The enzyme adds new dNTPs and dideoxyNTPs to the 3′ hydroxyl end of the primer, using the single stranded DNA as the template.

Figure 3.5

Schematic diagram showing the principle underlying dideoxy chain termination sequencing

Panel A shows the structural difference between normal dNTPs and dideoxynuleotides. Absence of a 3′ hydroxyl group in a dideoxyNTP stops chain elongation once it incorporates into a DNA strand. More recently dideoxyNTPs have been replaced with NTPs in which the 3′ end is blocked by chain-terminating fluorescent dyes. *Panel B* shows the pattern of bands that would be produced on a sequencing gel, and how they would be read to determine DNA sequence.

Originally, the four dideoxyNTP mixtures had to be separated in four different lanes of a sequencing gel. Then the sequence was read from bottom to top of the gel. Modern automated sequencing uses NTPs with four differently colored chain-terminating dyes (one per type of nucleotide). All four reactions are placed into a single lane of a gel and separated as usual. The four dyes are excited by a laser as the bands pass beneath a detector and the sequence is read automatically.

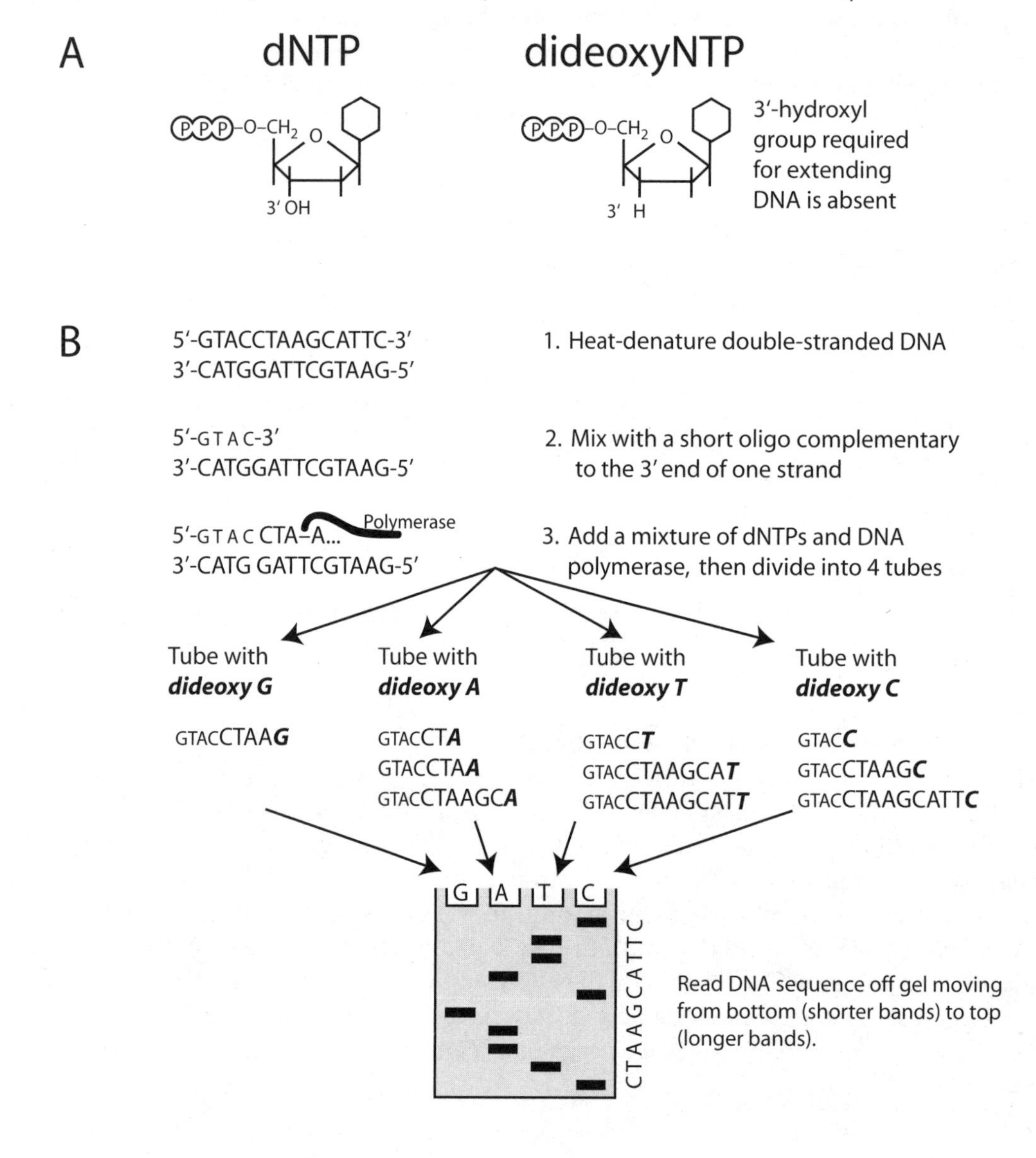

If only dNTPs are used, DNA polymerase replicates the complete template DNA strand. However, dideoxyNTPs are missing their 3′ hydroxyl group; when they are incorporated into a new DNA strand, the chain terminates. Dideoxy-NTPs are incorporated randomly, so DNA strands terminate at various distances downstream of the oligo. These new strands can be separated by size on an acrylamide gel.

Before the mid-1990s, individual labs usually sequenced their own DNA. Then the Human Genome Project provided an impetus for companies to develop automated sequencing instruments and more sensitive fluorescence dye labels for detecting the dideoxy-terminated DNA strands. DNAs of different sizes still are separated with an acrylamide gel, but a laser scanner and computer software reads the sequence automatically. Current instruments routinely read about 1000 bases from 100 samples per day.

An automated sequencer is expensive to purchase and maintain, but many research centers and private companies have a central facility that sequences DNA samples from other labs for a flat fee. Results are sent back to the investigator by e-mail. This is how the sequences you will be using were obtained. The pLac/WT and pLac/m# plasmids were sequenced in steps of ~1000 bases at a time. The individual sequence files were spliced together using the same software that you will use today, to create a complete sequence file for the DNA insert in plasmid pLac/WT and another for the insert in each pLac/m# mutant. To save time with today's exercises, you will be using these pre-assembled DNA sequences to analyze the inserted fragments of the *lac* operon in each plasmid.

Goals for Exercises 2 and 3

Your overall goal is to determine how the sequence of your pLac/m# mutant differs from the sequence of the normal promoter and LacZ coding region of pLac/WT. More specifically, you will do the following:

- Move the sequence files from a text document into Workbench, a Web-based bioinformatics software package, and learn to manipulate them.
- Align the DNA sequences for pLac/WT and your pLac/m# mutant, compare them to each other, and identify the mutations that make the fragment of the lac operon in pLac/m# different from the wild-type operon fragment contained in pLac/WT.

Exercise 2: Basic Sequence Manipulation

Creating a WorkBench Account and Starting a Session

Bio WorkBench 3.2 is a suite of bioinformatics programs maintained by the San Diego Supercomputing Center. Its web address is *http://workbench.sdsc.edu.*

1. Open a web browser, and go to the WorkBench website. The main page offers two options: entering the program or registering for a new account. Choose Register for a Free Account. Fill out the form and submit it.
2. Go back to the main page, log on, then click on the button that says Session Tools. The first screen you see will look like Figure 3.6 (you may need to scroll down).

Figure 3.6

Starting a WorkBench session

WorkBench stores your data in Sessions, spaces where you can store files containing individual or aligned sequences, the results of analyses, and other outputs. You can create more than one session if you want to store work related to different projects. When you signed up for an account, WorkBench created a dummy session called DEFAULT to hold a place for you in its servers. You cannot store data in this session, so you must either create a new one or resume one you created earlier.

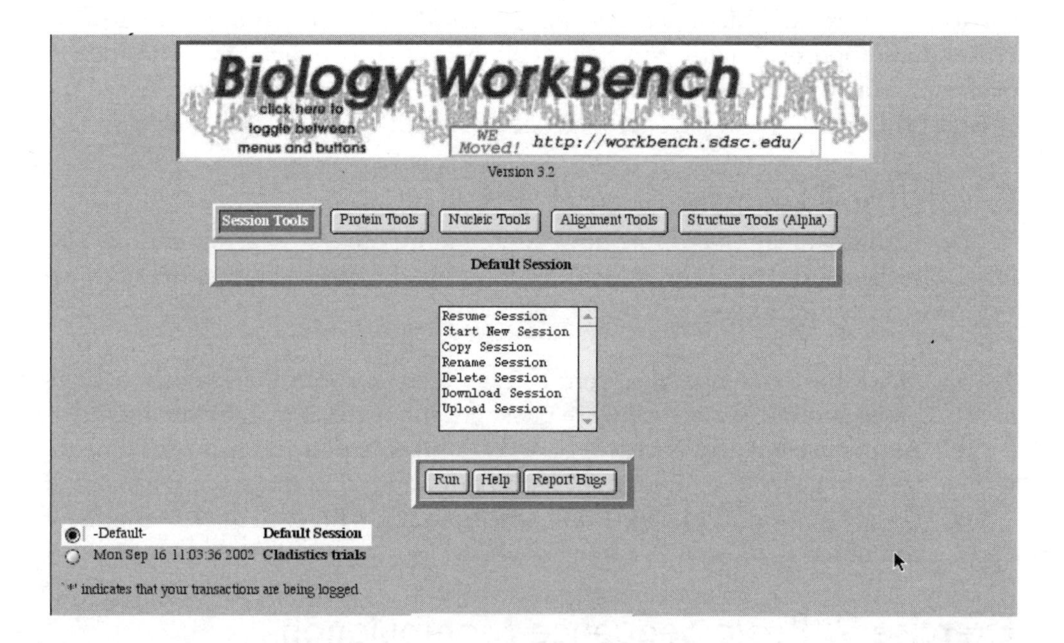

3. In the list of options, select Start New Session, then select Run. You will get the dialog screen shown in Figure 3.7. Give your new session a name and click on Start.

Figure 3.7

Creating a new work session

Anytime you need to return to an earlier Workbench session, just go to Session Tools and choose Resume Session.

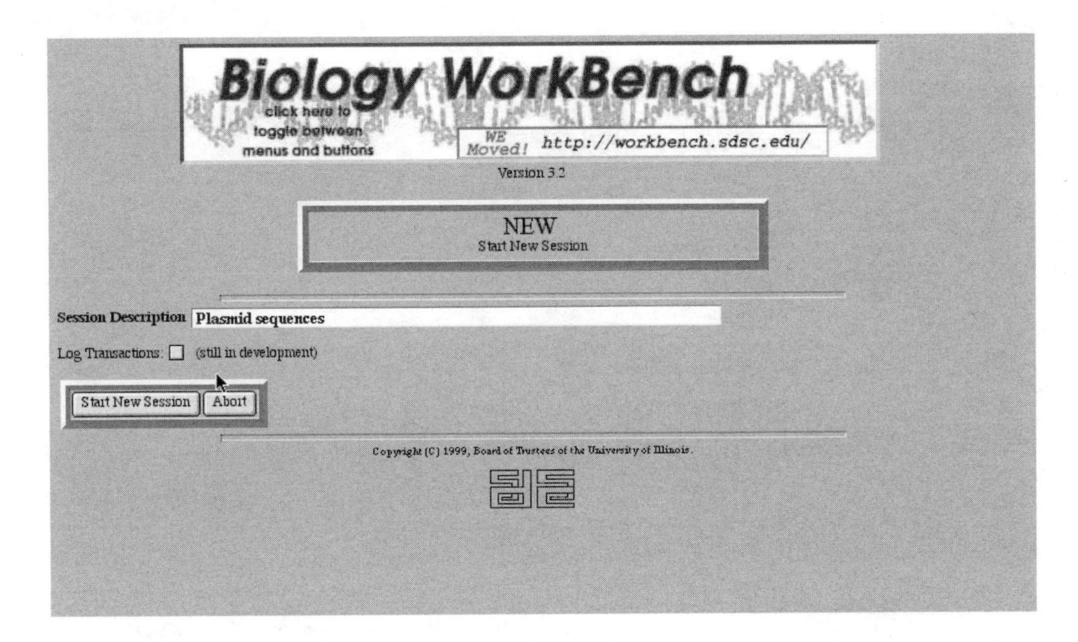

4. After you start or resume a session, you will return to the main work page. One view of this page is shown in Panel A of Figure 3.8 (p. 148). From this main work page, you can access dozens of search engines and sequence analysis programs, which are grouped into tool sets. For most of this unit, you will be using Nucleic Tools.

Figure 3.8

Creating a new sequence file

Panel A shows the dialog box that allows you to create a new sequence file. *Panel B* shows where to enter data. Type your name for the sequence file in the "Label" box. Paste actual nucleotide sequences in the larger "Sequence" box.

A

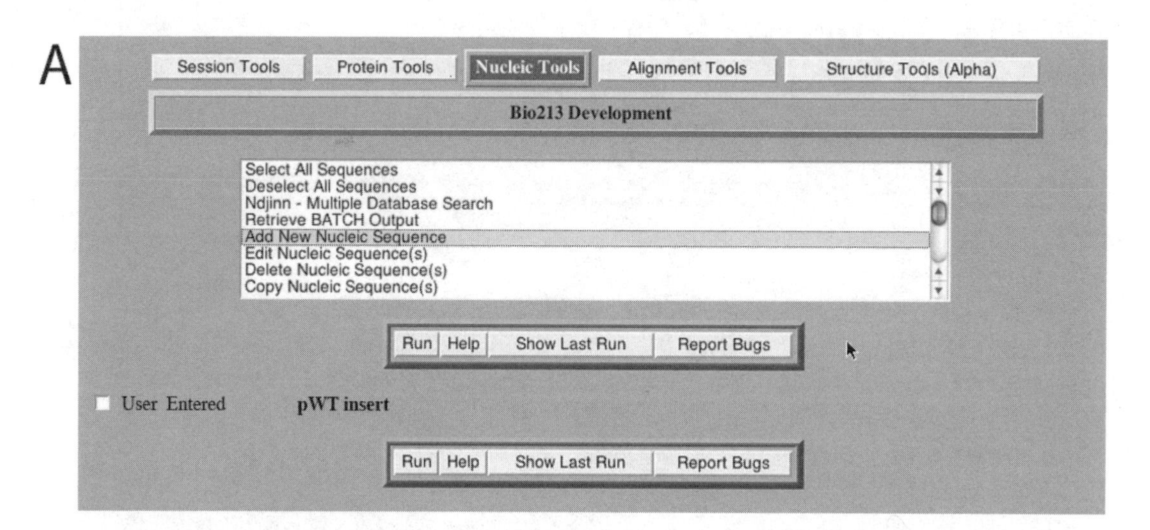

B

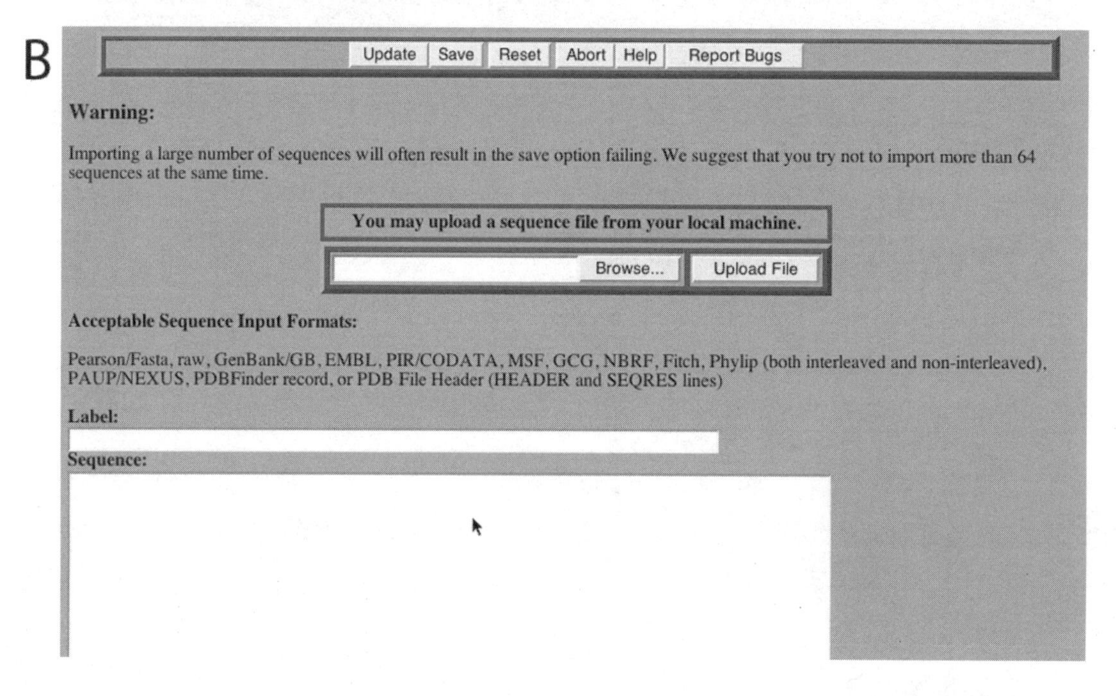

Entering Sequences

You can manually type sequences directly into a web form, but it is easier to cut and paste them. Your instructor will provide you with a text file containing the sequences of the pLac/WT and pLac/m# inserts and their backbone plasmid. Follow the instructions for opening the text file. Keep a copy of the text file on your computer as you continue with these steps.

1. To import your sequences to WorkBench, select Nucleic Tools on the main page (see Figure 3.8). Scroll down until you find the option Add New Nucleotide Sequence. Highlight this option, and click on Run. As you proceed, you will see the screen shown in Figure 3.8, Panel B.
2. When you reach the screen where you can enter data, minimize your browser window for a moment. Double-click on the file that contains your DNA sequences.
3. The text file contains several DNA sequences. Look for a sequence named "AmpR #1" and highlight just the nucleotide sequence; do not include any extraneous labels or numbers. From your program menu, select Copy.
4. Now click on the browser window to open it again. Point and click inside the box labeled "Sequence" to make it active. Go to the menu bar for the browser and choose Paste. The sequence for AmpR #1 that you just copied should appear in the window now.
5. Give your new sequence file a label name. At this point your screen should look something like Figure 3.9.

Figure 3.9

Sequence data from the MS Word file has been pasted into the WorkBench file and has been given a name

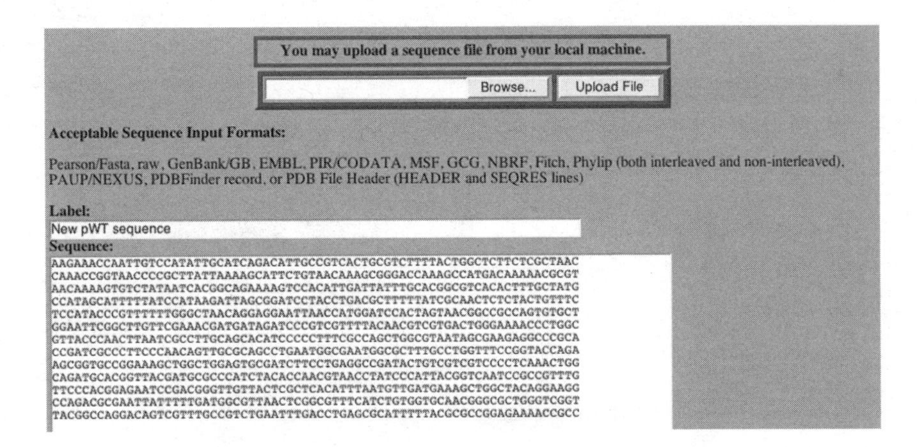

(*Hint:* Anytime you add or make changes to a sequence file, the sequence may extend beyond the right side of the boundary box. To make a file easier to view, you can click on Update.)

6. When you have pasted the AmpR #1 sequence and named the file, select Save from the menu bar at the bottom of the screen. A file with the sequence for AmpR #1 will be created in your WorkBench session and you will return to the main Nucleic Tools page.
7. Repeat these steps to import "AmpR #2" to your WorkBench session.
8. Repeat these steps to import the three files in the following list:

    - Plasmid backbone sequence
    - Insert sequence for pLac/WT
    - Insert sequence for the pLac/m# mutant you are analyzing

Splicing Sequences

If you can find where two sequences overlap, you can use Workbench to splice them together into a longer sequence. First, you need to locate where they overlap, using a search program called BL2SEQ.

1. Go back to Nucleic Tools. You will see your imported sequences and a list of several options for analysis. To run BL2SEQ, you must select two sequences you want to align. Check the boxes next to the two AmpR sequences.
2. Scroll down through the list of options. Look for the option BL2SEQ and click it once to select it. Then click Run (see Figure 3.10).

Figure 3.10

Starting a BL2SEQ alignment

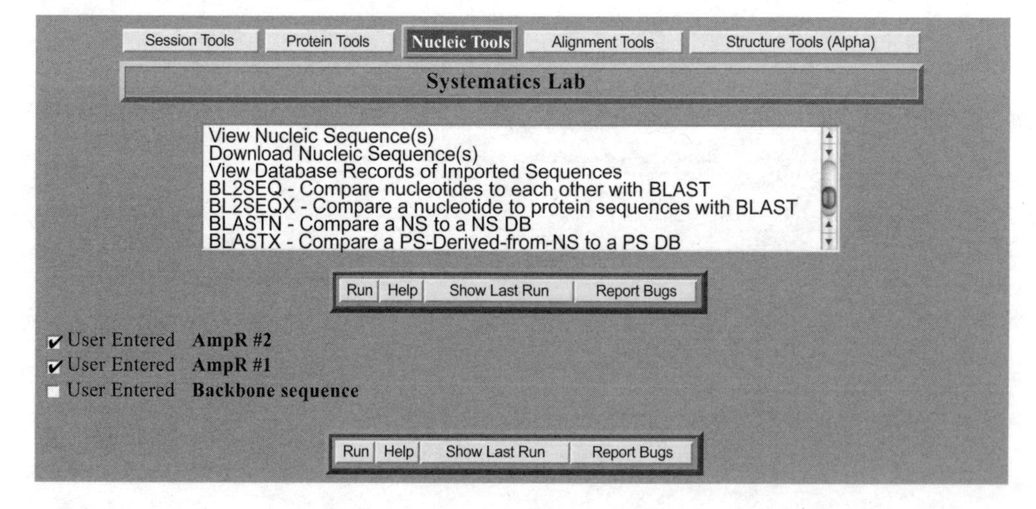

3. The next page lets you set several options to fine-tune how the program finds an alignment. For now, leave the options on their default values, and just click on the Submit button at the bottom of the page.

   Tip: In WorkBench, some steps use the term Run, and some use Submit. If this exercise tells you to click one and you see the other, do not panic. They do the same thing. Return always takes you back to the main tools page. Abort stops whatever program you started, cancels any changes, and sends you back to the main page.
4. The report will show one or more regions where the two sequences are similar. Find one that shows considerable overlap between the 3′ end of the AmpR #1 sequence and the 5′ end of AmpR #2.
5. You will use this alignment to pick where to join the two sequence files. Write down the numbers where the bases match perfectly between the two sequences and a string of six to eight bases that will help you recognize that region (see Figure 3.11). Then choose Return.

Figure 3.11

Output from a BL2SEQ alignment of the two AmpR sequences, showing a region of overlap

The top row of the alignment shows the 3′ end of AmpR #1. The bottom row shows the 5′ end of AmpR #2. The vertical lines show nucleotides that are identical in both sequences.

According to this analysis, there is a perfect overlap in the sequences between bases 446–481 of AmpR#1 and bases 1–36 of AmpR#2. Near the end of the overlapping region is the sequence "CCGC" followed by six "T"s in a row. This distinctive sequence should be easy to spot in both files.

AmpR #1 (User Entered)
- and -
AmpR #2 (User Entered)

```
Query:    >AmpR_#1
           Length = 481
Reference: Query= AmpR_#1
      (481 letters)

>AmpR_#2
       Length = 445

 Score = 71.9 bits (36), Expect = 5e-17
 Identities = 36/36 (100%)
 Strand = Plus / Plus

Query:446  gaaggagctaaccgctttttttgcacaacatgggga  481
           ||||||||||||||||||||||||||||||||||||
Sbjct: 1   gaaggagctaaccgctttttttgcacaacatgggga  36
```

6. Back on the main screen, check just the AmpR #2 sequence, then click View Nucleic Sequence. This lets you see the entire sequence at once.
7. Look through the sequence until you find the string of six to eight bases you wrote down earlier; it should be near the beginning of the sequence. This is the 5′ end of the sequence you are going to copy and paste on the end of another sequence.
8. Highlight all of the sequence that comes after the six to eight bases you recognized, then choose Copy from the browser's menu.
9. Choose Return to go back to the main Nucleic Tools menu.
10. Now check the box beside the AmpR #1sequence, and choose Edit Nucleic Sequence.
11. You will see a box that allows you to edit the AmpR #1 sequence. Look for the 6–8 bases that you wrote down; they should be near the end of the file this time. Highlight everything that comes after these bases and delete it using the backspace key.
12. Click just beside the last base, and choose Paste from the browser menu. This adds the sequence you just copied from the AmpR #2 file to the end of AmpR #1.
13. Once you have spliced together the two files, give the compiled sequence a new file name (e.g., AmpR #1+2, spliced), and select Save from the bottom of the screen.
14. You should return to the main screen, and a new file should be present.

Checking the Accuracy of a Spliced Sequence

To complete later exercises, you MUST be able to splice sequence files accurately. If you make a mistake, you may introduce a "mutation" into your sequence file that is not in the actual DNA. So before you move on to more complex analyses, you are going to check whether you correctly spliced together the AmpR practice sequences.

You were not told this before, but AmpR #1 and #2 are not random DNA sequences. When spliced properly, they form the coding sequence for **beta lactamase**, the enzyme that breaks down ampicillin. As was explained earlier, a gene in the backbone plasmid codes for ampicillin resistance (i.e., beta lactamase). If it was spliced correctly, your AmpR sequence should be a perfect copy of that part of the plasmid's sequence. Test this by using BL2SEQ again to compare the two files.

1. Go back to your main list of nucleic acid sequence files. Check your imported plasmid backbone file and your newly combined AmpR file. Run BL2SEQ, leaving the analysis options on the default settings.
2. Your spliced AmpR pieces and the plasmid backbone should match perfectly for more than 800 bases, with no gaps or mismatches. Even one mismatched base means you spliced the AmpR files together incorrectly. In that case, delete your spliced AmpR file, and try again.

3. Keep splicing your AmpR #1 and AmpR #2 files and checking them against the plasmid backbone until you get a perfect match.

Exercise 3: Identifying Differences Between DNA Inserts in pLac/WT and pLac/m#

Background

In Exercise 1 you imported the sequences of the pLac/WT and pLac/m# inserts. Now you will search your assigned pLac/m# DNA sequence for mutations. First, print out both sequence files. Look for six to eight bases near the beginning of both sequences that are identical. When you have found them, align the two printed files one beneath the other. Start comparing the two sequences, base by base. Each time you find a base that is different between the two sequences, circle the bases on both printed files. Continue searching for differences until you have compared all ~3000 bases. Count the number of bases that are different between the two sequences.

Once you are finished with the first pass, see if there is another sequence of six to eight bases that match in both sequences. Use these bases as a starting point, and, with another colored pen, mark every base that is different between the two sequences. When you are done, count up the number of differences between the two sequences again. Then compare that number to the number you got in the first pass. Keep repeating this process until you find the alignment that produces the fewest differences between the two sequences.

Does that seem like a very slow way to proceed? Believe it or not, in the bad old days before computers were readily available, this is how DNA sequence comparisons were done. Today, microprocessors do all the heavy lifting.

In Exercise 1 you looked for sequence regions that were identical so you could put sequences together accurately. This time, you are looking for differences between the wild type and mutant LacZ promoter and coding sequences.

Procedure

1. If you are not already there, go to the main page of WorkBench, and click on Nucleic Tools. Scroll down until you find BL2SEQ again.
2. Check the boxes next to the two sequence files you created for pLac/WT and pLac/m#, and click on Run.
3. The report will show the two aligned sequences stacked on top of each other. Bases that are identical are connected by vertical lines. Differences in the two sequences do not have connecting lines, and the bases may be highlighted in blue.
4. As you look through the alignment of the pLac/WT and pLac/m# inserts, note any differences between them in Table 3.2 (p. 154).

Table 3.2

Potential mutations in pLac/m# insert

Mutation #	Location in pLac/m#?	Type of Mutation (deletion, insert, frame-shift, or other) and Other Comments
1		
2		
3		
4		
5		
6		
7		
8		

5. You need a permanent record of your DNA sequence alignment of pLac/WT and pLac/m# to answer the summary questions. Click and drag the cursor so that the entire alignment is highlighted. Alternatively, use the browser menu to choose Edit > Select All.
6. Choose Edit > Copy. This copies the alignment data onto the clipboard of the computer.
7. Open a new blank document with a word processor or another text-based program, and paste in the alignment. Depending on your computer's settings, you may need to change the font of the pasted information to Courier, 9 pt.
8. The text file should look identical to what you originally had on the computer screen. If not, you may need to adjust the font size again.
9. On your text version of the alignment, find the lac promoter region in your two aligned DNAs. (*Hint:* The sequences of the promoter elements are listed in Figure 3.1 of this unit.) Highlight the locations of the CAP site, -35 site, -10 site, operator, and ATG start site. Also mark the locations where there are mutations in the pLac/m# DNA, compared to pLac/WT.
10. Save the annotated text document containing the alignment for printing out later.

Summary Questions for Exercises 2 and 3

When you turn in your answers to these questions, attach a copy of your aligned and marked-up DNA sequences.

1. Is the promoter identical in both pLac/WT and your pLac/m# mutant? If not, which elements differ between them? If you found differences, how do you expect them to affect gene expression? What is your reasoning for these predictions?
2. Look at the ATG that is the normal translation start site. Is this the only ATG in the DNA sequence? If there is more than one ATG, how does a ribosome select the proper one for starting translation?
3. Look at the rest of the aligned DNA sequences for pLac/WT and your pLac/m# mutant. What other differences did you find? How do you predict these mutations will affect expression, function, or regulation of the LacZ protein? What is your reasoning for these predictions?

DNA RESTRICTION MAPPING

Background: What Are Restriction Enzymes?

The pLac/WT plasmid you are using contains a fragment of the *lac* operon that was derived from bacterial genomic DNA. (If you need a reminder of what a plasmid is, refer to Figure 3.12, p. 156.) This fragment was cut from the genomic DNA at precise locations using **restriction endonucleases** (aka, **restriction enzymes**). Most species of bacteria produce restriction enzymes. These bind to a particular DNA sequence (called a restriction site or cut site) and cleave the phosphodiester backbone of both strands of the DNA double helix. Bacteria use these enzymes during DNA repair and as a defense against infection by bacteriophages (viruses that infect bacteria). Molecular biologists also use them to manipulate and analyze DNA molecules.

Figure 3.12

Schematic "mini-map" showing a piece of DNA being inserted into a plasmid vector

In this example, the DNA fragment being inserted in the plasmid backbone has two different cohesive ends, an EcoRI-end, and a BamHI-end; the plasmid DNA has the same two ends. If the two DNAs are mixed, then treated with DNA ligase, they will form a single, closed circle. Then the closed backbone plasmid with its new insert can be taken up and replicated by bacteria. The copies will be inherited and amplified by the daughter cells of the original bacterium.

Over 200 restriction enzymes are now available commercially. Most of them recognize a 4 to10 base pair (bp) palindrome, a DNA sequence that reads the same from either direction. Some restriction enzymes cut both DNA strands exactly in the center of the restriction site, creating **blunt ends**. Others cut the backbone in two places, so the pieces have short single-stranded **cohesive ends**.

The DNA sequence of four commonly used restriction enzymes are shown in Figure 3.13. The vertical lines indicate where the enzyme cuts the DNA palindrome.

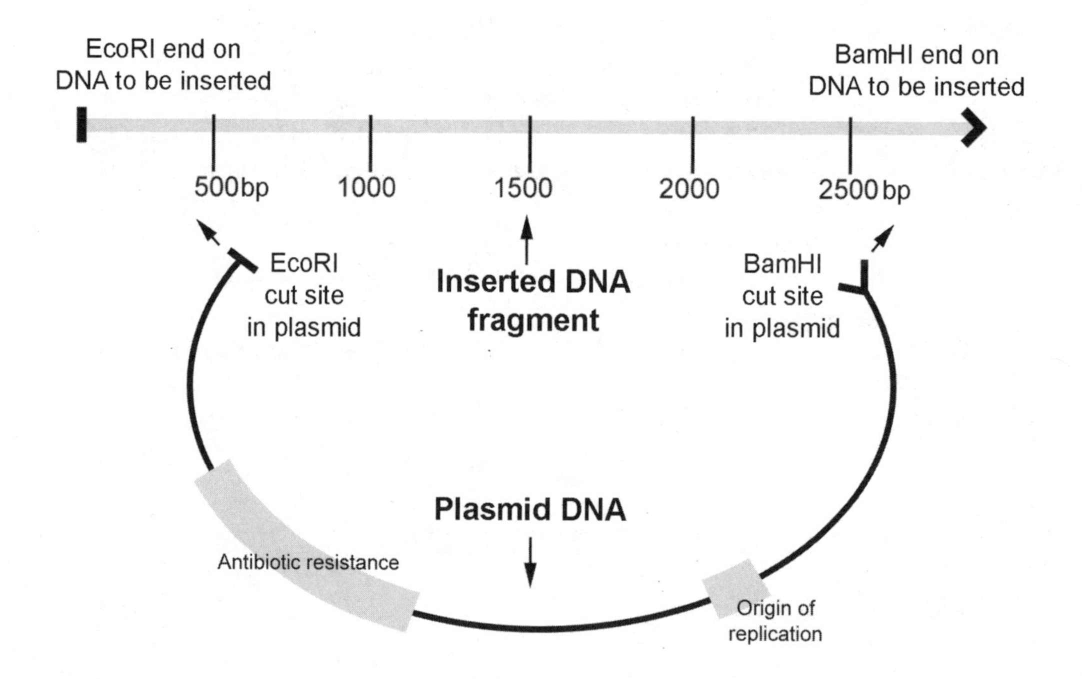

Figure 3.13

DNA sequences for the binding sites of several common restriction endonucleases (aka, restriction enzymes)

Enzymes That Create Cohesive Ends

Enzyme	Site		Fragment 1		Fragment 2
EcoRI	**5'–G|AATTC–3'** **3'–CTTAA|G–5'**	→	**5'–G** **3'–CTTAA**	**&**	**AATTC–3'** **G–5'**
BamHI	**5'–G|GTACC–3'** **3'–CCATG|G–5'**	→	**5'–G** **3'–CCATG**	**&**	**GTACC–3'** **G–5'**
Hind III	**5'–A|AGCTT–3'** **3'–TTCGA|A–5'**	→	**5'–A** **3'–TTCGA**	**&**	**AGCTT–3'** **A–5'**

An Enzyme That Creates Blunt Ends

Enzyme	Site		Fragment 1		Fragment 2
SmaI	**5'–CCC|GGG–3'** **3'–GGG|CCC–5'**	→	**5'–CCC** **3'–GGG**	**&**	**GGG–3'** **CCC–5'**

Each restriction enzyme is named for the bacterium that makes it. For example, EcoRI was the first enzyme purified from *Escherichia coli*, Strain R; Hind III was the third enzyme purified from *Haemophilus influenzae*, Strain D.

Two pieces of DNA that are cut with the same restriction enzyme usually have cohesive ends that can overlap and form base pairs. If DNA ligase is added, two pieces of DNA with the same cohesive ends can be covalently connected, even if they originated from different organisms. This ability to cut and reassemble DNA fragments from different sources to make new DNA sequences is the basis for the entire biotechnology industry.

Why Is Restriction Mapping Important?

Restriction enzymes also are essential tools for analyzing DNA structure. Restriction mapping is the process of cutting DNA at specific sequences with restriction enzymes, separating the resulting fragments by size using agarose gel electrophoresis, then estimating the size of those fragments. The size or number of DNA fragments produced tells us more about the structure of the original piece of DNA.

Here is a practical application of restriction mapping. The DNA sequence recognized by the restriction enzyme EcoRI is GAATTC. Imagine you have developed a rapid procedure for extracting a piece of genomic DNA from humans that is 2000 base pairs long and that contains the coding region for the essential enzyme "survivase." When you sequence the DNA fragment coding for normal (wild-type) survivase, you find an EcoRI recognition site located 1200 base pairs from the 5′ end. If you digest your isolated 2000bp DNA fragment with EcoRI, you get two smaller DNA fragments that are 1200bp and 800bp long. Using software to translate the DNA sequence into amino acids, you find that the palindrome is in frame, so that the "GAA" codes for the amino acid glutamic acid and "TTC" codes for phenylalanine.

Subsequently you learn of a seriously ill patient who does not make survivase. You extract and sequence the coding region for her version of survivase and find that in her DNA, a single nucleotide mutation has changed the palindrome from GAATTC to TAATTC. When you translate her DNA sequence, you find this mutation changes the codon for glutamic acid (GAA) to a "STOP" codon (TAA). This is probably why the patient has no functional survivase; she is only translating 1200 of the 2000 bases in the DNA code for the enzyme. The mutation also destroys the palindromic recognition site for EcoRI. As a result, EcoRI will not cut the mutant survivase DNA, and the DNA fragment you extracted will remain 2000 base pairs long.

Now you want to screen for other patients with this mutation. You could sequence each patient's DNA directly, but that would be extremely expensive and time-consuming. Instead, you can combine your rapid procedure for extracting a target region of genomic DNA with restriction mapping. If a screened patient's 2000bp extracted DNA forms two fragments when cut with EcoRI, the patient does not have the mutation; if the patient's extracted DNA is not cut by EcoRI, the patient likely has the mutation and is not likely to produce the enzyme survivase.

In the preceding example, a random mutation altered an important codon *and* changed a restriction site. Something so dramatic does not always happen though. Many mutations add or remove restriction enzyme sites, but do not significantly change the coding regions or functions of a protein. These "silent" mutations in the structure of DNA add up over time to create a unique genetic signature for each organism, which can be visualized by performing restriction mapping of that organism's DNA. Closely related organisms or species will share more restriction sites than distantly related ones. This principle has led to restriction mapping being used for paternity testing and in forensics to determine the origin of body fluids. Restriction mapping in combination with rapid sequence analysis also has helped biologists uncover close evolutionary relationships between species that were once thought to be only distantly related.

Restriction mapping is used for other purposes too, including the following:

- To confirm a sequence analysis, or prove that a mutation is present in one piece of DNA, but not another

- To determine the size of an unknown DNA fragment that was inserted into a plasmid
- To determine the size and structure of DNA elements in an organism's genome
- To compare the genomes of different species
- To compare DNA samples taken from different individuals within a population. This technique is widely used in forensics, systematics, evolutionary biology, and medicine.

It would take several semesters to experience every application of restriction mapping. For this exercise, you will use it to confirm the presence of any potential mutation(s) that you identified in your pLac/m# DNA sequence.

Goals for Exercises 4–7

In this set of exercises you will use WorkBench to align the 3′ ends of the pLac/WT and pLac/m# mutant inserts with the 5′ end of the sequence for the plasmid backbone, then splice the insert and backbone sequences together. Next you will search through these new sequence files for restriction enzyme recognition sites. Using that information, you will select one to two restriction enzymes that will cut pLac/WT and your pLac/m# mutant into different sets of fragments, then set up a restriction assay to test your predictions. You will separate the DNA fragments by gel electrophoresis, then determine their size and compare them to your predictions.

Exercise 4: Assembling Full-Length Sequences for pLac/WT and pLac/m#

So far you have focused on just the DNA inserts in pLac/WT and pLac/m# mutants. It was safe to ignore the sequence of the backbone because it is the same for both plasmids. To look for restriction sites in your plasmid DNAs, however, you must use the sequence of the *entire* plasmid (both insert and backbone.) Last week, you learned to splice together short DNA sequences. In this exercise, you must splice the 3′ end of the sequence for the inserted region in pLac/WT to the 5′ end of the sequence for the backbone plasmid. You will do the same for your assigned pLac/m# mutant. The process is no different than when you spliced together AmpR #1 and AmpR #2, except that you will be working with longer sequences.

Procedure

1. You should already have created a sequence file for the backbone plasmid backbone. If not, go back and do so now.
2. Following the same steps as in Exercise 2, use BL2SEQ to find the point where the 3′ end of the pLac/WT file overlaps the 5′ end of the plasmid backbone file.

3. Use View Sequence and Edit Sequence to connect the two files, just as you did in Exercise #2. Be sure that you place the sequence of the plasmid backbone on the 3′ end of the sequence for pLac/WT insert.
4. Save the combined sequences under a new file name.
5. Repeat the process, except this time add the sequence for plasmid backbone to the 3′ end of the pLac/m# mutant insert sequence. Save the combined sequences under a new name.
6. Use View Sequence to look at your two newly compiled sequences. They should be approximately 4000 to 6000 bases long. If they are significantly smaller, check that you have copied and pasted everything correctly.

Exercise 5: Locating Restriction Sites

WorkBench includes a program named TACG that can identify known restriction site palindromes in your DNA sequences and predict the sizes of the DNA fragments that would be produced if your pLac/WT and pLac/m# plasmids were actually digested with them.

Procedure

1. If you are not there already, open the session containing your pLac/WT and pLac/m# sequence files and go to Nucleic Tools.
2. Scroll through the list of programs until you find TACG (see Figure 3.14). Highlight it, then check the box next to the compiled sequence for pLac/WT. Click Run.

Figure 3.14

Starting a TACG search

This program also can translate DNA sequences into amino acids using both forward and reverse reading frames.

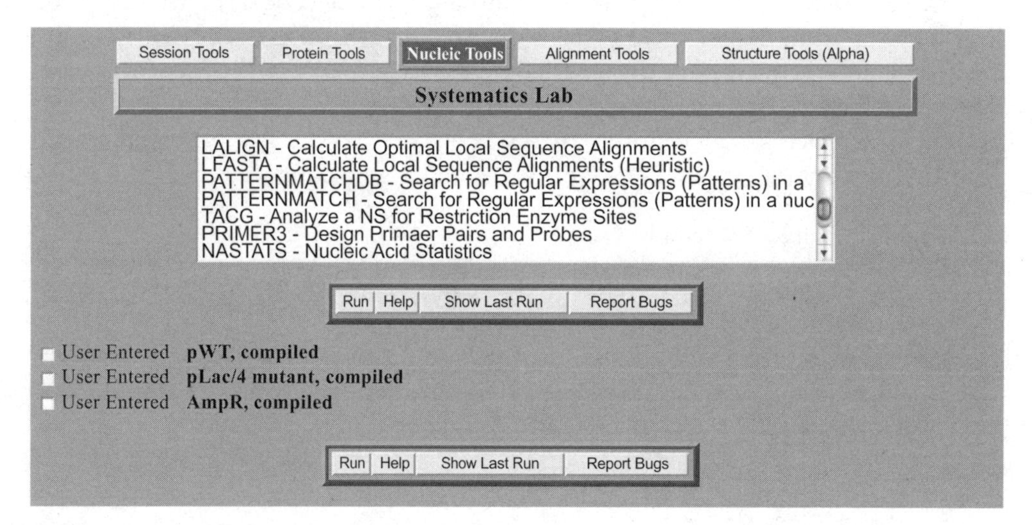

3. The dialog box for the TACG program (shown in Figure 3.15) allows you to pick the size of the restriction site, specific enzymes, and size of fragments or to set other parameters.

Figure 3.15

Dialog box for a TACG restriction site search

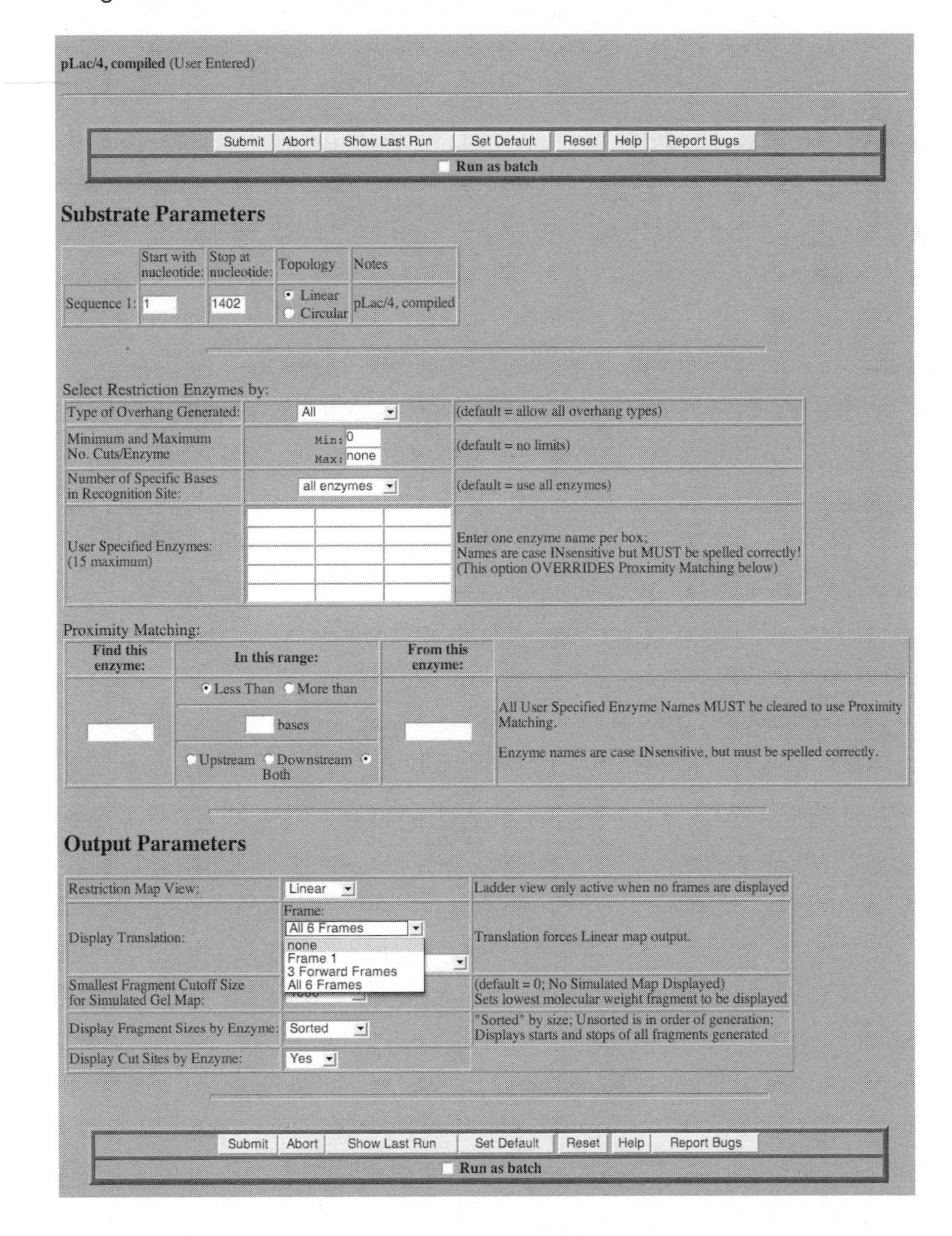

4. This time you do not want to use the default settings for the program, because they are set to search using all enzymes. There are many restriction enzymes that recognize common 4bp sequences. If you try to map out every one of these 4bp cut sites, you will have too many to look through. So in the dialog box, change the Number of Specific Bases option to "6+ bases." This will limit the search to enzymes that recognize a 6bp or larger palindrome.
5. Right now you are not concerned with how your DNA would be translated. Scroll down to the bottom of the dialog box. Change the option for Display Translation to "None." When you are done changing the defaults, click on Submit.
6. An output report from TACG is quite long, and often has a large amount of extraneous information at the top of the file. You are looking for the linear restriction map, which is near the end of the report (see Figure 3.16). The map shows 6+bp restriction sites for all enzymes that can cut the DNA sequence you specified.

Figure 3.16

A portion of a linear restriction map of a DNA sequence

All of the cut sites for a particular enzyme are listed under the enzyme's name, along with the sequence it recognizes. The name of a restriction enzyme over the sequence indicates that it will cut the DNA sequence at that particular position.

```
PvuII     CAG' CTG (0 Err) - 1 Cut(s)
      75

ScaI      AGT' ACT (0 Err) - 1 Cut(s)
     1039

SspI      AAT 'ATT (0 Err) - 2 Cut(s)
     585    715

XmnI       GAAnn' nnTTC (0 Err) - 1 Cut(s)
     920


== Linear Map of Sequence:
      DraI                      BmrI
      \                         \
 1   tttaaacgtcgtgactgggaaaaccctggcgttacccaacttaatcgccttgcagcacat  60
     aaatttgcagcactgaccctttgggaccgcaatgggttgaattagcggaacgtcgtgta
        ^    *    ^    *  ^    *    ^    *    ^    *    ^   *

                 PvuII        EarI                     PvuI
                 \            \                        \
61 ccccctttcgccagctggcgtaatagcgaagaggcccgcaccgatcgcccttcccaacag 120
   gggggaaagcggtcgaccgcattatcgcttctccgggcgtggctagcgggaagggttgtc
        ^    *    ^    *    ^   *    ^    *    ^    *    ^   *

        FspI          BglI
        \             \
121ttgcgcagcctgaatggcgaatgggacgcgccctgtagcggcgcattaagcgcggcggt 180
   aacgcgtcggacttaccgcttaccctgcgcgggacatcgccgcgtaattcgcgccgccca
        ^    *    ^    *    ^    *    ^    *    ^    *    ^   *
```

Using a TACG Report to Create a Restriction Site Mini-Map

TACG will show many more restriction sites than you can use. You must decide which ones will be useful and which ones you can ignore. Useful restriction enzyme sites will share two features. First, the enzyme is available. Some enzymes are very expensive, and all of them have a limited shelf life. So laboratories keep only certain ones in stock. You must take this into consideration when you plan a restriction mapping assay. Second, when pLac/WT and pLac/m# DNA are cut with the enzyme, the differences between the fragments released can be seen on an agarose gel. The gels you will use can accurately separate fragments from ~200 bp up to ~4000 bp long. Smaller fragments may run completely off of a gel or be too faint to see when stained, while larger fragments may not separate into distinct bands. Generally, you can see differences between two DNA bands if one is at least 10% smaller than the other.

1. Your instructor will provide a list of the restriction enzymes that are available to you. Write the list down in Table 3.3.
2. Begin searching for restriction sites for these enzymes in the TACG report for the pLac/WT sequence. Each time you find one, write down the position in the sequence in Table 3.3.
3. When you have tabulated all the useful restriction sites in pLac/WT, calculate the distance between the recognition sites for one restriction enzyme. After that, calculate the distances between the closest restriction sites for two different enzymes. Keep looking until you find two or three restriction enzymes that produce fragments that will separate well on an agarose gel. It is difficult to give more specific instructions on how to pick useful restriction sites. Your instructor will help you think through the process.
4. Once you have mapped out some useful restriction sites with pLac/WT, repeat the process for your assigned pLac/m# mutant construct. Try to find sites that are different in some way from pLac/WT. Record your data in Table 3.3.

Results

Table 3.3

Restriction enzyme sites

Restriction Enzyme	Position of Sites in pLac/WT, and Sizes of Fragments	Position of Sites in pLac/m#, and Sizes of Fragments

Mini-Maps of Your DNA Restriction Sites

The two mini-map templates in Figure 3.17 (p. 166) reflect how the *lac* operon fragment originally was inserted into the backbone plasmid to create the pLac/WT and pLac/m# plasmids you have been analyzing. Use the information in Table 3.3 to mark the approximate locations of restriction sites that you could use to confirm the presence and sequence of the inserted DNAs.

Figure 3.17

Two minimally annotated mini-maps for recording your DNA restriction sites

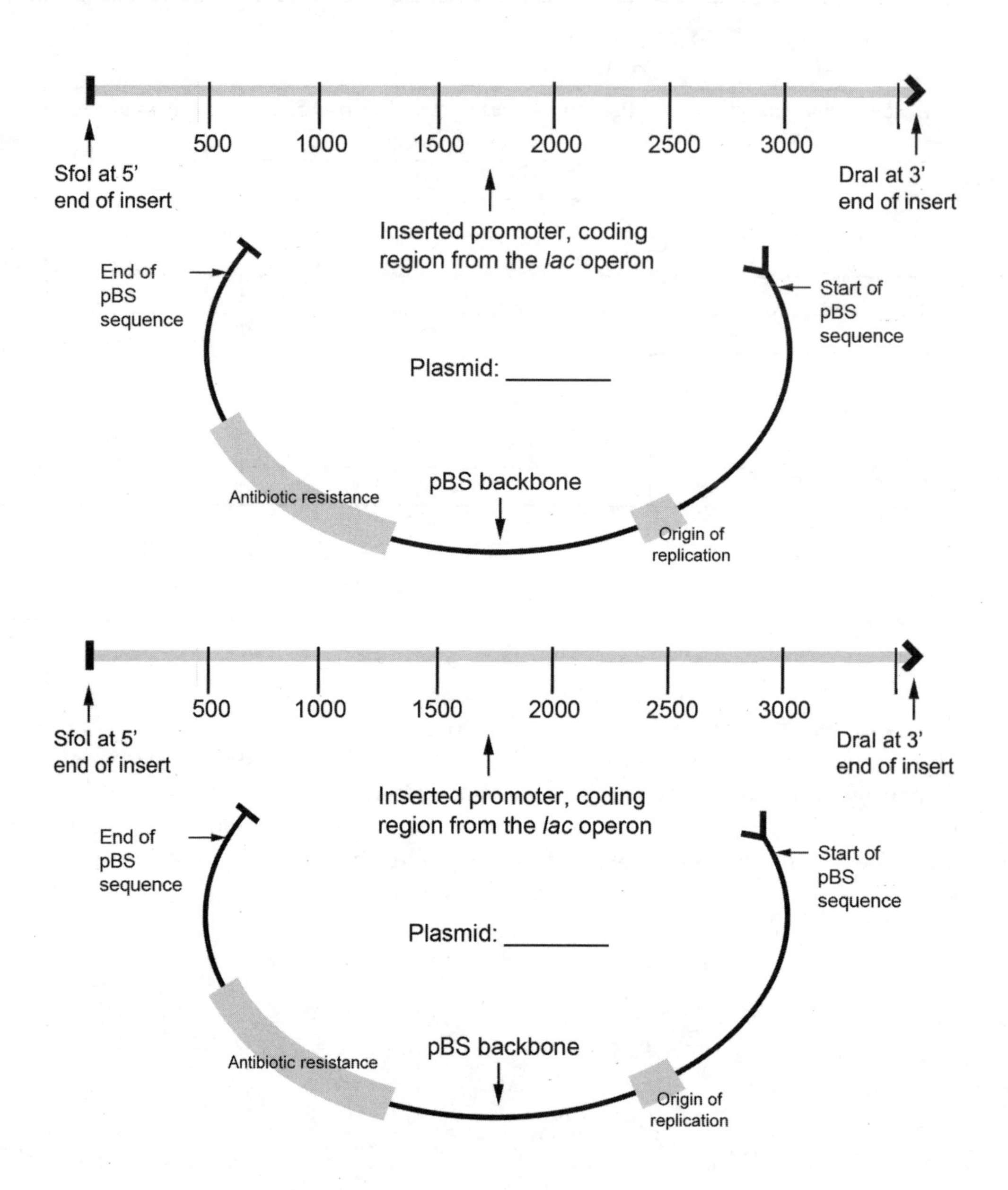

Summary Questions for Exercises 4 and 5

In addition to the questions below, you need to turn in your two mini-maps showing restriction enzyme sites for each of your two plasmids.

1. Based on the two maps you generated, which two enzymes can you use to demonstrate a difference in the sequences of pLac/WT and your pLac/m# mutant? For each enzyme, what will be the differences in sizes and numbers of fragments produced by digesting pLac/WT versus pLac/m#?
2. Where are the restriction enzyme cut sites you have chosen located in the operon fragment? In other words, are they in the promoter region, coding region for beta-galactosidase, or the plasmid backbone? How do you know?

Exercise 6: Restriction Digestion of DNA

Now you will actually digest the pLac/WT and pLac/m# mutant plasmids that you purified previously. As a control you will digest plasmid DNA of known size, that has one restriction site for the enzyme Hind III.

There are two warnings for this exercise. First, you will be pipeting very small volumes. When you add small volumes of solution to each tube, touch the pipet tip to the solution already in the tube. Once you push down on the plunger, look at the pipet tip to make sure all droplets of solution are left in the tube, not on the pipet tip. Second, restriction enzymes are stored in a solution containing 50% glycerol, which makes them very viscous and sticky. When you remove an enzyme from a stock tube, pay *close* attention. Never put a pipet tip all the way down in the tube of enzyme; you will pick up way more than you need. Just touch the TIP of the pipet to the surface of the enzyme plus glycerol, then withdraw what you need.

The two enzymes you will use are the ones you chose using Workbench in the preceding exercise. Individuals with different pLac/m# mutants may have chosen different enzymes, so the steps in the procedure will refer to your two selected enzymes simply as "Enzyme #1" and "Enzyme #2." It is up to you to keep them straight.

Procedure

1. Every restriction enzyme has a buffer and temperature at which it has the highest activity and specificity. Your instructor will provide the lab with a table that lists the specific buffer and incubation temperature for each enzyme that is available. In Table 3.4 (p. 168), enter the buffer and temperature of the two enzymes that you selected to use.

Table 3.4

Enzyme buffers and incubation temperatures

Name of Enzyme(s)	Buffer	Incubation Temp. (°C)
Enzyme #1:		
Enzyme #2:		
Hind III, for control DNA	Buffer #2	37°C

2. Your instructor will retrieve your pLac/WT and pLac/m# DNAs, and give you a tube of control DNA. Let them thaw, then mix each tube gently by flicking it with your finger.
3. Label eight small microfuge tubes with the numbers 1 through 8. Using Table 3.5 as a guide, add all the components except the enzyme. To help you keep track, mark off each item as you add it.

Table 3.5

Preparing restriction digests

Tube	Sample Description	DNA	Buffer	Water	Enzyme
1	Uncut pLac/WT	17 µL pLac/WT	none	3 µL	none
2	pLac/WT cut with Enzyme #1	17 µL pLac/WT	2 µL Bfr. ___	none	1 µL #1
3	pLac/WT cut with Enzyme #2	17 µL pLac/WT	2 µL Bfr. ___	none	1 µL #2
4	Uncut pLac/m#	17 µL pLac/m#	none	3 µL	none
5	pLac/m# cut with Enzyme #1	17 µL pLac/m#	2 µL Bfr. ___	none	1 µL #1
6	pLac/m# cut with Enzyme #2	17 µL pLac/m#	2 µL Bfr. ___	none	1 µL #2
7	Uncut control DNA	17 µL Control	none	3 µL	none
8	Control DNA cut w/Hind III	17 µL Control	2 µL Bfr. #2	none	1 µL Hind III

4. Once you have added everything but the enzyme, gently flick each tube to mix it.
5. Add 1µL of the correct enzyme to each tube. As you work, remember:

   - Keep all tubes of enzyme on ice, and immediately return them to the ice once you have removed what you need.
   - To prevent cross-contamination, use a fresh pipet tip each time you go back and take more enzyme from a tube.

6. Mix the enzymes with the DNAs by flicking each reaction tube with your finger.
7. When you are done, spin all eight tubes for one to two seconds in a microfuge to force all the liquid to the bottom of the tube.
8. Place the reactions at 37°C, or the temperature you wrote in Table 3.4.
9. To ensure complete digestion, your DNA samples will be allowed to incubate overnight, then frozen until you are ready to perform gel electrophoresis during a later lab period.

Exercise 7: Gel Electrophoresis of DNA

Background: Basic Principles of Electrophoresis

Gel electrophoresis separates charged molecules by how fast they migrate through a porous gel, under the influence of an electrical current. The gel consists of a shallow square tray filled with solidified agarose (a gelatin-like substance purified from seaweed) and containing pH 8.0 buffer as its liquid phase. A complex mixture of DNA fragments is placed in wells made at one end. When electrical current is passed through the gel, the negatively charged DNA molecules migrate toward the positively charged pole. The general process is shown in Figure 3.18 (p. 170).

Figure 3.18

Schematic diagram showing how DNA moves through a gel during electrophoresis

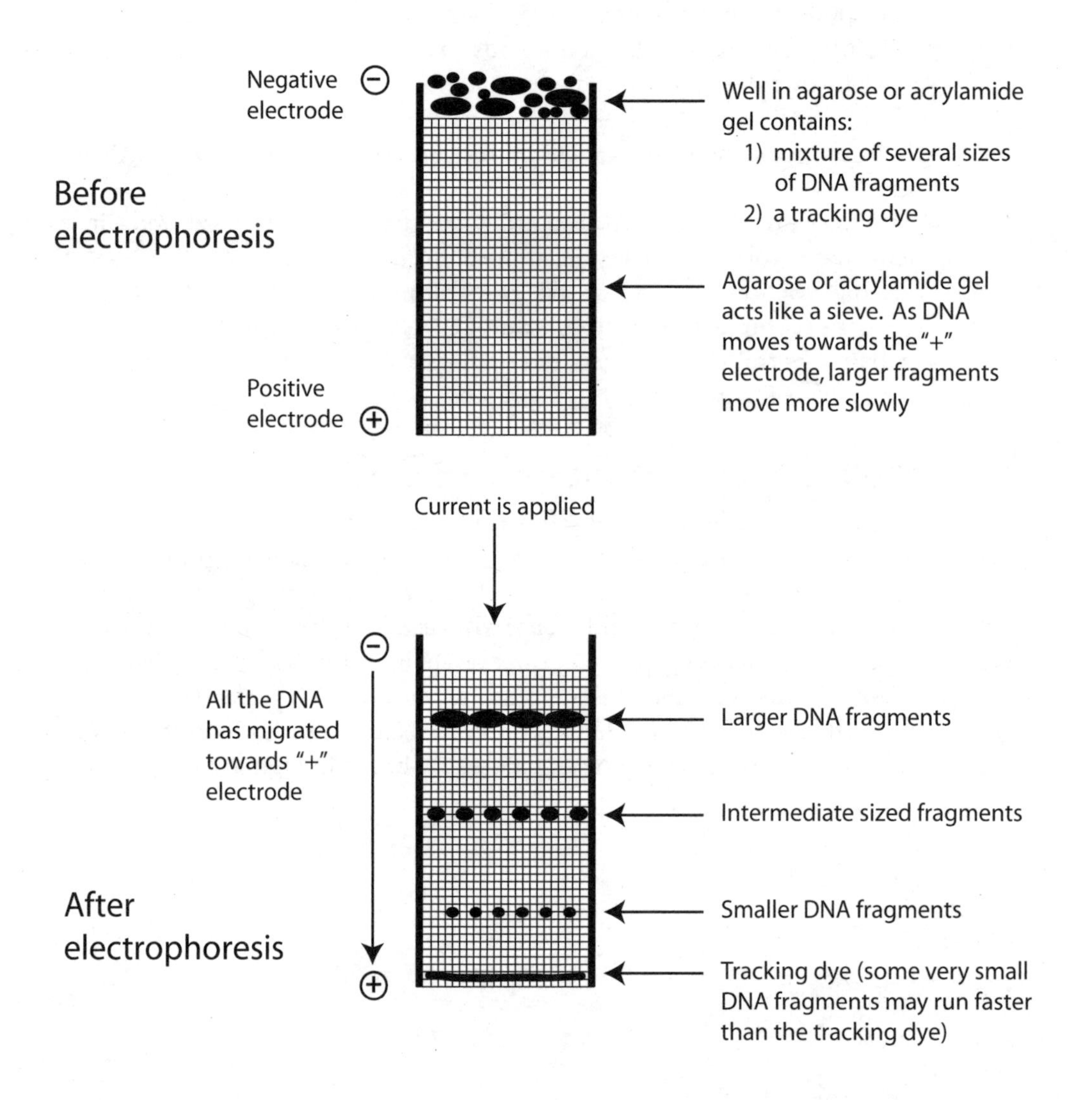

The overall rate at which a particular DNA molecule moves through the gel toward the oppositely charged pole depends on a number of factors:

- Size (molecular weight) of the DNA
- Shape (circular versus linearized) of the DNA molecules
- Porosity of the gel
- Buffer in the gel

Since the entire gel has the same buffer and porosity, DNA migrates at a rate that is proportional to the sizes of the molecules. By simultaneously electrophoresing a series of DNA standards of known size, you can estimate the size of every fragment that you generate in a restriction digest.

Safety Reminders

The gels used to separate your DNA contain ethidium bromide, which is a powerful mutagen. Never handle gels with your bare hands. Also, you cannot dispose of ethidium bromide-contaminated gels or other wastes in general trash. Put all pipet tips, gels, and other trash that come in contact with ethidium bromide into the proper labeled waste containers.

An electrophoresis unit can be deadly because it is filled with a highly conductive salt solution and uses DC current at high voltages. Always turn the main power supply switch "OFF" and wait 10 seconds before making any connection. Connect BOTH supply leads to the power supply just BEFORE turning on the power supply. After use, turn off the power supply, then disconnect both leads from the power supply. *Remember: Power supply on last...off first!* Also watch out for fluid leaks or spills! They can act as pathways for current and electrical shock or lead to fires.

You will view your stained gel using an ultraviolet light source. You must use safety goggles and a face shield while you are looking at or working anywhere near a UV light source. Otherwise, you risk blindness or a severe skin burn.

Procedure

1. Your instructor will prepare 1% agarose gels that you will use to separate your DNA fragments.
2. Thaw your completed restriction digests in your hand. Once thawed, add 5 μL of 6x DNA sample dye to each tube. Mix the reaction by flicking the tubes.
3. In the first well of your gel, load 18 μL of size marker mixture. This mixture contains DNA fragments ranging in size from 0.5 kb to 12 kb.
4. Write down the order in which you are going to load the remaining samples on your gel.
5. Open your first sample tube and carefully load all 25 μL of sample into the second well of the gel. Continue to load samples, in the same order that

you just wrote down. Be careful not to puncture the bottom of the well with the pipet tip, and avoid blowing air bubbles into the sample.

6. Connect the lid to the gel box and the electrical leads to the power supply. Remember that DNA is negatively charged, and thus it will run toward the positive electrode (red). (An easy way to remember this is "run toward red.")
7. Turn on the power supply, and set it to the voltage recommended by the gel box's manufacturer.
8. Let the gel run until the first dye front is about 2 cm from the lower end of the gel (~2 hours).
9. Turn off the power supply and disconnect the electrical leads.
10. Take a picture of the gel using an ultraviolet transilluminator box and a Polaroid camera, or use a video documentation system to create an electronic image of your DNA bands.
    Remember: Always wear gloves while handling DNA gels. You should also protect your eyes and face with a face shield and safety glasses while you are viewing your gel or working anywhere near a UV light box.

Estimating DNA Fragment Sizes

The size of an unknown DNA fragment is determined by comparing the distance it migrates on an agarose gel to the migration through the same gel of DNA fragments of known sizes. Figure 3.19 shows a previously completed mapping gel. This particular gel shows the separated fragments of a Hind III restriction digest. The DNA that was digested consisted of backbone plasmid that contained a 2 kb insert. The inserted DNA had Hind III sites on either side.

To determine the size of an unknown fragment of DNA on a gel, a standard curve for the gel must be constructed. The slope and intercept of the line you get allows you to estimate the size of any unknown DNA fragments that were separated on the same gel.

To learn how to create a standard curve from a gel photo, use Figure 3.19 and follow these steps.

1. Look at the lane with the DNA standard ladder. Use a millimeter ruler to measure the distance from the well at the top of the gel (where the samples were loaded) to the 1 kb DNA standard band near the bottom of the gel. Record your results in Table 3.6, page 174.
2. Measure the distance from the well to each of the other standard bands in your gel photograph. Record your measurements in Table 3.6.

Figure 3.19

Agarose gel of pBluescript containing a 2 kb insert

The standard ladder is on the left. The DNA digested with Hind III is in the right-most lane. The ~ 3 kb band is the pBluescript backbone vector DNA.

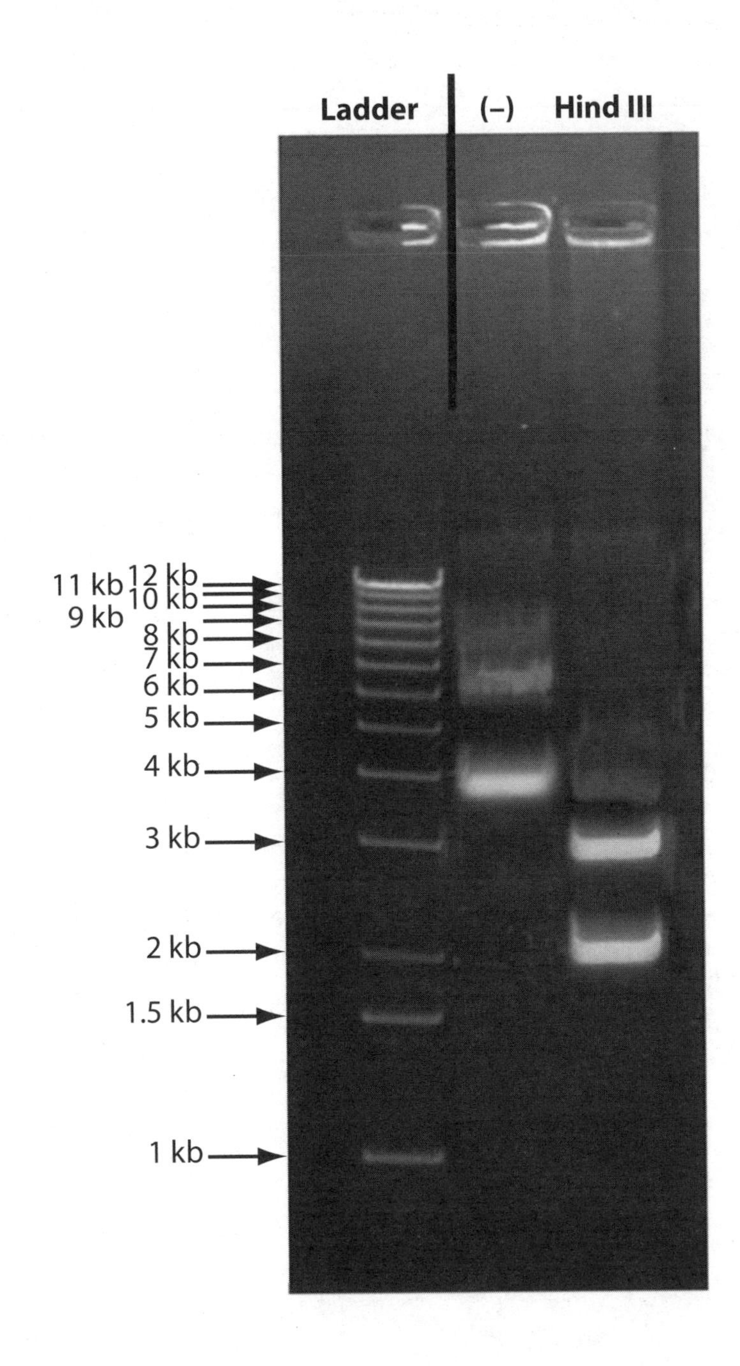

Table 3.6

Relative mobility of DNA standard bands and samples in Figure 3.19

Std. Band	Std. Band Size (base pairs)	Log10 of Size (bp) (*x* values)	Distance Migrated (mm) (*y* values)
#1			
#2			
#3			
#4			
#5			
#6			
#7			
#8			
#9			
#10			
	Hind III Sample Band #1		
	Hind III Sample Band #2		

3. Calculate the log10 of the number of base pairs in each standard band and enter it in the table.
4. Use a spreadsheet program or graph paper to plot the distance that each standard band migrated on the y axis, versus log10 of the number of base pairs on the x axis.
5. Determine the slope and y intercept for a line fitting the values you just plotted. Now you have a standard curve with which to determine the sizes of the two restriction fragments in the right-most lane.
6. In the lane marked "Hind III," measure and record the distances migrated from the well for each of the two bands.
7. Use the distances migrated as a y value and solve for x, using the slope and y intercept of the standard curve you just created.
8. Take the antilog of each x value you just calculated. These are the sizes of the bands in base pairs. The sizes that you calculated for the two bands in the restriction digest lane should be about 2000 bp and 3000 bp. If your calculations produced significantly different numbers (5% above or below the known sizes), go back and remeasure and recalculate the sizes of the unknowns.

9. Once you can calculate the sizes of the two unknown practice bands correctly for Figure 3.19, page 173, switch to the image of your own gel. Measure and record the migration distances of the standard bands in Table 3.7.

Table 3.7

Relative mobility of DNA standard bands in your gel

Std. Band	Std. Band Size (base pairs)	Log10 of Size (bp) (x-values)	Distance Migrated (mm) (y-values)
#1			
#2			
#3			
#4			
#5			
#6			
#7			
#8			
#9			
#10			

10. Next, measure the distances migrated by the DNA bands in each of your own restriction digests of pLac/WT and your pLac/mt# mutant plasmids. Record the migration distances for your digested pLac/WT and your pLac/m# mutant DNAs in Table 3.8, page 176.

Table 3.8

Sizes of DNAs from restriction digests in your gel

Sample and Major Band(s)	Distance Migrated (mm)	Calculated Size (in base pairs)
Your pLac/WT, digested with ________		
Band #1		
Band #2		
Band #3		
Band #4		
Your pLac/WT, digested with ________		
Band #1		
Band #2		
Band #3		
Band #4		
Your pLac/m#, digested with ________		
Band #1		
Band #2		
Band #3		
Band #4		
Your pLac/m#, digested with ________		
Band #1		
Band #2		
Band #3		
Band #4		
Control Plasmid, digested with Hind III		
Band #1		
Band #2		

11. Use the migration distances of the standard bands in Table 3.7 to create a standard curve. Calculate the sizes of the unknown bands in each of your restriction digests, then enter them in Table 3.8.

Summary Questions for Exercises 4–7

In addition to answering the following questions, you should turn in a copy of the restriction enzyme map you created for each of your two plasmids, copies of your data tables, and a photo or printout of your gel from the restriction mapping.

1. Did the sizes of the bands you obtained when you actually digested your pLac/WT and pLac/m# mutant plasmid DNAs match the sizes of the bands you predicted based on your TACG search? If not, what might account for the differences?
2. Were there any differences between the number and size of fragments produced by restriction digestion of pLac/WT and your particular pLac/m#? If so, did you predict them in advance? If not, where might they have come from?
3. Did your results confirm the presence of the mutations you identified before in your assigned pLac/m# mutant?

DNA Isolation and Analysis

Teacher Pages

INSTRUCTORS' NOTES

Background and Key Concepts

Nearly all students at the introductory level can define a mutation as a change in the sequence of DNA. If pressed to apply this knowledge though, the majority will have compartmentalized this as an isolated fact and be unable to explain how or why a change in DNA sequence actually leads to changes in phenotype and function for a cell. This unit was designed to help students bridge that intellectual gap. In it they isolate wild-type and mutant DNAs from bacteria, then make direct comparisons between their nucleotide sequences using bioinformatics software. Students also predict what effects the mutations they identify will have on the phenotype of the bacterial strains carrying them.

Students work with pLac/WT, a plasmid that contains the normal lac promoter and operator regions, and a modified version of the coding region for LacZ/bGAL. In pLac/WT, the ATG start site is in its proper place relative to the -10 site and operator. Just after the start site there is an in-frame, 24-nucleotide multiple cloning site. The bGAL enzyme that is encoded by pLac/WT has eight additional amino acids, but they do not affect enzyme activity. Moreover, students are unlikely to notice that there is a multiple cloning site that interrupts the normal sequence because they never align and compare the sequences for pLac/WT and the native lac operon in Genbank. The backbone of the plasmid is derived from pBlueScript II KS+. It contains a high copy number origin of replication and confers ampicillin resistance.

Using the pLac/WT as a starting point, specific mutations were generated in various regions of the promoter, operator, and coding sequences; each mutation was designed to add or remove at least one restriction site from pLac/WT. The various mutants and the wild-type plasmid were sequenced then transformed into *E. coli* strain 71-18, which lacks part of its endogenous *lac* operon.

As they progress through this unit, students learn how to

- isolate plasmid DNA by alkaline lysis,
- enter and edit DNA sequences using web-based bioinformatics software,
- identify mutations by aligning wild-type and mutant DNA sequences,
- predict the effects of DNA mutations on protein function,
- identify restriction sites in DNA,
- predict the sizes of DNA fragments that will be produced by a restriction digest, and
- determine the sizes of DNA fragments by comparing them to size markers.

By the end of the unit, students should be able to answer these questions.

- How is protein expression controlled in bacteria? What is a promoter? An operator?
- How is plasmid DNA different from genomic DNA? How are these differences exploited to extract plasmid DNA from bacteria?
- How can plasmids change the phenotype of a bacterial strain? Why is this important in nature and in the laboratory?
- What is the purpose and value of bioinformatics databases? Why is it useful to compare amino acid or DNA sequences?
- What are some common types of DNA mutations? Does a mutation always affect protein synthesis? Why or why not?

General Teaching Strategy and Common Problems

Like Unit 2: Mendelian Genetics, this is another multiweek series of exercises. It was originally designed for a sophomore majors course in genetics and molecular biology. Students will need four lab periods of about three hours each to complete the entire unit. The instructions in the manual and supplement are written assuming each lab section meets once a week, but the exercises can be completed in less time if students meet more often.

A week before the unit starts, students are given two LB-agar plates streaked with 71–18. One plate has bacteria carrying pLac/WT, and the other has bacteria carrying one of the mutant pLac/m# strains. The instructor demonstrates how to inoculate bacterial cultures, so that the students are responsible for starting their own 3 mL overnight cultures the day before their lab section meets. Alternatively, the instructor can start bulk liquid cultures and provide them to students on the first day.

In the first week, students isolate plasmid DNAs by alkaline lysis and, if there is time, quantify their DNAs with a check gel. They must obtain enough DNA to perform restriction digests, or they will have to repeat the procedure outside of normal lab time. Alternatively, instructors can supplement their material with DNA purified in bulk using a binding resin/column protocol. In the second week, students start their sequence analysis and, if not done beforehand, run their check gel. Students learn how to use WorkBench (a suite of bioinformatics computer programs) by splicing two training sequences called AmpR #1 and #2 and aligning them to a known standard. Next students align the sequences for pLac/WT and their pLac/m#, and identify the potential mutations present. In the third week they use WorkBench to find restriction sites that differ between pLac/WT and their pLac/m#. Based on their results they set up restriction digests using the DNAs they purified two weeks before, plus positive controls. The DNAs are digested overnight, then stored frozen. During the final week, students separate their plasmid DNA fragments using agarose gel electrophoresis and compare the restriction maps of their pLac/WT and assigned pLac/m# DNAs.

None of the individual steps of this unit are difficult, so most introductory students can complete them correctly. However, students may get lost in procedural details or fail to pay attention to why each step is necessary. The instructor should ask questions regularly of students to ensure they understand the purpose of each component. Stress to students that they are using relatively simple tools that can be applied to many tasks. It is the information they obtain by using these tools that is valuable. This unit is particularly good for encouraging students to use a lab notebook properly.

It is important to stress to students that DNA is simply an instruction set. The proteins and RNAs that are made using DNA as a template are what actually carry out most of the functions. Changing an RNA's nucleotide sequence or a protein's amino acid sequence is what changes their structure and functional properties. This in turn affects the overall organism or, in other words, it alters the organism's phenotype.

Recommended Prelab Skills

Students should know how to use serological and micropipetors. Basic microbiological methods are helpful, but not essential. They should be comfortable using a web browser and be able to perform basic "cut and paste" functions in word processing and spreadsheet programs. If students will make poster presentations, basic experience with a slide show or graphic design program is helpful.

Assessment

Given the difficulty of the material, prelab questions should not be a routine part of quizzes. Students are less likely to confuse the various elements in this unit if they take frequent short quizzes. Weekly postlab quizzes can be administered at the start of the next regular lab meeting. These should emphasize general knowledge and specific skills from the preceding week. More specifically, after completing Week 1 students should be able to describe the functions of plasmids in bacteria and how they are used in research. Upon completing Weeks 2 and 3, they should be able to describe the purpose and uses of the general classes of bioinformatics programs, including sequence alignment tools (BL2SEQ, CLUSTALW), and sequence analysis programs (TACG). They should be able to explain what restriction enzymes are, why bacteria make them, and in general how restriction enzymes are used in a research setting. Once they complete Week 4 students should be able to explain the general principles of agarose electrophoresis. They should be able to construct a semi-log curve and use it to estimate the size of a fragment of DNA of unknown size.

If an instructor chooses to give a final summary test at the end of the unit, the questions should emphasize higher-level thinking skills, not technical details. For example, students could be given a pair of novel sequences to align and compare. Another option is to provide them with a photograph of an agarose gel and ask them to interpret it. By the end of this unit it is also reasonable to ask students to explain how and why a mutation in a DNA sequence results in a change in phenotype of an organism.

In addition to regular postlab quizzes, students can be assessed based on personal notebooks and either a written report or a scientific poster. For posters, each student work group summarizes their analysis of a particular pLac/m# mutant, and presents their findings to the rest of their lab section. The poster should answer all of the following questions:

- What mutation(s) were found in the pLac/m# sequence?
- How are the mutations predicted to affect bGAL enzyme production or activity?
- What restriction enzymes were used to map pLac/WT and the mutant? What were the results?

Safety and Housekeeping

Ethidium bromide is a powerful mutagen. Students should collect all waste gels, staining solutions, and other bulk waste for decontamination and disposal. They must never put any ethidium-contaminated materials into general trash. Review basic electrical safety procedures with all students so they do not misuse a gel box or power supply and electrocute themselves. Students who are working around UV light sources must NEVER look at a UV source without safety glasses and a full face shield or other UV-blocking shield in place.

Other Tips

- Occasionally students cannot purify enough plasmid DNA for restriction digests, even after repeated attempts. As a backup, plasmids are isolated from 1 L cultures of bacteria carrying pLac/WT and each pLac/m# mutant. The DNAs are split into aliquots that are sufficient for one semester then stored frozen at -70°C, where they are stable for several years. Students can use the instructor's DNA or, alternatively, perform digests with the instructor's material alongside their own DNAs.
- If it is not practical to let students set their own overnight cultures, the instructor can grow enough of the bacteria containing pLac/WT and each pLac/m# plasmid for an entire week of lab, then break the cultures into 3 mL aliquots. These can be stored up to five days in the refrigerator before extracting plasmids without a significant reduction in yield.
- WorkBench has an integrated, context-sensitive Help menu. Selecting Help brings up a tutorial for the current program, which explains the default assumptions and parameters used. The instructor is strongly urged to take the time to review this information.
- When aligning sequences with BL2SEQ to find overlapping ends, the program may insert gaps in the sequence to get the best fit and not locate the overlap at the ends of the files. If this happens, use the browser's "Back" button to return to the program parameters, then click the selection box next to "Disable Gapped Alignment." This prevents sequence breaking, and the report should have just one or two alignments; one of them should show where the ends of the sequences overlap. Another alternative is to use Align instead. If you do, look under the option "Use End-Gap Weighting" and click on the button for "No."
- Instructors may want to supplement this unit with additional molecular methods. One option is to give students a small dilute aliquot of each plasmid, and let them transform and plate the 71-18 bacteria at the start of the unit. The instructor also might ask students to download the sequence of the endogenous *lac* operon, align it with pLac/WT, and describe the sequence changes between the native and cloned DNAs.
- Many institutions no longer permit use of ethidium bromide. Ethidium bromide can be replaced with any of the proprietary, less-toxic alternative dyes sold by science educational supply houses or molecular biology supply companies.
- Students also can test for production of functional bGAL enzyme by pLac/WT versus their pLac/m# mutant. There is a standard liquid cell lysate assay (see, for example, Goulding 1986), except chloroform is substituted for toluene in the lysis step. There is a problem in that pLac/WT behaves as if its promoter is constitutively active; apparently strain 71–18 does not produce enough LacI to block all copies of the plasmid, so the promoter leaks. There is a workaround strategy, and those who are interested in incorporating the liquid bGAL assay should contact the author for more details.

Supplemental Reference

Goulding, K. H. 1986. The time course of β-galactosidase induction in *Escherichia coli*. In *Experiments in molecular biology*, ed. R. J. Slater, 227–236. Clifton, NJ: Humana Press.

PREPARATORY NOTES

Quantities listed are for a lab section of 20 students working in pairs.

Before Week 1: Starting Overnight Cultures

Shared Materials

- 12 LB amp agar plates streaked with 71-18 *E. coli* bearing pLac/WT DNA
- 2 LB/amp agar plates of each mutant, streaked with 71-18 *E. coli* bearing pLac/# DNA
- 50 bacterial culture tubes (13 mL) (Falcon 2057, 2059, or equivalent) containing 3.5 mL of LB media supplemented with 50 µg/mL ampicillin (refrigerated)
- Shaking incubator with racks for culture tubes, preset to 37°C and 220 rpm.

Materials at Each Work Station

- Bunsen burner
- Striker
- Metal forceps
- 50 mL conical tube containing 10 to 12 autoclaved toothpicks
- 2 L waste bottle containing 200 mL of undiluted bleach (for bacterial waste)
- 10% bleach in a spray bottle (for cleaning up bacterial spills)
- 70% ethanol in a spray bottle (for general cleaning)

Week 1: DNA Purification, Check Gels: Shared Materials

- 2 aliquots (20 mL) of Solution 1
- 2 aliquots (20 mL) of Solution 2 (must be made fresh daily)
- 2 aliquots (20 mL) of Solution 3 (store refrigerated)
- 2 bottles (100 mL) of 100% ethanol (store in freezer)
- 2 bottles (100 mL) of 70% ethanol (store in freezer)
- 2 aliquots (20 mL) of sterile distilled water
- 2 to 3 microcentrifuges
- Disposable gloves
- ~6 g agarose, powder
- 10 L 1x TAE
- 100 microliters DNA standard (lambda DNA, digested with BstE II)
- Ethidium bromide solution, 10 mg/mL

Week 2: Materials at Each Work Station

- P-20, P-200, and P-1000 micropipets
- Boxes of yellow, blue tips, autoclaved
- 2 microfuge tubes (1.5 mL) in a 250 mL beaker, autoclaved
- 10 microfuge tubes (0.6 mL) in a 250 mL beaker, autoclaved
- Microfuge tube rack
- 200 mL 6x DNA sample loading buffer
- Agarose gel rack and box, with 10-well comb (for check gels, one box can be shared by two work groups)
- Power supply (shared by two work groups)
- 2 L waste bottle with 200 mL of bleach (for bacterial waste)
- 10% bleach in a spray bottle (for cleaning up bacterial spills)
- 70% ethanol in a spray bottle (for general cleaning)

Week 2: Bioinformatics (Part 1): Shared Materials

- Text file posted online or available on a disk, containing the following sequences:
  - ◊ AmpR #1
  - ◊ AmpR #2
  - ◊ Insert for pLac/WT
  - ◊ Insert for each pLac/m# mutant
  - ◊ Backbone plasmid

Week 2: Materials at Each Work Station

- Laptop or desktop computer with internet connection and web browser

Week 3: Bioinformatics (Part 2), Restriction Digests: Shared Materials

- 37°C water bath
- 50°C water bath (for incubating digests using BstX I)
- 60°C water bath (for incubating digests using BsrG I or BstE II)
- 2 microcentrifuges
- Floating microtube racks (alternative is sheets of closed cell foam)
- DNAs that students purified in Week 1
- Restriction enzymes (keep on ice)

  Aat II, Ase I, BamH I, Bgl II, BsrG I, BstE II, BstX I,
  Cla I, EcoR I, EcoR V, Hind III, Hpa I, Nco I, Pvu II
- 20 mL aliquots of reaction buffers for listed restriction enzymes
- 12 aliquots (40 μL each) of pBR322 (or other plasmid with a Hind III restriction site)

- 12 aliquots (40 µL each) of purified backup pLac/WT (optional)
- 3 aliquots (40 µL each) of purified backup pLac/m# (optional)

Week 3: Materials at Each Work Station

- P-20 and P-200 micropipets
- Boxes of yellow tips, autoclaved
- 20 microfuge tubes (0.6 mL) in a 250 mL beaker, autoclaved
- 1 mL microtube of sterile distilled water
- Microfuge tube rack

Week 4: Agarose Gel Electrophoresis: Shared Materials

- 12 aliquots (20 µL) of kilobase DNA standards ladder (store frozen)
- Rack or box containing students' digested DNAs from Week 3
- 2 to 3 microcentrifuges
- Disposable gloves
- ~6 g agarose, powder
- 10 L 1x TAE in a carboy
- Ethidium bromide solution, 10 µg/mL
- UV transilluminator
- Polaroid camera or other method for gel documentation
- Ruler that measures in millimeters

Week 4: Materials at Each Work Station

- 100 µL 6x DNA sample loading buffer
- Agarose gel rack and box, with 10-well comb
- Power supply (shared by two work groups)
- P-20 and P-200 micropipets
- Boxes of yellow tips, autoclaved
- 20 microfuge tubes (0.6 mL) in a 250 mL beaker, autoclaved
- Microfuge tube rack

Sources of Materials

Wild-type and mutant pLac/# plasmids, along with the sequence files, are available upon request from the author. Table 3.9 lists the plasmids by stock names; they should be assigned random numbers before being given to students, and the numbers should be changed each semester. Alternatively, new mutations can be generated using site-directed mutagenesis; this would be a good project for students in an advanced course.

Table 3.9

Characteristics of mutant plasmid DNAs

Name	Type, Location of Mutation	Major Changes in Restriction Map
pLac/WT	Wild type; no mutations	NA
pLac/Prom	A 1 bp mutation in the promoter blocks all bGAL activity	New BsrG I site added
pLac/Oper	A 2 bp mutation in the operator makes the operon constitutively active	New BsrG I site added
pLac/NcoA	A 1 bp insertion in 5' coding region creates frameshift and causes early termination	New Nco I site added
pLac/NcoB	Mutation creates a silent polymorphism in the early coding sequence	New Nco I site added
pLac/Hpa	Internal ~700 bp deletion within the coding region of the LacZ	Aat II, Hpa I, Cla I sites deleted; EcoR V fragment shortened
pLac/Trunc	Truncation of 3' end of coding region	Aat II, Hpa I, Cla I, EcoR V sites deleted
pLac/BAD	LacZ coding region translocated to araBAD promoter. This is a commercial plasmid (pBAD-Myc/His/Lac) that must be purchased from InVitrogen	Bgl II, BstE II, EcoR V, Nco I sites added Ase I, Cla I sites deleted

Plasmid pBR322 is the control plasmid for restriction digests, but other plasmids may be substituted for it. All plasmids are maintained in *E. coli* strain 71-18.

Stocks of pBR322, pLac/WT, and pLac/m# plasmid DNAs are prepared from 1 L liquid cultures of LB supplemented with ampicillin, using either CsCl purification or a bulk DNA prep kit. The kit is easier to use, but is more expensive. Stock DNAs are diluted to 0.1–0.15 µg/mL with distilled water, broken into 40 µL aliquots, and stored frozen. Stocks should be used within six months if stored at -20°C, but will keep indefinitely when stored at -80°C. If an ultracold freezer is unavailable, aliquots of DNA can be precipitated with sodium acetate/ethanol and stored as pelleted DNA under alcohol indefinitely. Pelleted DNAs are redissolved as needed in distilled water.

One lab section of 20 students working in pairs requires 75 µg of pBR322, 75µg of pLac/WT, and 20 µg of each of the pLac/m# mutants.

Maintaining Organisms

Saturated 1 mL cultures of 71–18 bacteria that are carrying plasmids can be mixed 1:1 with glycerol and stored indefinitely at -80°C. To start new cultures, scrape a sterile toothpick or yellow pipet tip across the frozen stock, then inoculate an LB-agar plate supplemented with ampicillin. Select well-isolated individual colonies for further amplification and to prepare students' stock plates.

Solutions, Reagents, Equipment

Ampicillin, 100x stock (50 mg/mL)

Dissolve 0.5 g ampicillin in 10 mL distilled water. Filter sterilize by forcing solution through a syringe filter with 0.22 µm pore size. Store frozen in 1–2 mL aliquots. Thawed aliquots can be stored for a week in the refrigerator. To inoculate liquid media, add 1 mL of ampicillin antibiotic per liter of media. Final concentration of antibiotic will be 50 µg/mL.

Luria-Bertani (LB) broth, LB agar

Prepare commercial LB mix according to instructions on the container. If commercial mix is unavailable, combine the following:

- 10 g Bacto-tryptone
- 5 g Bacto-yeast
- 5 g NaCl
- 2 mL 1N NaOH
- 1000 mL distilled water

Mix until dissolved. For pouring plates, add 15 g/L of microbiological grade agar. Split media into required volumes; if using Erlenmeyer flasks, select flasks that are 2–4 times larger than the volume of media. Cover the tops of flasks with foil, and autoclave 20 minutes on liquid cycle. Allow media to cool to <50°C before adding ampicillin or pouring plates. If required, add 1 mL of 1000x stock ampicillin per liter of LB media. If the LB media is not going to be inoculated immediately, wait and add the ampicillin just before using it.

Glucose, 1 M

Dissolve 18 g glucose in 100 mL distilled water. DO NOT autoclave.

Tris-Cl, 1 M, pH 8.0

Dissolve 12.11 g Tris base in 60 mL distilled water. Adjust to pH 8.0 with concentrated HCl. Bring to 100 mL with water.

EDTA, 500 mM

Mix 93.1 g ethylenediamine tetraacetic acid-disodium salt with 400 mL distilled water. Begin stirring, then add 10 M NaOH until pH is stable at 8.0 (~25 mL); the powder will not go into solution until the pH is adjusted. When dissolved, bring volume to 500 mL.

NaOH, 10 M

Dissolve 40 g sodium hydroxide in 50 mL water. Bring to 100 mL.

SDS, 10%

Dissolve 10 g sodium dodecyl sulfate in 90 mL distilled water. Heat gently to dissolve, then adjust volume to 100 mL.

Potassium acetate, 5 M

Dissolve 4 g potassium acetate in 50 mL distilled water, then bring to 100 mL final volume.

Solution 1

Mix:

5 mL 1 M glucose
2 mL 500 mM EDTA
2.5 mL 1 M Tris-Cl pH 8

Bring to 100 mL with water. Use within 24 hours. For longer storage, refrigerate between labs, or autoclave aliquots. If you see bacteria growing in the solution, discard it.

Solution 2

Mix:

8.8 mL water
1 mL 10% SDS
0.2 mL 10M NaOH

This solution must be made on the day of the lab. Mix the SDS and water, then add NaOH last. If SDS precipitates, mix for several minutes to re-dissolve.

Solution 3

Mix:

60 mL 5 M potassium acetate
11.5 mL glacial acetic acid
28.5 mL distilled water

The final solution is 3 M potassium, and 5 M acetate. Store refrigerated.

6x DNA sample buffer

Mix:

22.5 mL distilled water
0.125 g xylene cyanol
0.125 g bromophenol blue
2.5 mL 500 mM EDTA
25 mL glycerol

Dissolve dyes in water, then add EDTA and glycerol. Divide into aliquots of 1 mL or less. Unused buffer may be refrigerated for three months or frozen for one year.

TAE buffer, 10x stock

Mix:

48.4 g Tris base
11.4 mL glacial acetic acid
40 mL 500 mM EDTA
900 mL distilled water

Check the pH; if between 8.2 and 8.6, bring up to 1 L with water. To use, dilute 1:10 with distilled water.

Ethidium Bromide, 10 mg/mL

To avoid handling powder, purchase premade solutions.

Agarose, 1%

Mix 6 g of agarose with 600 mL 1x TAE buffer in a liter flask. Microwave on high, mixing occasionally, until agarose melts. Let cool to ~50°C (you should be able to hold your hand on the bottom of the flask for more than two seconds). Add 10 μL of ethidium bromide (10 mg/mL), fill gel racks, and set sample combs in place.

Lambda and Kilobase DNA Standards Ladder

Dilute BstE II-digested lambda DNA standard solution to a FINAL concentration of 0.4 mg/mL with TE (10 mM Tris pH 8.0, 1 mM EDTA). Add 1/6 volume of 6x DNA sample buffer. Split into 20 μL aliquots in 0.6 mL microtubes and store frozen.

Prepare kilobase ladder according to manufacturer's instructions. Add enough TE so that the final volume to be loaded is 15 μL/lane. Add one-sixth volume of 6x DNA sample buffer and mix well. Split standard into 20 μL aliquots in 0.6 mL microtubes. Store frozen.

Restriction Enzymes and Buffers

Do not purchase more enzymes than can be used in one year. Store enzymes in non–frost free freezers. If one is not available, store enzymes in a closed plastic box with ice packs to insulate them from rapid changes in temperature during defrost cycles. Enzymes may lose activity if split into small tubes. It is better if the instructor removes an aliquot of each enzyme from the stock vial and gives it to students during the lab, rather than preparing small aliquots beforehand. Reaction buffers can be split in 20 μl aliquots and refrozen prior to lab.

Properties of Enzymes

Student Pages*

BACKGROUND

Enzymes are specialized proteins that catalyze chemical reactions in biological systems without being permanently changed or used up themselves. Many biologically important molecules are quite stable. Without enzymes, cells and organisms could not break down or construct molecules quickly, and metabolic reaction would proceed *very* slowly.

Metabolic reactions (the chemical reactions in cells and organisms) do not always run at the same speed. The rate of any reaction in a cell can be turned up or down by changing the activity of the enzyme responsible for that reaction. These exercises will let you discover firsthand how cells might regulate their enzymes.

How Do Enzymes Work?

Enzymes catalyze reactions by lowering the **activation energy (E_A)** so that the reaction starts more easily. Most enzymes are proteins that have folded into specific three-dimensional structures. The shape includes one or more active sites that help stabilize the intermediate steps in a chemical reaction and make it more likely to occur.

To give you an idea of how much an enzyme can lower the energy of activation, consider what happens when cellulose (wood fibers) is broken down. We routinely do this when we burn wood. There is a noncatalyzed reaction that releases heat, water, and carbon dioxide. Overall, burning is an exothermic reaction, but we must put considerable heat into the system to start it. Wood must be heated to over 400°F to ignite; if it is wet, the wood must be heated long enough to drive off most of the water. When wood burns, most of the energy released by the reaction is lost as heat.

*Teacher Pages begin on page 207.

Now think about the mushrooms that spring up after a summer rainstorm. Most are decomposers that use the enzyme **cellulase** to break down cellulose for energy. The reaction occurs in wet wood at ambient temperatures. In addition, mushrooms convert the cellulose into water and carbon dioxide via small stepwise reactions so that the mushrooms can capture, store, and reuse a large part of the energy in the cellulose molecules for other metabolic processes, rather than releasing it all as heat.

Factors That Affect Enzyme Activity

Anything that changes the shape of a binding site on an enzyme will affect the enzyme's activity. The shape can be affected by pH, ion concentrations, or temperature. Molecules other than the main substrate may also compete for the same active site and inhibit the enzymatic reaction. To further complicate things, coenzymes like metal ions may be required for the active site to function properly. In this laboratory, you will examine how some of these factors can affect the activity of a model enzyme.

Our Model Enzyme

There are many forms of sugars. Simple sugars (monosaccharides) are molecules like glucose or fructose. Simple sugar molecules can be covalently bonded together into groups, ranging from two (disaccharides) to thousands of repeated units (polysaccharides). You are familiar with many of the complex sugars already. Both starch and cellulose are polysaccharides made up of repeated glucose molecules. Among the disaccharides you already know about is sucrose (the chemical name for table sugar), but there are many others as well. Another you may have heard of before is lactose (milk sugar), which is a combination of glucose and another simple sugar, galactose.

Almost all living organisms can use sugars to provide energy for their cellular processes. However, cells can only import simple sugars; unless di- and polysaccharides are broken down into their single sugar components, most cells cannot use them for energy. The enzyme that you will study in these exercises is **β-galactosidase** (bGAL for short), the enzyme that most organisms use to split lactose into the monosaccharides galactose and glucose. A schematic of the reaction catalyzed by bGAL is shown in Figure 4.1.

Figure 4.1

Schematic diagram of how the enzyme β-galactosidase (abbreviated bGAL) cleaves the two-ring sugar lactose

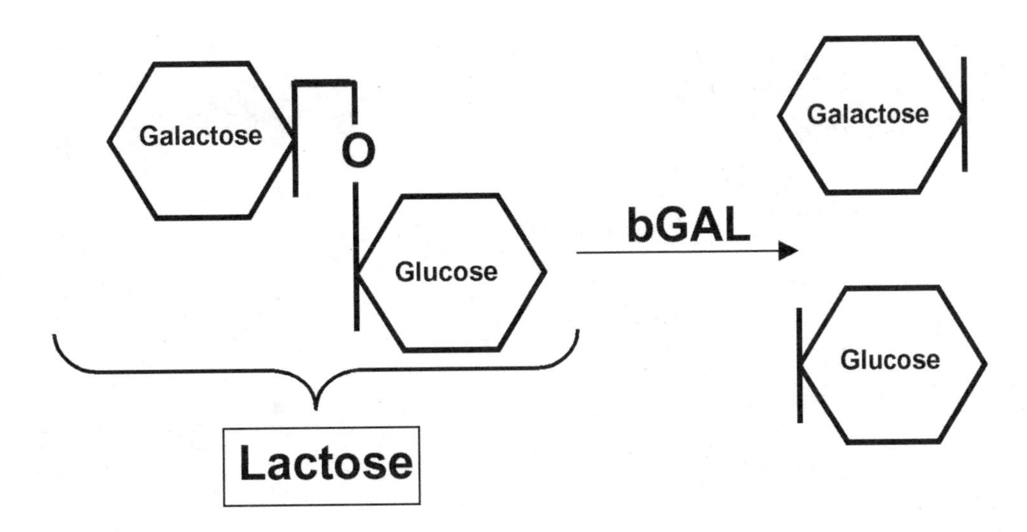

In case you think bGAL is not a very interesting model enzyme, remember that millions of people either are born without the ability to make bGAL or lose the ability to make it once they become adults. When the cells in their pancreas stop producing the enzyme, they cannot digest lactose, so the sugar passes into their large intestines. The sugar prevents water from being reabsorbed, and the afflicted persons develop gas and frequent diarrhea. This condition is called **lactose intolerance**, and as a result millions of people cannot comfortably eat foods that contain lactose, such as milk, butter, cheese, yogurt, ice cream, or other dairy products.

The bGAL activity in a sample can be measured using **ONPG**, a modified substrate that resembles lactose but has a colorless, unstable dye molecule attached to the galactose rather than the normal glucose. If bGAL is present, it hydrolyzes the chemical bond and releases the dye from galactose. The unstable dye molecule rearranges itself to form a bright yellow product. The reaction can be summarized schematically as shown in Figure 4.2, page 196.

Figure 4.2

Schematic diagram of cleavage of the test substrate ONPG by β-galactosidase
ONPG = O-nitrophenyl-β-D-galactopyranoside (which is colorless); ONP = O-nitrophenol (the bright yellow product); bGAL = enzyme

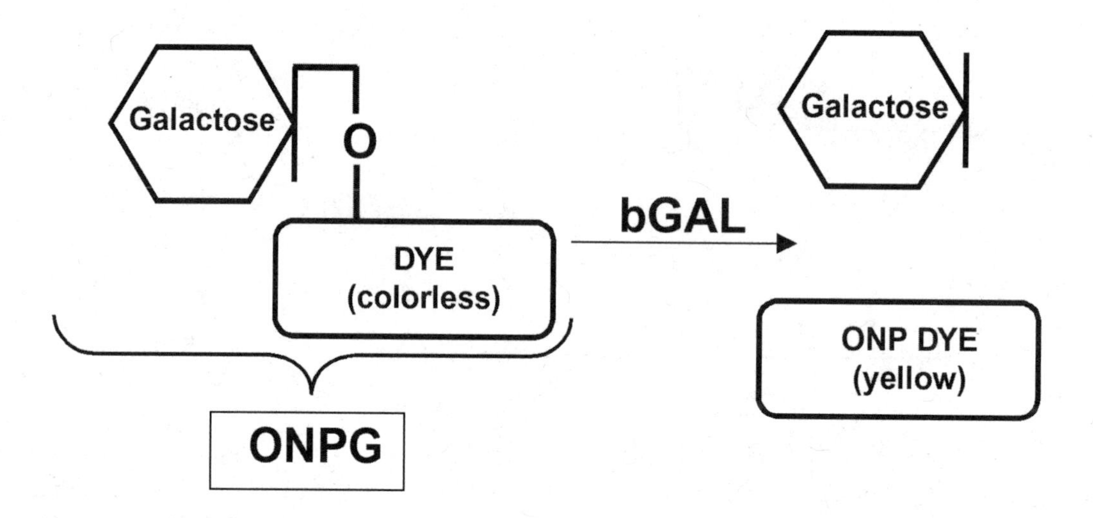

The amount of yellow product produced by the above reaction is directly proportional to the amount of enzyme present. The amount of yellow product formed can be measured using an instrument called a spectrophotometer.

Goals for This Unit

You will use β-galactosidase purified from a cereal mold (*Aspergillus oryzae*) to answer the following questions:

How does substrate concentration affect enzyme activity?

What is the effect of enzyme concentration on the rate of a reaction?

What is the effect of temperature on activity of bGAL?

Once you have completed the three exercises, you will develop an experiment of your own that you will carry out. You can use purified enzyme from mold or other sources to determine the effects of pH, or of various cofactors and possible inhibitors, on enzyme activity. Alternatively, you may decide to test something other than these variables.

General Safety Precautions

The enzyme assay uses an acetate buffer that has a low pH of 4.5. This is about the same acidity as white vinegar. It may irritate your skin, so immediately wash off any buffer that spills onto your skin. Both ONPG substrate and its cleaved yellow ONP product are irritants, especially if splashed into your eyes. Be careful when mixing the solutions. In addition, ONPG and ONP reaction products produced during the enzyme assays are toxic wastes that must be collected for destruction. Place all waste solutions into the designated waste containers; *do not* pour them down the drain.

UNIT EXERCISES

General Procedure

In each exercise, you will set up and perform most of your enzymatic reactions in the same way.

1. You will make reaction mixtures by adding colorless OPNG substrate to a pH 4.5 buffer in large test tubes. If needed, you will then add any necessary alternative substrates or inhibitors and enough additional buffer to bring the final volume of each tube to 5 mL.
2. To start the reactions, you will add 1 mL of bGAL enzyme that has been diluted to the proper concentration to each tube. Then mix the tube. Pay close attention because timing is crucial. The assays work best if you keep the tubes in order and add the enzyme to one tube at a time at regular intervals (say, every 15 seconds.)
3. Reaction mixtures will be incubated for 10 minutes at 22°C (room temperature) or the temperature indicated in the specific instructions for the assay.
4. After exactly 10 minutes, you will add 1.5 mL of Stop solution (1 M sodium carbonate) to each reaction. Again, the assay works best if you add Stop solution to the tubes in the same order and at regular time intervals, as in Step 2.
5. Once the reaction has stopped, you will read the absorbance of the sample using a spectrophotometer.

Exercise 1: How Does Substrate Concentration Affect a Catalyzed Reaction?

In this exercise, you will test concentrations of the substrate from 1 mM to 20 mM ONPG, while the concentration of the bGAL enzyme stays the same in all of the tubes.

Remember that bGAL is a catalyst. What do you predict will happen to the amount of product formed as the concentration of substrate increases?

Procedure

1. See Table 4.1 for what to mix in eight large, clean test tubes. Tube #1 is the Blank; it is used to set the baseline on the spectrophotometer. A blank is treated identically to the samples, except that enzyme is not added.

Table 4.1

Volumes for testing effects of substrate concentration

Tube	Buffer	Volume and Concentration of ONPG
1 (Blank)	5 mL	1 mL of 10 mM
2	4 mL	1 mL of 1mM
3	4 mL	1 mL of 2.5 mM
4	4 mL	1 mL of 5 mM
5	4 mL	1 mL of 7.5 mM
6	4 mL	1 mL of 10 mM
7	4 mL	1 mL of 15 mM
8	4 mL	1 mL of 20 mM

2. Add 1 mL of undiluted enzyme to Tubes #2–#8, and mix it thoroughly. Remember, try to add the enzyme to one tube about every 15 seconds.
3. Place all eight tubes in a rack on the bench top for 10 minutes.
4. Add 1.5 mL of Stop solution to each tube (including #1) and have your lab partner immediately mix the tube thoroughly. Once again, try to add the Stop solution to one tube every 15 seconds.
5. Transfer each reaction mixture to a separate, clean cuvet. Your instructor will demonstrate how to use the spectrophotometer. Read the absorbance of each sample, then record the values in Table 4.2.

Results

Remember you will not record an absorbance for Tube #1, because it is the blank used to set the baseline absorbance to zero.

Table 4.2

Results of Exercise 1

Tube Number	Concentration of Substrate	OD at 421 nm	Calculated Activity (nM/min)
2	1 mM		
3	2.5 mM		
4	5 mM		
5	7.5 mM		
6	10 mM		
7	15 mM		
8	20 mM		

Calculating Enzyme Activity From Absorbance

Activity is a measure of the rate at which the enzyme cleaves ONPG to a colored product. It is given in units of nanomoles per liter (nM) of ONPG substrate that is converted per minute. The absorbance (also called the optical density or OD, which is measured by nanometers [nm]) can be used to calculate the amount of enzyme activity in a sample. Activity is determined using the following formula:

$$\text{Activity} = \frac{\text{OD}}{[1.36 \times 10^{-4}\ (/\text{nM}) * \text{T}\ ]}$$

Where: T = time of incubation (usually 10 minutes). The other factor in the denominator is a constant that describes the absorbance of ONP in solution at a given concentration.

Summary Questions for Exercise 1

Plot the calculated enzyme activity against the substrate concentration (use either graph paper or a spreadsheet program).

1. Based on your data, is there a substrate concentration at which enzyme activity reaches a plateau—that is, becomes saturated? What is your evidence for this conclusion? What do you think your results mean?
2. Earlier you predicted what would happen as the concentration of substrate increased. Did the results you obtained match your predictions? If not, what did you observe that was different from what you predicted? What do you think it means?

3. Remember that many cells and organisms control the rate of their metabolic reactions by controlling the enzymes that catalyze those reactions. How could cells or organisms use differences in substrate concentration to control metabolic rate?

Exercise 2: How Does Enzyme Concentration Affect a Catalyzed Reaction?

Concentrations of various enzymes in cells and organs is not static. Most enzymes are quickly made or broken down as the needs of a cell or organ change. So how does adding or removing an enzyme affect the rate of a reaction it catalyzes? In this exercise, you will answer this question. You will change the concentration of bGAL enzyme, while the concentration of ONPG substrate stays the same in all of the tubes. This time, the amount of substrate is constant, but the amount of enzyme changes.

What do you predict will happen to the amount of product formed as the concentration of enzyme increases?

Procedure

1. In five large, clean test tubes, mix the amounts of buffer and 5 mM ONPG as shown in Table 4.3. Do not add the enzyme yet.

Table 4.3

Volumes for testing effects of enzyme concentration

Tube	Buffer	Volume of 5 mM ONPG	Volume and Concentration of Enzyme
1 (Blank)	5 mL	1 mL	1 mL of 0% (buffer)
2	4 mL	1 mL	1 mL of 25% solution
3	4 mL	1 mL	1 mL of 50% solution
4	4 mL	1 mL	1 mL of 75% solution
5	4 mL	1 mL	1 mL of 100% solution

2. When you are ready to start timing the reactions, add 1 mL enzyme to Tubes #2–#5 and mix each thoroughly. Be sure to add the correct dilution of the enzyme to each tube. As before, try to add the enzyme to one tube every 15 seconds.
3. Place all five tubes in a rack on the bench top for 10 minutes.

4. After 10 minutes, add 1.5 mL of Stop solution to each tube (including #1) and have your lab partner immediately mix the tube thoroughly. Once again, try to add the Stop solution to one tube every 15 seconds.
5. As in Exercise 1, transfer each reaction mixture to a separate, clean cuvet. Read the absorbance of each tube and record it in Table 4.4.

Results

Table 4.4

Results of Exercise 2

Tube Number	Concentration of Substrate	OD at 421 nm	Calculated Activity (nM/min)
2	25%		
3	50%		
4	75%		
5	100%		

Summary Questions for Exercise 2

1. Plot activity (y axis) against enzyme concentration (x axis). Based on your results, what is the effect of enzyme concentration on a catalyzed assay? What is your evidence for this conclusion?
2. Suppose the enzyme concentration was 20-fold higher than was used in the assay; can you predict or estimate how much product would be formed per minute? What if the enzyme was diluted 100-fold; can you predict the amount of product that would be formed per minute?
3. Many cells and organisms control the rate of their metabolic reactions by controlling the enzymes that catalyze those reactions. How could cells or organisms use differences in enzyme concentration to control metabolic rate?

Exercise 3: How Does Temperature Affect an Enzyme-Catalyzed Reaction?

In this exercise, you will incubate identical reaction mixtures at five different temperatures (4° [refrigerator or ice water]), 22°[room temperature], 37°, 50°, and 90°C) for 10 minutes, then measure the amount of product formed.

What do you predict will happen to the amount of product formed as temperature increases?

Procedure

1. In six large, clean test tubes, mix the amounts of buffer and 5 mM ONPG as shown in Table 4.5. Do not add the enzyme yet.

Table 4.5

Volumes for testing effects of temperature

Tube	Buffer	Volume of 5 mM ONPG
1 (Blank)	5 mL	1 mL
2	4 mL	1 mL
3	4 mL	1 mL
4	4 mL	1 mL
5	4 mL	1 mL
6	4 mL	1 mL

2. When you are ready to start timing the reactions, add 1 mL of the same concentration of enzyme to Tubes #2–#6, and mix each tube thoroughly. As you did before, try to add the enzyme to one tube every 15 seconds.
3. Immediately place each tube in a rack in the water bath at the appropriate temperature for 10 minutes. You can place the blank on the bench top.
4. After 10 minutes, add 1.5 mL of Stop solution to each tube (including #1), and have your lab partner immediately mix the tube thoroughly. Once again, try to add the Stop solution to one tube every 15 seconds.
5. Transfer each reaction mixture to a separate, clean cuvet. Read the absorbances, and record them in Table 4.6.

Results

Table 4.6

Results of Exercise 3

Tube Number	Concentration of Substrate	OD at 421 nm	Calculated Activity (nM/ min)
2	4°C		
3	22°C		
4	37°C		
5	50°C		
6	90°C		

Summary Questions for Exercise 3

1. Plot enzyme activity against temperature. Based on your results, what is the effect of temperature on a reaction that is being catalyzed with an enzyme? What is your evidence for that conclusion?
2. If the enzyme did not work above a certain temperature, what has happened to the enzyme? Do you think this is true for all enzymes? How could you find out?
3. Remember that many cells and organisms control the rate of their metabolic reactions by controlling the enzymes that catalyze those reactions. Can cells or organisms use differences in temperature to control metabolic rate? If not, why not; if so, how?

Unit Summary Questions

1. Look back at the results of Exercises 1–3. Now think about this problem. Amylase is another enzyme that our body uses to process sugars. Amylase in our small intestine cleaves starch in foods into individual glucose molecules that are burned for energy. However, the enzyme is not needed if there is no starch to digest.

   What is the simplest way for the body to regulate amylase activity?

   What might be some other ways for the body to regulate amylase activity?

   How can the body turn amylase activity OFF quickly?

2. Look back at the results of your temperature assay in Exercise 3. Now think about this problem. All organisms use enzymes to catalyze their metabolic reactions. Yet not all animals have a stable body temperature as humans do. Ectotherms like frogs or toads are found in a range of habitats, from icy mountain stream water to sunny gardens. They do not generate their own body heat but must absorb heat from their environments.

   Do you think the enzymes used by ectotherms are also sensitive to temperature? If you say yes, how do you think ectotherms make sure their metabolic reactions occur fast enough to keep them alive? If you say no, what might be different about the enzymes of ectotherms that makes them less temperature sensitive?

Your Own Experiment

Now you have the opportunity to design and execute an experiment to test a hypothesis of your own about enzyme activity. You can work with just your lab partners or the entire class can design and carry out a single, larger experiment. In the latter case, four to six lab groups of two to three people each can test a different factor within the larger experiment. Each student in a group must participate in carrying out the assays and will be responsible for understanding what was done.

You may ask, Why should anyone care about how environmental conditions affect enzyme activity or whether an enzyme can or cannot catalyze a reaction involving molecules other than its primary substrate? There are several reasons. Very often diseases are caused when a genetic defect, pathogenic organism, or other event causes body enzymes to become over-active or inactivated. Therefore, high or low enzyme activity can be used to diagnose disease. Understanding how the enzyme activity has been changed often points to effective treatments too. For example, remember from the Introduction that insufficient bGAL enzyme is the cause of lactose intolerance in human adults. The simplest treatment for this condition is to take a supplement that provides the missing enzyme; there are several over-the-counter medications that do just that. However, the enzyme also is sensitive to pH; in stomach acid, the enzyme can break down and lose activity. To develop an effective commercial product, it was essential to find a version of the bGAL enzyme that could remain active after passing through the stomach. Other factors also might affect bGAL activity. The question is, Which ones?

As you discuss what to do for your experiment, consider the following:

- Think about the range of conditions found in the cells and tissues of organisms, as well as in the world outside of organisms. Does bGAL continue to function across the entire range of conditions found in organisms?
- How else (besides what you observed already) might organisms control the activity of their enzymes?
- Is there any reason why one organism might want to block the actions of enzymes in another? If so, how might they do it?

Next, decide what you would have to do to the bGAL enzyme to test the parameter on which you focused. State it as a testable hypothesis. Plan out precisely what you will need to do and what data you will need to collect from your experiment. Be sure to plan for proper negative controls and a sufficient number of replicates.

Materials Available for Experiments

The reagents in the following list are routinely available. If you'd like to use something different, ask your instructor whether the appropriate materials are available.

- Different pH buffer solutions, ranging from pH 2–10
- Alternative substrates for bGAL in which the sugar has been changed or removed: ONP-xylose, ONP-glucose, ONP-phosphate

- Water baths at different temperatures: refrigerator or ice bucket (4°C) room temperature several heated water baths
- Competitors: pure lactose, galactose, glucose, pectin
- Complex solution of other proteins and sugars
- Purified bGAL extracted from other sources

UNIT 4

Properties of Enzymes

Teacher Pages

INSTRUCTORS' NOTES

Background and Key Concepts

Most undergraduates know that enzymes are biological catalysts that accelerate chemical reactions. Most also understand that enzymes have optimal conditions under which they operate and that enzyme activity can be affected by temperature, pH, and other environmental conditions. However, many students do not realize organisms routinely control the rates at which enzyme-catalyzed reactions occur—by changing the amount of substrate available to an enzyme, by changing the amount of enzyme available, or by using inhibitors and activators. Others fail to understand how enzymes can become saturated or that cells and organs must compensate by maintaining substrates well below the Michaelis constant (K_M) for the enzyme. This unit was developed to help students develop a more functional, intuitive understanding of how enzymes work, what factors affect enzyme activity, and how enzyme activity can be regulated by organisms. For nonmajors, this is a sufficient level of understanding; among majors, developing an intuitive understanding first can help them understand formal enzyme kinetics more deeply.

The model enzyme for this unit, β-galactosidase (bGAL), was selected for several reasons. Purified enzyme is inexpensive and relatively stable. The assay for bGAL activity is simple for students to perform. Nonmajor students find bGAL interesting because it is the enzyme responsible for lactose intolerance, so it is relevant to human biology. The enzyme is historically important to the field of molecular biology in that bGAL is the product of one of the first bacterial operons to be explained in detail; today, bGAL still is a widely used marker enzyme. Finally, the enzyme is highly specific, so it is ideal for demonstrating substrate selectivity under a range of conditions.

It is also possible to demonstrate connections between cell biology and environmental biology or ecology with this enzyme. The bGAL that students use is extracted from an industrial strain of the fungus *Aspergillus oryzae*. The enzyme's pH optimum is ~5.5, much lower than the pH optimum for bGAL produced by bacteria but very near the optimal pH for fungal growth in culture. It is very likely that the biochemical differences in bacterial versus fungal bGAL reflect differences in the microenvironmental conditions for which each domain of organism is adapted.

General Teaching Strategy and Common Problems

This two-week unit was designed for a nonmajors course. In Week 1 students learn how substrate concentration affects enzyme activity and how an organism can control metabolic reactions by controlling enzyme concentration. They also see how external environmental conditions alter enzyme activity. For their experiments in Week 2, students are encouraged to look in more depth at effects of external conditions on activity or, alternatively, to explore the substrate specificity of enzymes.

For this unit specialized vocabulary and equations have been reduced to the bare minimum. Some instructors may choose to alter the exercises to include more specific terms and concepts from classical enzymology. In the author's experience, though, this approach is counterproductive; it encourages students to focus more on learning factual details and not on developing a functional understanding of enzyme regulation.

At first, the Unit Summary Questions at the end of each exercise may seem overly difficult. However, they have been designed to reinforce the principal goals of the unit. Students must interpret their results for each exercise, not just summarize them or perform a basic calculation. They are asked to predict what would happen if the experimental procedure were modified, and to explain how an organism could use the properties they have just observed to regulate enzymatic processes.

Students with minimal science background may be intimidated by what appears to be a very technical assay. Yet most students can complete the exercises without difficulty. Encourage students to work carefully and to take the time needed to understand one small step thoroughly before they tackle the next one.

Other students may struggle to understand what a spectrophotometer measures. Avoid discussing the optics or mechanics of spectrophotometry. Instead remind students that the color they see is an *indirect* indicator of an enzymatic process. The spectrophotometer merely acts as a mechanical "eye" that reports color intensity as a number. Past instructors have demonstrated the relationship of color to concentration by making serial 1:2 dilutions of a stock solution of red food coloring in water. If a stock solution is sufficiently concentrated, 1:2 and 1:4 dilutions will not look significantly different. Once the stock solution is diluted 1:8 or more, each step in the sequence looks different, until the color change is too small to see. The 1:2 and 1:4 dilutions represent the upper plateau range of the instrument (i.e., their eyes), while the colorless dilutions represent the lower limit of sensitivity.

Another common problem arises as students design experiments of their own. Unless they have some prior experience with experimental design, most decide to test a chemical or condition at random. One way to encourage deeper thinking is to require them to submit an experimental proposal. They must write down a reasonable, testable hypothesis and their rationale, which must be based on background observations or prior knowledge. They cannot start their own experiment otherwise. A Sample Form for Students' Experimental Outline is found in Appendix B.

Recommended Prelab Skills

Some prior experience performing calculations and graphing data with a spreadsheet program is helpful but not essential. Students must be able to measure solution volumes using serological pipets.

Assessment

If pre- and postlab quizzes are used as part of formal assessment, the prelab quiz should focus only on basic concepts such as the general functions of enzymes. After they complete the unit, students should be able to explain at least two mechanisms by which organisms can regulate enzyme activity. Good practical questions include asking students to blank a spectrophotometer and take a reading of an unknown.

Safety and Housekeeping

The yellow reaction product of the enzyme assay is 2, 4-dinitrophenol (DNP), which is an EPA-regulated hazardous waste. Unused nitrophenyl-derivatized substrates and all of the liquid reaction products should be collected and pooled for destruction by photocatalysis. Dilute the pooled DNP wastes to less than 1% wt./vol. in a clear glass container. Add 1 g/L titanium dioxide (TiO_2) and expose solution to sunlight or a strong UV light source. When the reaction is complete and the solution is decolorized, the solution can be flushed down the drain with copious amounts of water.

Other Tips

If spectrophotometers fail or are unavailable, prepare a stock solution of 50 mM dinitrophenol in water, and make multiple dilutions to set up a colorimetric curve. The curve should include ~21 points (100% (50 mM), 95%, 90%…10%, 5%, zero). Put samples of the standard curve solutions into the same tubes that the students use for their assay, and label each tube with the concentration of dinitrophenol it contains. Students simply match their tubes to the various standards and record the concentration.

PREPARATORY NOTES

Quantities listed are for a lab section of 20 students working in groups of three or four.

Week 1: Shared Materials

- 5 mL and 10 mL serological pipets
- Pipet pumps
- 125 mL flask
- Small refrigerator or ice bucket
- Small water baths set at 37°, 50°, and 90°C
- Paper towels
- Waste container for spent reaction mixtures
- 1 L potassium acetate buffer, pH 4.5
- Preweighed powder for 50 mL of 20 mM ONPG (store in freezer)
- Preweighed powder for 50 mL of 5 mM ONPG (store in freezer)
- 50 mL stock beta galactosidase enzyme (store in freezer)
- 250 mL, 1 M sodium carbonate

Week 1: Materials at Each Work Station

- Spectrophotometer
- Paper towels or tissues
- Test tube rack with 20 large (15 × 150 mm) test tubes
- Test tube rack with 12 cuvets for spectrophotometer
- 95% ethanol in wash bottle

Week 2: Shared Materials

- 5 mL and 10 mL serological pipets
- Pipet pumps
- Refrigerator or buckets containing ice water
- Small water baths set at requested temperatures
- Paper towels
- Waste container for spent reaction mixtures
- 1 L potassium acetate buffer, pH 4.5
- Preweighed powder for 2 × 50 mL of 5 mM ONPG (store in freezer)
- 100 mL stock beta galactosidase (store in freezer)
- 500 mL, 1 M sodium carbonate

Week 2: Materials at Each Work Station

- Spectrophotometer
- Paper towels or tissues

- Test tube rack with 20 large (15 × 150 mm) test tubes
- Test tube rack with 12 cuvets for spectrophotometer
- 95% ethanol in wash bottle

Optional Materials for Experiments (make only if requested)

- 100 mL each of pH 2, 4, 5, 6, 7, 8, and 10 reaction buffers
- Preweighed powder for 5 mL of alternative substrates (store in freezer)
  - ◊ 10 mM ONP-xylose
  - ◊ 10 mM ONP-glucose
  - ◊ 10 mM ONP-phosphate
- 25 mL each of 100 mM sugar solutions in pH 4.5 acetate buffer
  - ◊ Lactose
  - ◊ Glucose
  - ◊ Galactose
  - ◊ Pectin
- 25 mL, 1% normal goat, bovine, or horse serum diluted in pH 4.5 buffer (used to make a complex mixture of proteins, bGAL, and substrate)

Sources of Materials

Enzyme

Powdered bGAL enzyme from *A. oryzae* can be purchased from a research chemicals supplier. The most common unit definition is "one unit hydrolyzes 1.0 mmole of ONPG to o-nitrophenol and D-galactose per minute at pH 4.5 and 30ºC." One vial containing 10,000 units costs ~$15–$25.

Every semester, prepare a new enzyme stock solution (10 units/mL); stored properly, this solution can be used for up to 30 days. Dissolve 500 units (~50 mg) of enzyme in 50 ml of distilled water. If precipitate remains, filter or centrifuge the solution to clarify it. Split the enzyme stock solution into individual, labeled 1mL aliquots in microfuge tubes and store them frozen in non–frost free conditions. Do not refreeze thawed aliquots.

To check enzyme activity, thaw one aliquot of enzyme and dilute it to 50 mL with distilled water to a working concentration of 0.2 units/mL. Prepare 20 mM ONPG solution and pH 4.5 acetate buffer (instructions are below), and perform Exercise 1. A standard reaction mixture containing 0.2 U enzyme should reach maximum activity between 5 and 10 mM ONPG.

After 10 minutes at room temperature, the OD421 should be ~1.0 for a 1 cm path length. If the OD is less than 0.7, extend the reactions to 20 minutes. If the OD is higher than 1.0, dilute each enzyme aliquot to 100 mL instead of 50 mL; do not try to shorten the incubation time.

Solutions, Reagents, Equipment

Potassium Acetate Buffer, pH 4.5

Dissolve 1.9 g potassium acetate in 1200 mL distilled water. Add 3 mL glacial acetic acid. Check pH. If needed, adjust pH with 10% acetic acid or 1 M KOH. Store at room temperature.

20 mM, 5 mM ONP Stock Solutions

Weigh out 0.30 g of o-nitrophenyl-β-D-galactopyranoside and place it in a 50 mL conical tube labeled "20 mM ONPG."

Weigh out 0.075 g of ONPG and place it in a second 50 mL conical tube labeled "5 mM ONPG." Label with "Add 50 mL pH 4.5 acetate buffer." Store both tubes in the freezer until just before lab.

1 M Sodium Carbonate

Dissolve 106 g Na_2CO_3 in 1000 mL distilled water.

Alternative Substrates, 10 mM Solutions

Weigh out each of the following into separate, labeled 15 mL conical tubes.

- 0.03 g ONP-xylose
- 0.03 g ONP-glucose
- 0.04 g ONP-phosphate, cyclohexylammonium salt

Label each tube with "Add 10 mL pH 4.5 acetate buffer." Store tubes in the freezer until just before lab.

Sugar Competitors, 200 mM Solutions

Weigh out each of the following into separate, labeled 50 mL conical tubes.

- 0.8 g lactose (competitive inhibitor)
- 0.4 g glucose (product inhibitor)
- 0.4 g galactose (product inhibitor)
- 1.0 g liquid pectin (complex soluble polysaccharide)

Add 50 mL distilled water to dissolve. Store at room temperature.

Prelab Preparatory Instructions

1. Thaw two aliquots of enzyme stock solution. Mix 2 mL enzyme with 100 mL of pH 4.5 acetate buffer in an Erlenmeyer flask. This is the "Working Enzyme Solution" at 0.2 units/mL. Divide in half and place at the two work stations on the side benches.
2. Prepare ONPG solutions for Exercise 1 by adding 50 mL of pH 4.5 acetate buffer to one of the premade tubes labeled "20 mM." Once the ONPG has gone into solution, make the dilutions shown in Table 4.7 in labeled 50 mL conical tubes.
3. Prepare "Diluted Enzyme Solutions" for Exercise 2 as shown in Table 4.8.

Table 4.7

Preparing substrate solutions

Label	Vol. of 20 mM Stock	Vol. of pH 4.5 Buffer
1 mM ONPG	1 mL	19 mL
2.5 mM ONPG	2 mL	14 mL
5 mM ONPG	4 mL	12 mL
7.5 mM ONPG	6 mL	10 mL
10 mM ONPG	8 mL	8 mL
15 mM ONPG	12 mL	4 mL
20 mM ONPG	Use balance of stock substrate	

Table 4.8

Preparing enzyme dilutions

Label	Volume of Working Enzyme	Volume of pH 4.5 Buffer
25% Enzyme	5 mL	15 mL
50% Enzyme	10 mL	10 mL
75% Enzyme	15 mL	5 mL
100% Enzyme	20 mL	none

4. Make the substrate for Exercises 2 and 3 by adding 50 mL of pH 4.5 acetate buffer to one of the premade tubes in freezer labeled "5 mM ONPG." Warm to 37°C to dissolve.
5. For the second week only, if students have requested alternate substrates, add 10 mL of pH 4.5 acetate buffer to one 15 mL conical tube each of ONP-xylose, ONP-glucose, and ONP-phosphate.

UNIT 5

Applied Enzymology

Student Pages*

BACKGROUND

All organisms use enzymes to catalyze metabolic reactions and regulate body processes. Some reactions are unique to certain cell types, and only those cells will produce the requisite enzymes. For example, plant cells contain **RUBISCO**, the chloroplast enzyme complex that fixes carbon dioxide during photosynthesis. Since animal cells do not undergo photosynthesis, it would be extremely unusual for an animal cell to make rubisco. Conversely, other cell processes are ubiquitous. For example, all prokaryotic and eukaryotic cells can break down glucose to pyruvate via glycolysis. So nearly all plant, animal, fungal, and bacterial cells contain the same general sequence of glycolytic enzymes.

Even when two different organisms produce the same enzyme for the same function, the enzyme molecules are not identical. The DNA that encodes the enzyme in each species has undergone different mutation and natural selection events. As a result, the two enzymes often function slightly differently. Another source of variation in enzymes is gene duplication. Occasionally the DNA coding for an enzyme is duplicated, and the duplicate is inserted in a second place in the chromosomes of a cell. If the duplicated DNA is not harmful, the cell passes that duplication along to all daughter cells. There is one working copy of the DNA for the enzyme, so the second copy can be mutated and altered without killing the cell. Over time, enough changes may accumulate in the DNA sequence to create an enzyme that resembles the original one but has different functional properties. This is the underlying mechanism leading to cells having families of functionally and structurally similar enzymes.

*Teacher Pages begin on page 233.

Phosphatases Are Our Model Enzymes

Enzymes can be divided into major classes based on their general activity. For example, **phosphatases** cleave the diester bond between an hydroxyl group on various molecules and an inorganic phosphate (PO_4^{-3}) group. Every living organism uses phosphorylation and dephosphorylation to regulate cellular processes, so all cells possess several different phosphatases. The general reaction is shown in Figure 5.1.

Figure 5.1

The generalized phosphatase reaction

In cells, "R" can be any number of molecules, including nucleotides (e.g., when a phosphatase cleaves ATP to ADP and P_i), amino acid side groups (serines, threonines, and tyrosines in particular), and various phosphorylated sugars (such as those used in gluconeogenesis).

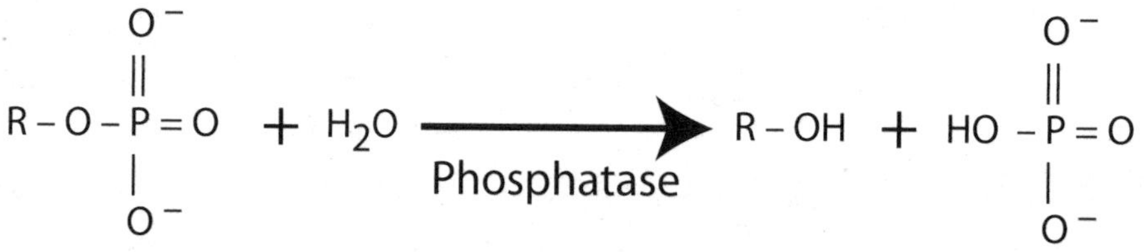

$$R-O-\overset{\overset{O^-}{||}}{\underset{\underset{O^-}{|}}{P}}=O + H_2O \xrightarrow[\text{Phosphatase}]{} R-OH + HO-\overset{\overset{O^-}{||}}{\underset{\underset{O^-}{|}}{P}}=O$$

Phosphatases do not cleave indiscriminately. Each enzyme has specific substrate molecules on which it can act, as well as an optimum set of working conditions. Most can be classified as either **acid phosphatases**, or **alkaline phosphatases**, depending on the pH at which they are most active. In addition to having a pH optimum, some phosphatases require cofactors, while others are sensitive to certain chemical inhibitors.

Differences between enzymes from different organisms are used routinely in medicine as a diagnostic tool. Imagine this scenario: A male patient, age 45, comes into a hospital emergency room at 2 a.m. He has been ill with a fever for three days, but has no idea why. You draw blood; the test results show abnormally high levels of phosphatase activity. There are many potential causes for elevated phosphatase levels in blood, including (but not limited to) the following:

- A systemic bacterial infection
- A fungal infection (candidiasis, coccidiosis)
- Metastatic prostate cancer
- Early stages of cirrhosis of the liver
- Kidney disease
- Colorectal cancer

Each of these conditions requires a different treatment. Identifying the particular type of phosphatase present in the blood sample would help with the diagnosis and ensure proper treatment. Fortunately, the phosphatases from each of these cellular sources have very different functional properties. So how would you distinguish between these possibilities for this particular patient? To answer that question, you will need to first learn more about the basic chemistry of phosphatases and about how to measure phosphatase activity.

Goals of This Unit

During the first part of this unit, you will learn how to perform a basic enzyme assay, using acid phosphatase purified from wheat germ. By the end of the first week, you should be able to answer the following questions:

How does enzyme concentration affect the rate of a reaction?

What is the effect of pH and other local conditions on reaction rate?

Subsequently you will be given a sample of mock serum from a normal patient and a sample from the patient described above, who has one of the conditions listed. You must use the assay methods you learn, plus a summary of the functional characteristics of the phosphatases from each possible source, to determine the specific type and concentration of phosphatase that is present in the sample you are assigned.

General Safety Precautions

The mock serum samples you are working with were prepared especially for this laboratory and have no hazardous components. Some sensitive individuals may react to the proteins in the solution, so it is recommended you wear gloves when handling the mock serum samples.

The phosphatase assay employs two strong buffers and two nitrogenous organic salts. All of them are irritants. Immediately wash off any chemicals or solutions that splash on you, and wipe up any spills.

The sodium hydroxide stop solution is a strong corrosive. Be extremely careful not to spill or splash it. If you spill it on yourself, *immediately* flush the area with bicarbonate neutralizer or body flush solution (kept at the front of the lab), followed by ample water. If the skin blisters, notify your instructor. If anyone gets sodium hydroxide splashed into their eyes, flush their eyes for 10 minutes using the eyewash station at the sink in the lab and have someone call the campus emergency service.

The yellow nitrophenol reaction product of the assay is classified by the U.S. Environmental Protection Agency (EPA) as a hazardous chemical. Do not pour it down the drain. All enzyme substrate solutions and yellow end products should be placed in the labeled waste containers for proper disposal.

UNIT EXERCISES

Exercise 1: How Does Enzyme Concentration Affect Reaction Rate?

Background: Chemistry of the Phosphatase Assay

Para-nitrophenyl phosphate (pNPP), a small cyclic organic molecule with an attached phosphate, is used to quantify phosphatase activity. At the appropriate pH, most acid and alkaline phosphatases can hydrolyze pNPP substrate to produce nitrophenol, a bright yellow product. The reaction can be summarized schematically as shown in Figure 5.2.

Figure 5.2

Schematic of the reaction used to detect phosphatases

In solution, para-nitrophenyl phosphate (pNPP) is pale or colorless, but once the inorganic phosphate is cleaved, nitrophenol is formed. Stopping the reaction with sodium hydroxide (NaOH) raises the pH and the nitrophenol becomes bright yellow.

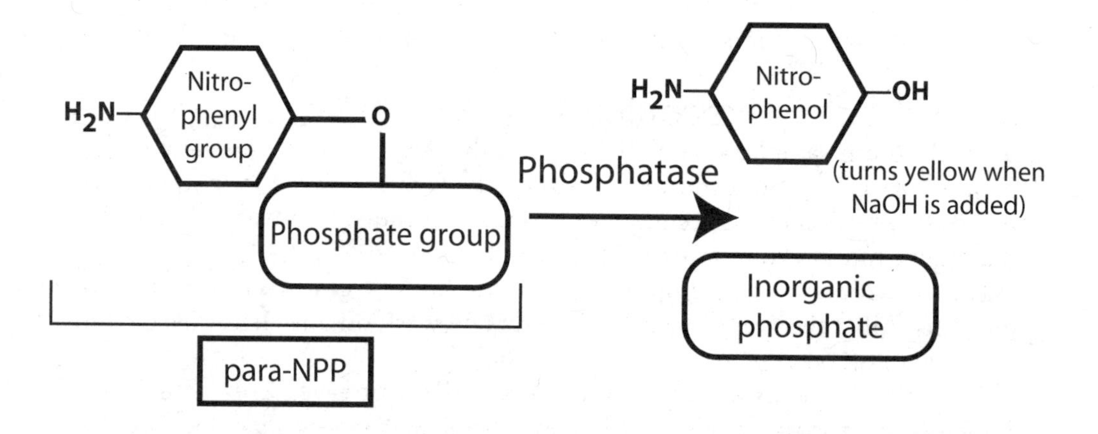

A known concentration of pNPP is mixed with the phosphatase enzyme in an appropriate buffer and incubated at room temperature for 10 minutes. The reaction is stopped using sodium hydroxide, and the amount of nitrophenol produced is measured using a **spectrophotometer**. The more yellow the sample, the more nitrophenol there is in solution and the higher its absorbance at 420 nM.

The basic assay can be used to generate a **standard curve** that in turn can be used to determine the enzyme concentration in experimental or diagnostic samples. Samples are prepared with known concentrations of enzyme, and allowed to react with a known quantity of pNPP substrate. The reactions are stopped, and the intensity of yellow color (which comes from nitrophenol product) measured. The absorbance of each sample (*y* values) are graphed against the known concentration of enzyme in each sample (*x* values). With this curve, it is possible to calculate the amount of enzyme in an unknown sample, based on how much product it generates when mixed with a known amount of pNPP.

Procedure

You will use several known concentrations of wheat germ acid phosphatase to generate a standard curve, with which you will determine the concentration of enzyme in two "unknown" samples.

Before you begin, what do you predict will happen to the amount of product formed as the concentration of enzyme increases?

1. In 1.5 mL microtubes, prepare the following enzyme dilutions for the standard curve (see Table 5.1).

Table 5.1

Preparing diluted enzymes

Tube	Volume of Buffer	Volume of Stock Enzyme	Final Concentration of Enzyme
1	990 µL	10 µL	0.01 U/mL
2	950 µL	50 µL	0.05 U/mL
3	900 µL	100 µL	0.1 U/mL
4	800 µL	200 µL	0.2 U/mL
5	700 µL	300 µL	0.3 U/mL
6	600 µL	400 µL	0.4 U/mL
7	500 µL	500 µL	0.5 U/mL

2. Label each microtube with the concentration of enzyme it contains, and place them on ice.
3. Wash and dry 11 test tubes, and label them "Blank," then "1–10." See Table 5.2, page 220, for what to place in each tube.

   The Blank will be used to set the baseline absorbance on the spectrophotometer. It is treated identically to the samples, except that enzyme is not added.

Table 5.2

Preparing tubes for Exercise 1

Tube	pH 5 Buffer	Volume of Diluted Enzyme	Concentration of Enzyme
Blank	4.5 mL	none	—
1	4 mL	0.5 mL	0.01 U/mL
2	4 mL	0.5 mL	0.05 U/mL
3	4 mL	0.5 mL	0.1 U/mL
4	4 mL	0.5 mL	0.2 U/mL
5	4 mL	0.5 mL	0.3 U/mL
6	4 mL	0.5 mL	0.4 U/mL
7	4 mL	0.5 mL	0.5 U /mL
8	4 mL	0.5 mL	1.0 U/mL
9	4 mL	0.5 mL	Unknown A
10	4 mL	0.5 mL	Unknown B

4. To start the assay, quickly add 0.5 mL of 100 mM pNPP substrate to each tube, and mix it thoroughly.
5. Place all 11 tubes in a rack at room temperature.
6. Allow the assay to incubate for 10 minutes.
7. After 10 minutes, remove the rack, and immediately add 0.5 mL of Stop solution (2 M NaOH) to each tube. Carefully mix each tube thoroughly.
   Be very careful with the NaOH. It is extremely corrosive! If you spill it on yourself, immediately flush the injury with the bicarbonate solution at the front of the lab.
8. Record the absorbances for each sample in Table 5.3.

Results

Table 5.3

Results of Exercise 1

Tube	Sample	Absorbance (420 nm)
1	0.01 U/mL	
2	0.05 U/mL	
3	0.1 U/mL	
4	0.2 U/mL	
5	0.3 U/mL	
6	0.4 U/mL	
7	0.5 U/mL	
8	1.0 U/mL	
9	Unknown A	
10	Unknown B	

Remember, you do not record an absorbance for the blank used to set the baseline absorbance.

Determining Enzyme Activity of an Unknown

To determine how much acid phosphatase activity is in Unknown A and Unknown B:

1. Use graph paper or a spreadsheet program to make an xy graph of the data above. On the y axis, place absorbance at 420 nm for samples 1–8, and on the x axis, the concentration of enzyme in each sample.
2. Determine the slope (m) and y-intercept (b) of a line through your data points. If there is a plateau in your data, calculate the slope and intercept using just the data points that form a straight line.
3. Use the recorded absorbance for each unknown as y, then solve the equation $y = mx + b$. The resulting values for x are the enzyme activity in the two unknowns.
4. Record the calculated activities as part of Question #2 on page 220.

Summary Questions for Exercise 1

1. Did the results you observed match what you predicted before you started? If not, what did you observe that was different from what you predicted? What do you think it means?
2. How much phosphatase activity was present in Unknown Sample A? Unknown Sample B?
3. Go back to the enzyme standards you analyzed. Imagine the absorbance values you obtained for 0.4 U, 0.5 U, and 1.0 U/mL enzyme standards were nearly identical. If you plotted absorbance versus enzyme concentration for these standards, there would be a plateau in the line connecting the data points. (If this scenario does not make sense, go back and actually plot it out.)

   When you calculated the standard curve, you were instructed to ignore any data points in a plateau region of the graph and use only data points that formed a straight line. Why can't you use the data points in the plateau region?

Exercise 2: How Does pH Affect Phosphatase Activity?

As was stated in the background section of this unit (pp. 216–217), phosphatases can be divided into acid and alkaline types, based on the pH at which they are maximally active. But how much effect does pH have on overall activity? Does activity change only a little (say, +/- 10%) or dramatically (50% or more)?

In this exercise, you will answer the question yourself by performing the same assay as in Exercise 1, except using three different pH buffers. So that you have a range of enzyme concentrations to test, you will assay the stock enzyme solution, plus Unknowns A and B.

Before you continue, what do you predict will happen to the amount of product formed as the pH changes?

Procedure

1. Clean 11 test tubes, label them. See Table 5.4 for what to place in each tube.

Table 5.4

Preparing tubes for Exercise 2

Tube	Volume, pH of Buffer	Volume, Source of Enzyme
Blank	4.5 mL of pH 5	None
1	4 mL, pH 5	0.5 mL of stock (0.1 U/mL)
2	4 mL, pH 4.0	0.5 mL of stock (0.1 U/mL)
3	4 mL, pH 7.0	0.5 mL of stock (0.1 U/mL)
4	4 mL, pH 10.0	0.5 mL of stock (0.1 U/mL)
5	4 mL, pH 4.0	0.5 mL of Unknown A
6	4 mL, pH 7.0	0.5 mL of Unknown A
7	4 mL, pH 10.0	0.5 mL of Unknown A
8	4 mL, pH 4.0	0.5 mL of Unknown B
9	4 mL, pH 7.0	0.5 mL of Unknown B
10	4 mL, pH 10.0	0.5 mL of Unknown B

2. To start the assay, add 0.5 mL of 100 mM pNPP substrate to each tube, and mix it thoroughly.
3. Place the tubes in a rack at room temperature. Allow the assay to incubate for 10 minutes.
4. Add 0.5 mL of Stop solution (2 M NaOH) to each tube. Carefully mix each tube thoroughly.
5. Record the absorbances for each sample in Table 5.5, page 224.

Results

Table 5.5

Results of Exercise 2

Enzyme	pH of Buffer	Absorbance (420 nm)	Calculated Enzyme Activity (U/mL)
Positive control	5.0		
Stock enzyme (0.1 U/mL)	4.0		
	7.0		
	10.0		
Unknown A	4.0		
	7.0		
	10.0		
Unknown B	4.0		
	7.0		
	10.0		

Using the absorbances you recorded and the standard curve you created for Exercise 1, calculate the phosphatase activity in each sample and enter the results in the last column of the table. Then answer the following questions.

Summary Questions for Exercise 2

1. Did the results you observed for the stock enzyme (0.1 U/mL) match what you predicted before starting? If not, what did you observe that was different from what you predicted? What do you think it means?
2. Did pH affect the relative phosphatase activity in Unknown A and Unknown B in the same way? If not, how were they different? Was this the result you expected? Why or why not? If it is not, how could you explain the observed results?

Exercise 3: How Do Potential Inhibitors Affect Phosphatase Activity?

Background

Every enzyme has a specific three-dimensional structure that it must maintain in order to function properly. Besides pH, temperature and other local conditions or factors can affect the overall shape of an enzyme and thereby affect its activity.

Some enzymes will not function without **cofactors**, which are small molecules or ions needed to form the active site. Many enzymes use metal ions such as copper, nickel, iron, calcium, or magnesium as cofactors.

Some compounds or ions are called **inhibitors** of an enzyme because they reduce enzyme activity in a specific manner. Specific enzyme inhibitors can be separated into different subcategories. Some are **irreversible** inhibitors; once an enzyme has come into contact with them, it cannot become active again. Other enzyme inhibitors are **reversible**, and if these reversible inhibitors are removed from the system, the enzyme is active once more. Inhibitors can be further subdivided based on the primary mechanism by which they inhibit the enzyme.

The vast majority of enzyme inhibitors and activators are small molecules. However, many enzymes inside cells are additionally regulated by their associations with larger proteins or macromolecules. Like the small molecules, these larger complexes affect the overall three-dimensional structure of the protein backbone of the enzyme. Two common mechanisms by which they affect enzyme function are summarized in Figure 5.3.

Figure 5.3

Schematic diagram of two common mechanisms by which inhibitors affect enzyme activity

Some bind to the same site as the normal substrate, while others change the shape and activity of the enzyme by binding to another location outside of the active site.

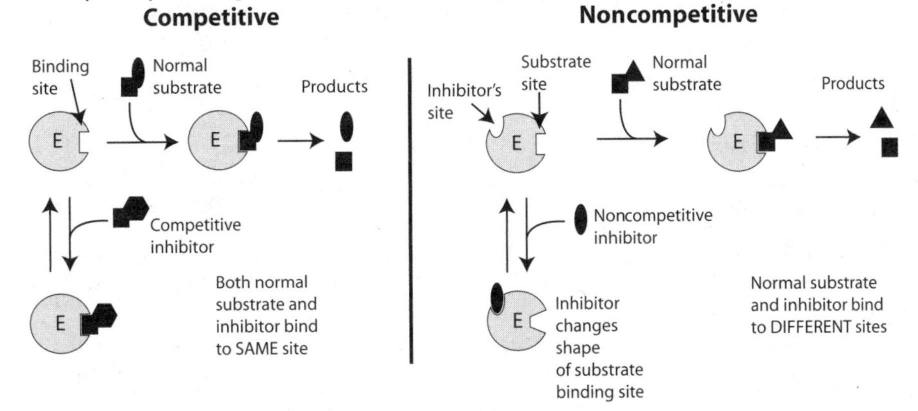

In this exercise you will test the effects of four different inhibitors on the activity of wheat germ acid phosphatase:

Potassium phosphate: Inorganic phosphate is a product released by any phosphatase-catalyzed reaction.

Sodium fluoride: The Fl- ion binds irreversibly to a reactive serine that is part of the active site of the phosphatase enzymes.

Molybdic acid: The exact action is still unknown; current evidence suggests Mo^{+6} ions forces phosphatase enzymes into a three-dimensional state that cannot bind substrates.

Phosphonoacetic acid: This molecule competes with pNPP for binding to the phosphatase. However, it cannot be cleaved, so it stays within the binding site.

Before you proceed, answer the following questions.

Based on the chemical laws of equilibrium, how do you think an excess amount of phosphate product will affect enzyme activity? What is your reasoning behind this conclusion?

Since fluoride ion irreversibly changes the shape of the active site in acid phosphatase, what do you predict will happen to overall enzyme activity? Why?

Phosphonoacetic acid and molybdate both bind reversibly to the phosphatase enzyme. However, they bind at two different sites on the enzyme. What do you predict that each chemical will do to enzyme activity? What is your rationale?

Procedure

1. You will be provided with 20 mM stock solutions of each potential inhibitor to be tested. Label 1.5 mL microtubes and prepare the dilutions as indicated in Table 5.6.

Table 5.6

Preparing Inhibitors for Exercise 3

Tube	Volume of Water	Volume of Inhibitor
2 mM phosphate	900 μL	100 μL of 20 mM phosphate
10 mM phosphate	500 μL	500 μL of 20 mM phosphate
2 mM fluoride	900 μL	100 μL of 20 mM NaF
10 mM fluoride	500 μL	500 μL of 20 mM NaF
2 mM molybdate	900 μL	100 μL of 20 mM molybdate
10 mM molybdate	500 μL	500 μL of 20 mM molybdate
2 mM phosphonoacetate	900 μL	100 μL of 20 mM phosphonoacetate
10 mM phosphonoacetate	500 μL	500 μL of 20 mM phosphonoacetate

2. Clean 10 test tubes and label them. See Table 5.7 for what to put in each test tube.

Table 5.7

Preparing tubes for Exercise 3

Tube	Buffer	Potential Inhibitor	Enzyme (0.1 U/mL)
Blank	4.5 mL	None	None
1	4 mL	None	0.5 mL
2	3.5 mL	0.5 mL of 2 mM phosphate	0.5 mL
3	3.5 mL	0.5 mL of 10 mM phosphate	0.5 mL
4	3.5 mL	0.5 mL of 2 mM fluoride	0.5 mL
5	3.5 mL	0.5 mL of 10 mM fluoride	0.5 mL
6	3.5 mL	0.5 mL of 2 mM molybdate	0.5 mL
7	3.5 mL	0.5 mL of 10 mM molybdate	0.5 mL
8	3.5 mL	0.5 mL of 2 mM phosphonoacetate	0.5 mL
9	3.5 mL	0.5 mL of 10 mM phosphonoacetate	0.5 mL

3. To start the assay, add 0.5 mL of 100 mM pNPP substrate to each tube, and mix it thoroughly.
4. Incubate the tubes at room temperature for 10 minutes, stop the reactions with 0.5 mL of Stop solution (2 M NaOH), and record the absorbances for each sample in Table 5.8, page 228.

Results

Table 5.8

Results of Exercise 3

Sample	Absorbance (420 nm)	Calculated Enzyme Activity (U/mL)
Positive control (no additive)		
0.5 mL of 2 mM phosphate		
0.5 mL of 10 mM phosphate		
0.5 mL of 2 mM fluoride		
0.5 mL of 10 mM fluoride		
0.5 mL of 2 mM molybdate		
0.5 mL of 10 mM molybdate		
0.5 mL of 2 mM phosphonate		
0.5 mL of 10 mM phosphonate		

Using the absorbances you recorded and the graph you created in Exercise #1, calculate the phosphatase activity in each sample, and enter the results in the last column of the table. Then answer the following questions.

Summary Questions for Exercise 3

1. Briefly, how did each of the additives affect enzyme activity? Did your results match your earlier predictions? If not, what did you observe that was different from what you predicted? What do you think it means?
2. Suppose the concentration of substrate in the reaction mixture was increased 50-fold. Based on your observations and the descriptions of how each compound interacts with the phosphatase enzyme, do you predict that enzyme activity will be higher, lower, or the same? What is your reasoning behind this conclusion?
3. As you will learn next week, human cells produce many different alkaline phosphatases. Two very rich sources are intestine and placenta. Activity of placental alkaline phosphatase is completely inhibited by 10 mM levamisole (a small amine-containing molecule), but the intestinal form of the enzyme is not affected. How is it possible for two different isoforms of an enzyme purified from the same organism to respond differently to the same inhibitor?

Exercise 4: Characterizing Unknown Phosphatases in a Serum Sample

Now you must work with your lab partners to determine the source of the excess phosphatase in a mock serum sample. You will be given two tubes. One has a solution with the same phosphatase levels as normal human serum. The material in the second tube has phosphatase levels consistent with one of the disease states described in the next section. Both samples have already been diluted 1:10 with 0.9% saline.

While planning your assays, keep these suggestions in mind:

- You can get confused very quickly by all the different characteristics of the various phosphatases. Try arranging the major features in a table. Then see if you can identify one or two characteristics that allow you to limit the possibilities.
- Do not plan an overly elaborate assay. You will have to test several different sets of conditions and inhibitors or activators. So do not make each assay too complicated.
- Another reason to plan on smaller assays is that you may have to repeat some sample points that fail to work properly. It is better to do several smaller assays and get usable data than to design a huge assay that you have to repeat over and over before it works.
- Plan to assay replicate samples of both the normal and "unknown disease" sera.
- Write down your group's thoughts along the way and document all of your results.

Selected Phosphatase Isoforms Found in Serum

Acid phosphatases (most active at low pH)

Prostate-specific: Prostate gland cells produce high levels of the enzyme so patients with advanced prostate cancer typically have very high levels of the enzyme in their blood. This enzyme is inhibited by 20 mM tartrate and partially denatured by heating it to 65°C for 10 minutes.

Fungal: Found in patients with a widespread infection such as candidiasis or coccidiosis. These are common in immunocompromised patients, those who are undergoing chemotherapy, and patients with active AIDS. Depending on the species, the fungal enzyme may be heat stable, but it is never inhibited by tartrate.

Hepatic: The liver produces large amounts of both acid and alkaline phosphatases. In cirrhosis and some other liver diseases, high levels of both are present in serum. Hepatic acid phosphatase is harder to characterize; it is partially inhibited by 20 mM tartrate and is not always inactivated by heat. However, hepatic acid phosphatase is almost always present at the same time as very high levels of alkaline phosphatase.

Alkaline Phosphatases (most active at high pH)

Bacterial: Bacteria release this enzyme as they are killed by the immune system. The bacterial isoform requires Zn^{+2} and Mg^{+2} ions as cofactors, so it is inhibited by 10 mM EDTA (a chelating agent that removes divalent cations from solutions). Bacterial alkaline phosphatase also remains active after one hour at 65°C. Unlike mammalian forms, which have maximum activity at pH 10.0 and 37°C, the bacterial enzyme is active at pH 8.0 and works well at 25°C.

Intestinal: This enzyme is elevated in patients with colorectal cancer, necrotic bowel, and certain other intestinal diseases. This isoform is most active at pH 10, and remains active after 1 hour at 65°C. It is relatively unaffected by 10 mM NaF and not inhibited by 1 mM levamisole. It requires Zn^{+2} and Mg^{+2} ions as cofactors so is inhibited by 10 mM EDTA.

Hepatic: This enzyme is released in large amounts along with acid phosphatase during cirrhosis and some other liver diseases. The alkaline phosphatase released is most active at pH 10, but is inactivated after just 20 minutes at 65°C. It is also partially to completely inhibited by 1 mM levamisole, and 10 mM EDTA.

Renal: The properties of the enzyme released during kidney failure are almost identical to the intestinal isoform. However, renal alkaline phosphatase activity is completely inhibited by 1 mM levamisole.

Handling and Storing Mock Serum Samples Safely

- Keep all mock serum samples on ice, except when you are removing aliquots for an assay.
- Do not freeze mock serum samples. Store them in the refrigerator.
- Always wear gloves when handling mock serum or samples containing any of the mock serum. Wipe up any spills, and disinfect the area with 70% alcohol.
- Dispose of mock serum only in the labeled waste receptacles. Do not place it in the general trash or pour it down the sink.

Materials Available

These items will be available for your assays. You may not need to use everything on this list. You may also request other items that are not on this list if you identify another reagent during your research. When you request it, be prepared to show your instructor your literature source.

- 1 M diethanolamine buffer, pH 10.0
- 10 mM Tris buffer, pH 8.0
- 150 mM sodium acetate buffer, pH 5.0
- 100 mM potassium phosphate in water
- 10 mM $MgCl_2$ in water
- 1 mM $ZnCl_2$
- 100 mM EDTA, pH 8.0
- 10 mM levamisole
- 200 mM L-tartaric acid
- 100 mM sodium fluoride (NaF)
- 100 mM molybdic acid
- 37°C water bath
- 65°C water bath
- Hot plate, large beaker of water, and thermometer (for other incubation temperatures)

UNIT 5

Applied Enzymology

Teacher Pages

INSTRUCTORS' NOTES

Background and Key Concepts

Like Unit 4, this unit focuses on properties and functions of enzymes. Instructors should read the background material for that unit since much of it applies here. Students once again discover for themselves how enzyme concentration, environmental conditions, and potential inhibitors affect enzymes. What makes this unit more challenging is that students must apply their knowledge to the practical problem of clinical diagnosis. This set of exercises is particularly appealing for preprofessional students.

The phosphatase enzyme family is subdivided into acid, neutral, or alkaline types, based on the pH at which they exhibit optimum activity. Most organisms and tissues produce specific phosphatases, and their presence or absence is used routinely for diagnostic purposes. For example, raw cow's milk contains very little acid phosphatase but ~500 mU/mL of a thermosensitive isoform of alkaline phosphatase. When properly pasteurized, cow's milk has <0.35 mU/mL of alkaline phosphatase. Batches of milk are checked routinely for phosphatase activity to determine if they have been properly pasteurized. Alkaline phosphatase levels greater than 0.35 mU/mL indicate insufficient pasteurization, while high acid or neutral phosphatase levels indicate the milk is contaminated with microorganisms or other foreign matter.

For this unit, seven mock serum formulations were developed; one serves as a normal control and six mimic different disease states. All of them are formulated with purified enzymes, so they do not contain potentially pathogenic materials. The phosphatase assay is very simple to perform, which limits the potential technical difficulties students might have. As they work through this unit, students learn how to

- use a spectrophotometer to measure optical density,
- calculate enzyme activity from absorbance,
- use proper controls and blanks in an assay, and
- construct a standard curve and use it to determine the concentration of enzymes in an unknown sample.

General Teaching Strategy and Common Problems

This two-week unit was designed for a sophomore-level-majors cell biology course. In the first week, students work primarily with wheat germ acid phosphatase. They determine the effects of enzyme concentration, pH, and four known inhibitors on phosphatase activity. Wheat germ phosphatase is minimally active at pH 7 or 10 and nicely demonstrates how pH can be used to distinguish enzyme subtypes. Two of the inhibitors tested (potassium phosphate and phosphonoacetate) do not significantly affect wheat germ acid phosphatase, while the other two (sodium fluoride and molybdic acid) potently inhibit it. This allows students to see that not all inhibitors work on all isoforms of these enzymes.

For Week 2, students are given samples of mock serum that were "drawn" from a patient in a hospital emergency room. The patient's symptoms point to several clinical possibilities but none can be ruled out. The students' task is to use the information they learned about phosphatases in Week 1 and the background data on pages 229–230 to determine the most likely source of the phosphatases the mock serum contains and, thus, the most likely cause of the patient's illness.

Students need one period of about three hours to complete the general exercises and a second block of three hours to perform mock serum analysis. Ideally, students should have several days between the two blocks to look at one or more of the background papers listed in Supplemental References (p. 236) and plan their strategy. It is not necessary to provide students with mock sera until the start of the second lab meeting.

Our goal in this unit is not for students to make the "right" diagnosis, but rather for students to learn to state claims and defend them with reasonable evidence that has been interpreted correctly. In the author's experience, not all of the serum samples perform exactly as formulated. From time to time, components in a mock serum mixture have interacted to give surprising enzyme properties, or a particular batch of phosphatase has been unaffected by its normal inhibitors. When grading students, focus on their ability to formulate a reasonable strategy for identifying the phosphatase isoform and how well they interpret the data they obtain.

Relatively few mechanical problems arise during this unit. However, there are conceptual areas with which some students struggle. In the first week, students may not understand why they use o-nitrophenyl phosphate for an enzyme substrate. Take the opportunity to introduce general concepts and principles of indirect measurements and reporter assays. In this case, phosphate ions are ubiquitous in biological systems so it would be almost impossible to measure changes in them directly. It might be possible to measure the concentration of the phosphorylated and unphosphorylated native substrates, but since phosphatases do not have just one substrate, which substrate should be used? Scientists and clinicians use an alternative substrate that is recognized by nearly all phosphatases. The covalent bond connecting the phosphate to the nitrophenyl group has the same chemical structure as the normal covalent bond between phosphate and the hydroxyl groups of serines, threonines, tyrosines, and various sugars. Freed nitrophenyl groups rearrange to form a bright yellow dye molecule. Such indirect measurements are generally accepted, as long as the substrate and conditions under which the alternative is used are documented and the proper controls are used.

Students may claim, "the serum unknowns are too complex for me to figure out." Reiterate that they should work slowly and methodically through their options. Let them struggle some but do not let them become hopelessly frustrated. If a group of students cannot decide how to proceed, ask them to identify just one variable that eliminates some of the possible conditions; nearly all will choose pH. Have them perform the pH assay alone, then discuss the next step once they have that data in hand.

Other groups will design overly complex experiments with dozens of tubes, increasing the chance for error. Have these groups break their complex assay into two or more smaller ones. They should have ample time to complete at least three different assay trials in a single three-hour lab period.

Students may not understand why they are given a normal mock serum control. Remind them that phosphatases are present in normal tissues as well. Enzyme activity is not defined as high or low just because the spectrophotometer reading is larger or smaller. It is always relative to the normal levels found in that tissue or a known reference standard.

Recommended Prelab Skills

Students should know how to use serological and micropipets accurately, and how to make serial dilutions. Experience in graphing X-Y data with a spreadsheet program is useful.

Assessment

If pre- and postlab quizzes are used as part of formal assessment, the prelab quiz should focus mainly on general functions of enzymes and the general functions of phosphatases. A postlab quiz can be given at the start of the second week of this unit. Students should be able to explain at least two mechanisms by which organisms can regulate enzyme activity. Exercise 3 describes the mechanisms by which four inhibitors block phosphatase activity, but does not define them as competitive, noncompetitive, or product inhibitors; if students have learned these concepts in lecture, it would be appropriate to ask them to name the specific modes of action. Include a practical question for which students must blank a spectrophotometer and take a reading of an unknown. It is not reasonable to expect them to know from memory which diseases have which phosphatase profiles.

Students also can write a three-to-four-page paper describing their analysis of the mock serum. It is difficult to fit this assignment into the format of a traditional scientific paper. Instead, consider having students complete a "Clinical Report" in which they state the Patient ID#, describe the analysis methods they used and their rationale, then state and defend the diagnosis. Another option is to use a directed-writing format similar to the Science Writing Heuristic (Keys et al. 1999).

Safety and Housekeeping

While mock serum does not contain hazardous blood products, students still should practice *universal precautions*. This term is used frequently in hospitals and by regulatory agencies, and this lab provides an opportunity for introducing it. Universal precautions include regular hand washing, routine use of disposable gloves and safety glasses during procedures, and proper disposal of contaminated materials like pipets and dishes.

Students must be very careful with the sodium hydroxide solution. It is caustic and will leave a scar if spilled onto skin and not washed off quickly. Keep a 4 L liter bottle of 100 mM sodium bicarbonate in water on hand to neutralize any sodium hydroxide that gets spilled onto skin or clothing.

The yellow reaction product of the enzyme assay is an EPA-regulated hazardous waste. The unused nitrophenyl-derivatized substrates and liquid reaction products should be collected and pooled for destruction by photocatalysis. A destruction process is described in detail on page 209; check with your institution's environmental health office for other options.

Other Tips

Try to have each group in a lab section analyze a different patient's serum.

Occasionally the enzyme activity increases in wheat germ phosphatase solutions during storage. As a result, the enzyme solutions prepared for Exercise 1 may be too potent. If this occurs, simply dilute the enzyme stock solution 1:10.

When students heat samples of phosphatases to test thermal stability, make sure they heat their enzymes and cool them again before adding substrates. If they try to heat inactivate in the presence of substrate, the reaction will go to completion before the enzyme denatures. Never heat mock serum solutions above 85°C, as the albumin in the solution cooks like egg white.

Supplemental References

Bull, H., P. G. Murray, D. Thoma, A. M. Fraser, and P. N. Nelson. 2002. Acid phosphatases. *Journal of Clinical Pathology: Molecular Pathology* 55:65–72.

Matsushita, M., T. Irino, T. Kawaguchi, and T. Komoda. 2002. The effect of different buffers and amounts of intestinal alkaline phosphatase isoforms on total alkaline phosphatase activity. *Clinica et Chimica Acta* 319: 49–55.

Keys, C. W., B. Hand, V. Prain, and S. Collins. 1999. Using the science writing heuristic as a tool for learning from laboratory investigations in secondary science. *Journal of Research in Science Teaching* 36: 1065–1089.

PREPARATORY NOTES

Quantities listed are for a lab section of 20 students working in groups of three or four.

Week 1: Shared Materials

- 5 L distilled water (stored in a carboy)
- 100 mL of 100 mM para-nitrophenyl phosphate in water
- 50 mL wheat germ acid phosphatase, diluted to 1.0 U /mL in water
- 1 L pH 5.0 assay reaction buffer
- 500 mL pH 4.0 assay reaction buffer
- 500 mL pH 7.0 assay reaction buffer
- 500 mL pH 10.0 assay reaction buffer
- 20 mL Unknown Sample A
- 20 mL Unknown Sample B
- 100 mL of 2 M sodium hydroxide in water
- 20 mL of 20 mM sodium phosphate
- 20 mL of 20 mM sodium fluoride
- 20 mL of 20 mM molybdate
- 20 mL of 20 mM phosphonoacetate
- 5 serological pipets (10 mL)
- Pipet pumps
- Paper towels
- Waste container for spent reaction mixtures

Week 1: Materials at Each Work Station

- Spectrophotometer
- Paper towels or tissues
- Test tube rack with 20 large (15 × 150 mm) test tubes
- Test tube rack with 12 cuvets for spectrophotometer
- Water bath at 37°C
- 20 microtubes (1.5 mL) in a 250 mL beaker, autoclaved
- P-1000 and P-200 micropipets
- Boxes of blue and yellow tips, autoclaved
- Microtube rack

Week 2: Shared Materials

- 7 aliquots (15 mL) of Control mock serum
- 2 aliquots (15 mL) of mock serum from Patient A
- 2 aliquots (15 mL) of mock serum from Patient B
- 2 aliquots (15 mL) of mock serum from Patient C
- 2 aliquots (15 mL) of mock serum from Patient D
- 2 aliquots (15 mL) of mock serum from Patient E
- 2 aliquots (15 mL) of mock serum from Patient F
  (Store all mock sera frozen until ready to use)
- 5 L distilled water (stored in a carboy)
- 100 mL of 100 mM para-nitrophenyl phosphate
- 2 L pH 5.0 assay reaction buffer
- 2 L pH 4.0 assay reaction buffer
- 2 L pH 7.0 assay reaction buffer
- 2 L pH 8.0 assay reaction buffer
- 2 L pH 10.0 assay reaction buffer
- 250 mL of 2 M sodium hydroxide in water
- 500 mL of normal saline
- 100 mL of 10 mM magnesium chloride
- 25 mL of 10 mM levamisole
- 25 mL of 200 mM tartrate
- 25 mL of 100 mM EDTA, pH 8.0
- Hot blocks or water baths at 65°C and 75°C
- Serological pipets (10 mL)
- Pipet pumps
- Paper towels
- Waste container for spent reaction mixtures

Week 2: Materials at Each Work Station

- Spectrophotometer
- Paper towels or tissues
- Test tube rack with 20 large (15 × 150 mm) test tubes
- Test tube rack with 12 cuvets for spectrophotometer
- Water bath at 37°C
- 20 microtubes (1.5 mL) in a 250 mL beaker, autoclaved
- P-1000 and P-200 micropipets
- Boxes of blue and yellow tips, autoclaved
- Microtube rack

Solutions, Reagents, Equipment

1 U/mL Wheat Germ Acid Phosphatase

Make a stock of ~5 units/mL of Type I wheat germ phosphatase (Sigma #P3627, ~0.4 units/mg solid) in distilled water. Dilute the stock enzyme 1:5 with water, and perform Exercise 1 from the unit. The enzyme is at proper concentration if maximum OD420 for Exercise 1 is 1.0-1.3 for a cuvet with a 1 cm path length. If the maximum absorbance values are lower than 0.7, try using a 2:5 dilution of enzyme.

Dilute sufficient enzyme for one week, then divide it into 50 mL aliquots (one aliquot per lab, plus a couple of extras). Store diluted aliquots frozen. Do not try to keep the aliquots of diluted enzyme solution longer than one week. Make fresh ones.

100 mM para-Nitrophenyl Phosphate

Weigh out 4.0 g p-nitrophenyl phosphate hexahydrate, disodium salt (Sigma # 104-0) and place in 15 mL conical tubes. Label with "Mix into 100 mL distilled water." Store in the freezer until just before lab. Other salts are commercially available and can be substituted for the disodium salt form.

pH 4.0, 5.0 Assay Reaction Buffers

Prepare Solution 1 by dissolving 40.8 g sodium acetate trihydrate in 2000 mL distilled water. Prepare Solution 2 by diluting 5 mL glacial acetic acid in 600 mL distilled water.

For pH 5.0 buffer, mix 1500 mL of Solution 1 with 500 mL of Solution 2. Check the pH. It should be near 5.0. If needed, adjust the pH higher by adding Solution 1; lower pH with Solution 2.

For pH 4.0 buffer, make 1200 mL of diluted acetic acid. Mix 1000 mL of acetic acid with 1000 mL of sodium acetate. Check pH, and adjust as before.

Final concentration is 150 mM acetate in both buffers.

pH 7.0 Assay Reaction Buffer

Dissolve 20 g MOPS free acid (e.g., Sigma #M1254) in 1500 mL distilled water. Adjust pH with 1 or 10 M HCl or NaOH, then bring to 2 L with distilled water. Final concentration is 50 mM MOPS.

pH 8.0 Assay Reaction Buffer

Dissolve 1.2 g Tris base in 600 mL distilled water. Adjust to pH 8.0 with concentrated HCl. Bring to 1000 mL with water. Final concentration is 10 mM Tris.

pH 10.0 Assay Reaction Buffer

Add 200 mL diethanolamine to 1700 mL distilled water. Mix, then adjust pH with 10 M HCl or NaOH as needed to 9.8 to 10.0. Bring to final volume of 2000 mL with distilled water.

Unknown Sample A

Dilute 1 mL of 5× stock wheat germ acid phosphatase into 100 mL distilled water. Final concentration will be ~0.05 units /mL, which should be in the linear range of the students' assays.

Unknown Sample B

Dilute a commercial glycerol solution of molecular grade calf *alkaline* phosphatase (e.g., Promega #M1821, 1000 U/mL) to 0.1 units/mL with distilled water. In Exercise 2, as pH rises, activity in this unknown suddenly appears at pH 9 or higher. Below pH 8, there should be no activity.

2 M Sodium Hydroxide

Dissolve 80 g sodium hydroxide in 900 mL distilled water. Bring to 1000 mL final volume. WARNING: Exothermic solution! Highly corrosive.

Enzyme Inhibitors (see Table 5.9)

Table 5.9

Making stocks of inhibitor solutions

Inhibitor	**Add to 100 mL water**
20 mM sodium phosphate	0.24 g
20 mM sodium fluoride	0.084 g
20 mM molybdic acid ammonium salt, tetrahydrate	2.5 g
20 mM phosphonoacetic acid	0.28 g
10 mM levamisole	0.24 g
200 mM L-tartrate	3 g
100 mM EDTA	Dilute 500 mM EDTA stock solution 1:5. Instructions for preparation are in Unit 3, p. 189.

Normal Saline

Dissolve 9 g NaCl in 1 L of distilled water.

10 mM Magnesium Chloride

Dissolve 0.095 g of anhydrous $MgCl_2$ in 90 mL of water, then bring to 100 mL.

Preparing Mock Serum Samples

1. Purchase the following stock enzymes or equivalents. Store at -20°C in frost-free conditions.

    - Potato acid phosphatase (e.g., Sigma #P3752)
    - Bovine intestinal alkaline phosphatase (e.g., Sigma #P3681)
    - Bacterial alkaline phosphatase (e.g., Sigma #P4069)
    - Pig kidney alkaline phosphatase (e.g., Sigma #P4439)
    - Bovine prostatic acid phosphatase (e.g., Sigma #P640)
    - Shrimp alkaline phosphatase (e.g., Promega #M8201)

2. Prepare baseline diluent solution (BDS).

    - 9 g NaCl
    - 8 g bovine serum albumin, Fraction V (Sigma #7906)
    - 1000 mL distilled water

3. Prepare 1 U/mL stock solutions of each phosphatase by dissolving powdered enzymes or diluting liquid forms directly in baseline diluent.
4. For every 20 students, prepare 100 mL of Control mock serum and 30 mL of mock serum for Patients A–F. Quantities of enzyme to add are given in Table 5.10, page 242.

Table 5.10

Preparing the mock sera samples

Serum	Condition	Typical Clinical Phosphatase Levels	Diagnostic Feature	Composition
Control	Normal	0.03 U/mL alkaline; 0.005 U/mL acid	Not applicable	3 units of calf intestinal phosphatase and 0.5 units of potato acid phosphatase in 100 mL of BDS
Patient A	Bacterial infection	0.2 U/mL alkaline	Active at pH 8.0; inhibited by EDTA	6 units of bacterial alkaline phosphatase in 30 mL BDS
Patient B	Fungal infection	0.1 U/mL acid	Thermostable; tartrate insensitive	3 units of potato acid phosphatase in 30 mL BDS
Patient C	Prostate cancer	0.2 U/mL acid	Unstable to heat; tartrate sensitive	6 units of bovine prostatic acid phosphatase in 30 mL BDS
Patient D	Hepatic disease	0.2 U/mL acid; 0.5 U/mL alkaline	Acid, alkaline forms together; alkaline inhibited by EDTA, levamisole	6 units wheat germ acid phosphatase AND 15 units of shrimp alkaline phosphatase in 30 mL BDS
Patient E	Kidney disease	0.5 U/mL alkaline	Inhibited by levamisole	15 units of pig kidney alkaline phosphatase in 30 mL BDS
Patient F	Colon cancer	0.5 U/mL alkaline	Not affected by levamisole	15 units of bovine intestinal alkaline phosphatase in 30 mL BDS

Note: Potato, arctic shrimp, and wheat germ phosphatases are used to prepare the control and two of the mock sera. They are much less expensive and for instructional purposes they have chemical characteristics nearly identical to the human enzymes. If these substitutions are not acceptable to other instructors, the corresponding enzymes should be purchased instead.

5. Divide the mock sera into ~15 mL aliquots. Label tubes and store them frozen for up to one week.

UNIT 6

Energetics and Photosynthesis

Student Pages*

BACKGROUND

You probably know already that enzymes catalyze metabolic and other chemical reactions in biological systems. However, many cellular reactions also require a source of energy to proceed forward. Living organisms must obtain the required energy from external sources and then store the energy they obtain as one or more of these forms:

- Chemical energy in covalent bonds
- An **oxidation-reduction** (**redox**) potential
- An electrical or chemical gradient across a lipid bilayer

Plants, some protists, and certain bacteria contain chlorophyll that allows them to undergo **photosynthesis**, in which they convert and store solar energy in high-energy chemical bonds. This stored chemical energy is later released by the processes of **glycolysis** and **respiration**, which occur in cytoplasm and mitochondria, respectively. Animals and other organisms that cannot undergo photosynthesis must consume other organisms to obtain sufficient chemical energy, which they also extract by glycolysis and respiration. Ultimately, photosynthesis provides most of the energy required for the fundamental metabolic processes of nearly all living things on this planet.

*Teacher Pages begin on page 255.

Photosynthesis and respiration are similar in that they are multistep oxidation-reduction (aka, redox) reactions. High-energy electrons are passed through **electron transport chains** from one molecular complex to another. As energy leaves the electrons, both systems use it to pump protons (H+ ions) across a lipid bilayer and generate a combined chemical and charge gradient that drives ATP synthesis.

There are significant differences too. In photosynthesis, absorbed solar energy powers a chemical reaction that cleaves water to form oxygen, and releases high-energy electrons that are stored as NADPH. The electrons on NADPH and the ATP are then used to drive anabolic reactions that synthesize glucose from carbon dioxide. Conversely, glycolysis and respiration are catabolic reactions that break glucose apart again, releasing carbon dioxide and several high-energy electrons. The electrons are transferred back to oxygen molecules, forming water.

Photosynthesis Is a Multistep Process

The summary equation for photosynthesis is shown in Figure 6.1.

Figure 6.1

Summary of the general chemical reactions of photosynthesis

$$6\,CO_2 + 12\,H_2O \xrightarrow{\text{Light}} C_6H_{12}O_6 + 6\,H_2O + 6\,O_2$$

Carbon dioxide + Water → Glucose + Water + Oxygen

While this equation is accurate, it does not indicate the complexity of the process. Photosynthesis actually occurs in three distinct stages, depicted in more detail in Figure 6.2. All of these processes are occurring within the specialized organelles called **chloroplasts**. The first two stages, called the **light-dependent reactions**, use solar energy to generate high-energy NADPH and ATP molecules. The light-dependent reaction requires several cofactors and components that are not consumed: the green pigment **chlorophyll**, Mn^{+2} ions, several protein-rich electron transport complexes, and an intact phospholipid bilayer for producing a proton gradient. Subsequently, **light-independent reactions** (also called the Calvin cycle) use the high-energy products of the light-dependent reactions to synthesize glucose. During this cycle, the enzyme complex RUBISCO combines six 1-carbon CO_2 molecules with six 5-carbon ribulose phosphate molecules, then rearranges them to create twelve 3-carbon glyceraldehyde-3-phosphate (G3P) molecules. Ten of the G3P molecules, plus the ATP and NADPH from the light-dependent reactions, are used to regenerate ribulose phosphate. The two remaining G3P molecules are used to synthesize one 6-carbon glucose molecule.

Figure 6.2

Schematic summary of the major stages of photosynthesis

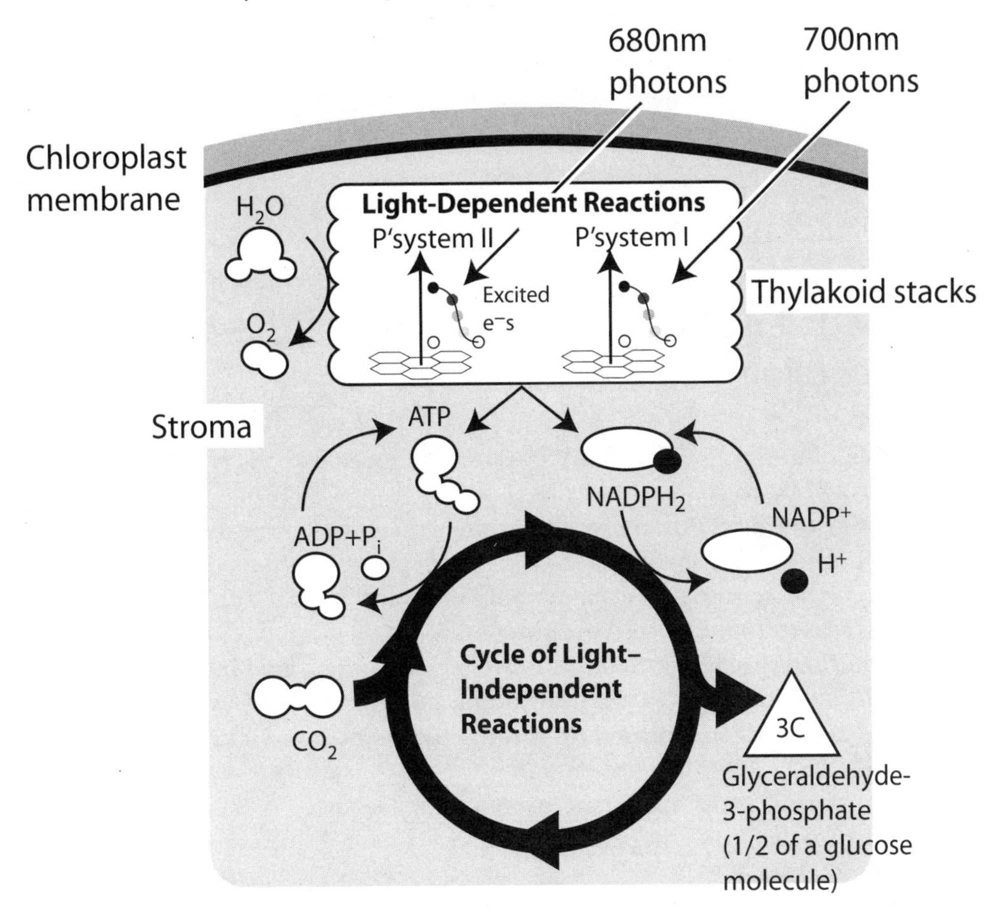

Goals for This Unit

The process of photosynthesis is the key to life on Earth. Anything that alters photosynthetic rates in plants can have dramatic effects on plant productivity. It is technically difficult to measure glucose production directly. However, relative photosynthetic output can be estimated indirectly by measuring the amount of electron transfer occurring as part of the light-dependent reactions.

In the first part of this unit, you will learn to make an enriched chloroplast preparation from spinach leaves. Next you will measure their relative redox activity by mixing them with DCIP, an indicator dye that changes color when it becomes reduced. You will also determine whether certain herbicides (chemicals used to kill plants) affect electron transport. After completing Exercises 1 and 2, you will design and execute an experiment to test a hypothesis of your own.

General Safety Precautions

You will be using a tabletop centrifuge; pay close attention to any safety procedures for properly loading and operating this instrument.

DCIP and some other compounds you might use for your experiment can be very toxic if swallowed or absorbed through your skin. Cover tubes when you mix them, and wear gloves if necessary. Wash your hands thoroughly before leaving lab or if any of these chemicals spill or splash onto your skin.

UNIT EXERCISES

Exercise 1: Enriching Chloroplasts by Differential Centrifugation

Procedure

1. Pack a mortar, pestle, 50 mL beaker, and three 50 mL centrifuge tubes in crushed ice to cool.
2. While the above items cool, remove the large central vein from four medium-sized spinach leaves.
3. When the mortar is cool, empty out the ice and wipe it dry. Add the spinach leaves and 10 mL of cold grinding buffer.
4. Grind the spinach until it has completely disintegrated into a paste.
5. Add an additional 10 mL of grinding buffer, mix with the spinach paste, and strain the green mixture through three layers of cheesecloth into the chilled 50 mL beaker.
6. Transfer 200 μL of this liquid to a microtube labeled "A,"and store it on ice.
7. Transfer the remaining green liquid to a chilled 50 mL centrifuge tube, and spin it at 200 × G (200 times the force of gravity) for three minutes at 4°C. Be sure to include a balance tube in the centrifuge.
8. Pour the supernatant into the second chilled 50 mL centrifuge tube. This time centrifuge the chloroplast suspension at 1000 × G for 10 minutes at 4°C.
9. Pour the supernatant into a clean 50 mL tube marked "B." Be careful not to disturb the pellet. Place the tube on ice.
10. Add 20 mL of grinding buffer to the pellet and suspend it by stirring gently with a transfer pipet. Label the tube "C," and place it on ice.
11. Load 10 μL aliquots of A, B and C onto a hemacytometer, and examine them under phase optics. Look for chloroplasts in each of the three samples; they will be small, oval to egg-shaped, and shine brightly under phase optics. They are bright green, but are so thin and small that it may be difficult to see their color.

Estimate the relative numbers of chloroplasts in the three samples. It is not necessary to count them; just rank the three samples from most to fewest chloroplasts. Write down your observations in Table 6.1.

Results

Table 6.1

Comparison of chloroplasts in different fractions

Fraction	Amt. of Random Debris	Relative No. of Chloroplasts
A		
B		
C		

Summary Questions for Exercise 1

1. Which aliquot had the largest number of chloroplasts—A, B, or C? Which sample had the smallest number of chloroplasts? Which aliquot had the largest fraction of randomly sized debris and tissue fragments?
2. The raw homogenate you made in the first steps already contained active chloroplasts. So why is it necessary or useful to perform the subsequent steps?
3. Plants regularly transport large quantities of water through the interstitial spaces of leaves. Yet when you isolated the chloroplasts, you used a buffered grinding solution of 400 mM sucrose plus NaCl. Why is it necessary to use a buffer solution, rather than just water? (*Hint:* This buffer is approximately iso-osmotic with the internal fluid within the chloroplasts. Now go back and look at what is required in order for photosynthesis to occur.)

Exercise 2: Quantifying Photoreduction Activity With the Hill Reaction

Background

The light-dependent reactions are summarized schematically in Figure 6.3, page 248.

Figure 6.3

Redox reactions in the thylakoid membrane of a chloroplast

Dashed lines represent electron transfer, solid lines represent proton movement. PS-I = Photosystem I; PS-II = Photosystem II; PQ = plastiquinone; PC = plastocyanin; Fd = ferredoxin; NR = NADP reductase. (Taiz, L. and E. Zeiger. 1998. *Plant physiology*. 2nd ed. Sunderland, MA: Sinauer Press, page 173. Redrawn from original with permission.)

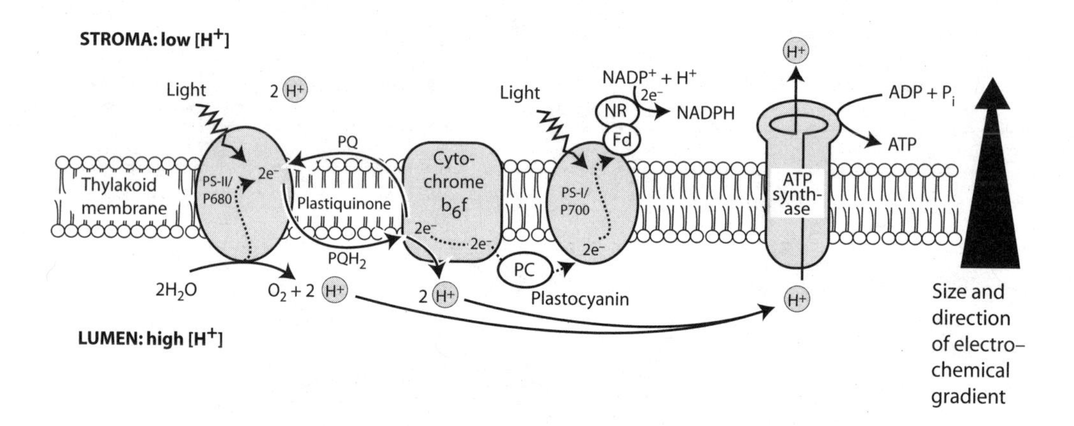

In Photosystem II, energy from absorbed photons of light causes electrons to leave chlorophyll and move to plastiquinone. Oxidized chlorophyll indirectly stimulates cleavage of water to produce molecular oxygen (O_2), plus electrons that restore the ones lost from chlorophyll. The free protons (H^+ ions) from water are then pumped out across the thylakoid membrane. Reduced plastiquinone transfers its electrons to cytochrome b_6f, which in turn transfers the electrons to plastocyanin. At the same time, the cytochrome b_6f complex also pumps protons across the thylakoid membrane, adding to the proton gradient.

In Photosystem I, energy carried by incoming photons of light is again used to excite electrons on chlorophyll molecules. This time the excited electrons leave the chlorophyll and travel through a multi-subunit complex to NADP+, to produce NADPH. In Photosystem I, electrons carried by plastocyanin from Photosystem II replace the electrons that leave the chlorophyll molecules.

Accumulation of excess protons in the lumen creates an electrochemical gradient that is pushing protons back toward the **stroma** (innermost space) of the chloroplast. However, the only path by which protons can traverse the thylakoid membrane is through an **ATP synthase complex**. When protons move down their combined electrical and diffusion gradient, new ATP molecules are synthesized by this complex.

How Are Electron Transfer Reactions in Chloroplasts Measured?

The general process of electron transfer in the presence of active chloroplasts plus light is called the Hill reaction (after its discoverer) and can be summarized by the equation in Figure 6.4.

Figure 6.4

The general equation for the Hill reaction

$$2\,H_2O + 2\,A \xrightarrow{\text{Light, chloroplasts}} 2\,AH_2 + O_2$$

Water + Electron acceptor → Reduced electron acceptor + Oxygen

Electrons pass between several very specific donors and acceptors in the two photosystems. However, in a redox reaction, electrons can pass from the reduced donor molecule to any acceptor molecule that has a lower energy level and that comes close enough for the transfer to occur. Early experiments on photosynthesis showed that the electron carrier molecules in chloroplasts could pass electrons to many other electron acceptors besides their normal ones. Furthermore, some of these alternative electron acceptors were much easier to measure because they changed color when they gained or lost electrons.

One alternative electron acceptor is two **6-dichlorophenol indophenol** (DCIP). DCIP can accept electrons from reduced plastiquinone molecules in Photosystem II. Oxidized DCIP is a dark blue compound with an aromatic ring structure that absorbs light maximally at 600 nm. In contrast, reduced DCIP has conjugated C=C bonds and is colorless. Because of this color change, DCIP reduction can be measured "on the fly," as a *decrease* in light absorbance at 600 nm (see Figure 6.5), p. 250.

Figure 6.5

Behavior of the electron accepting dye DCIP during the Hill reaction

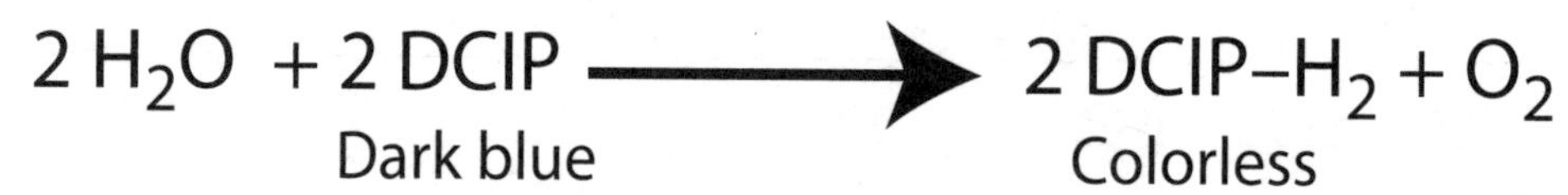

The assay you will use today uses loss of color of DCIP to measure Photosystem II activity. If a chemical, environmental condition, or other factor slows down the photosystem, then chloroplasts will not decolorize a solution of DCIP.

Procedure

1. Label 13 large test tubes as shown in Table 6.2 and place them in a tube rack. Then prepare three replicates of each of the reaction mixes. You only need one replicate of the blank.

Table 6.2

Setup for Hill reactions in Exercise 2

	Tube 0 Blank (make 1)	Tubes 1–3 No Chloroplasts	Tubes 4–6 + C'plasts, in Dark	Tubes 7–9 +C'plasts, in Light	Tubes 10–12 +C'plasts, +Herbicide
Water	8 mL	6 mL	6 mL	6 mL	5 mL
130 μM DCIP	none	2 mL	2 mL	2 mL	2 mL
Reaction buffer (100 mM phosphate, pH6.5)	2 mL	2 mL	2 mL	2 mL	2 mL
Herbicide:	none	none	none	none	1 mL

2. Wrap the three tubes to be kept in the dark (#4–6) in foil to block out incident light.
3. When you are ready to begin the assay, add 100 μL of the enriched chloroplasts you prepared in Exercise 1 to the blank tube and to tubes #4–12.

4. Cover with Parafilm and invert all tubes three to four times to mix well.
5. Place tubes #0, #1–3, and #7–12 in a rack on a sunny windowsill or under a grow light. Place tubes #4–6 in darkness.
6. After 10 minutes, blank the spectrophotometer with tube #0. Then pour each sample into a spectrophotometer cuvet, and read the absorbance at 600 nm. After recording absorbance, pour each sample back into its respective large tube. Be sure to put the three dark tubes back in foil after each measurement.
7. Repeat the previous step again at 20 minutes, and once more at 30 minutes.
8. Record your results in Table 6.3.

Results

Table 6.3

Results of Hill reactions in Exercise 2

		Absorbance		
Tube #	**Sample**	**10 min.**	**20 min.**	**30 min.**
1	No chloroplasts added			
2				
3				
Mean				
4	Plus chloroplasts, in dark			
5				
6				
Mean				
7	Plus chloroplasts, in light			
8				
9				
Mean				
10	Plus chloroplasts and herbicide, in light			
11				
12				
Mean				

1. Using a spreadsheet program, plot the mean absorbance against time for each of the four treatments. Format your data as an XY graph, not a bar graph.
2. Use your XY graph to answer the following Summary Questions for Exercise 2.

Summary Questions for Exercise 2

1. Compare the line you obtained for samples with no chloroplasts with that of chloroplasts in the light. What happened to the absorbance at 600 nm in the tubes with chloroplasts? What specifically is occurring at the level of individual molecules to produce the effects you have observed?
2. Compare the lines for samples with chloroplasts kept in light versus those kept in darkness. Explain what is happening at the molecular level and how it produced the observed results.
3. Each lab group tested a different herbicide to determine if it inhibited photosynthetic redox reactions. Consult with the other groups in the lab: Which compound(s) appear to inhibit electron transport? Which do not? If more than one altered electron transport, rank them in relative order of inhibition.
4. What chemical structural characteristics would you predict you would find in a compound that acts as an electron transport blocker? Write down your predictions. Now look up the chemical structure of each herbicide molecule tested; did your predictions match the actual structures?
5. Some of the herbicides you tested might *not* have affected electron transport. What are some other potential targets for herbicides? (*Hint:* Most commercially available herbicides are not particularly toxic to animals because they interfere with processes that are unique to plants.)
6. The herbicide triazine prevents electrons from passing through the electron transport chain from plastiquinone to cytochrome b_6f. Triazine has been used widely for weed control, but several weed species now are resistant to triazine. Electron transport is essential for photosynthesis, so the weeds cannot just eliminate the pathway and gain resistance. How then might a weed become resistant to triazine? (*Hint:* Work with your lab group on this question. First decide what triazine would have to do to block electron transport. Then ask, how could a plant prevent triazine from having that effect?)
7. What other environmental factors or conditions do you think could increase or decrease the rate of redox reactions by chloroplasts? What is your rationale?

Your Own Experiment: Environmental Cell Biology

Nearly all organisms depend directly or indirectly upon the process of photosynthesis to obtain usable energy. The amount of energy converted into usable forms by photosynthetic organisms in an ecosystem is termed **primary productivity**.

Environmental changes can affect photosynthetic rates, in turn changing primary productivity and ultimately the ecosystem that depends on it. You have heard innumerable times that global warming, acid rain, nuclear winter, soil salinization, and other insults can alter plant growth. But why? More specifically, do these environmental insults affect photosynthesis directly, or do they change primary productivity by some other means? What about other abiotic factors in the environment: If they change, will photosynthesis also be affected? If so, how much? Conversely, are there environmental factors that can *increase* the rate of photosynthesis?

Using the principles and assays you learned today, you can identify and test many different environmental conditions to see how they might affect vascular plants, or phytoplankton in an aqueous environment. Unlike measuring overall plant growth rates, this assay allows you to focus on and quickly measure just the redox reactions that drive photosynthesis.

Ideas for Experiments

There are many directions in which you can go. The following are just general suggestions to stimulate thinking and discussion by your group. Check with your instructor when you have a tentative plan of what you want to test, to make sure the materials you need are available, and you can complete your assays in a reasonable amount of time.

- What are the typical contaminants in storm water or agricultural waste-water runoff? Could these affect photosynthesis? What is the maximal amount that can be present in the environment before a contaminant reduces photosynthesis significantly?
- Several crop varieties have been bred or genetically engineered to be more productive in low versus high light conditions. These varieties are marketed heavily to farms in the northern tier of states and to Canadian provinces. However, chloroplasts can still undergo photosynthesis even when light levels are quite low. So is light availability really a problem? How low does the light level have to be to reduce photosynthetic rates?
- Agro-scientists use the Hill reaction (among other assays) to characterize potential herbicidal compounds. They compare the effectiveness of new versus existing chemicals by determining a 50% inhibitory dose (ID50), that is, the concentration of a compound needed to inhibit activity by 50%. What is the ID50 for one of the commercially available herbicides? Are there other compounds that might work just as well as commercial herbicides?

Available Materials

These items will be routinely available. You may not need to use everything on this list. You may also request other items that are not on this list. If you identify another reagent as you are doing research. When you request it, be prepared to show your instructor your literature source.

- Spinach for making extracts
- All of the buffers and reagents for the Hill assay
- Shade cloth and colored acetate to produce different wavelengths or intensities of light
- Concentrated stock solutions of various commercial herbicides
- Temperature-controlled chambers
- Other leafy crops: lettuce, kale, or mustard greens
- Various salts and buffer solutions (please ask if you need a specific item)

UNIT 6

Energetics and Photosynthesis

Teacher Pages

INSTRUCTORS' NOTES

Background and Key Concepts

In preparing students for this unit, emphasize these concepts:

- Each organelle has distinct properties that are different from other structures in the cell.
- Specific organelles can be isolated or enriched by using differences in their size, weight, or density relative to other cellular materials.
- Redox reactions can change the three-dimensional shape of a molecule by changing the organization of its covalent bonds. This change causes molecular movement; in chloroplasts such movements can drive transport of protons across the membrane bilayer.
- Environmental conditions can dramatically affect cellular chemical processes. This is one way abiotic factors affect the rate of photosynthesis and, thereby, affect primary productivity of plants. This in turn can affect an entire ecosystem.

General Teaching Strategy and Common Problems

This two-week unit was designed for a sophomore-level majors cell biology course. In the first week students learn to use differential centrifugation to enrich chloroplasts extracted from spinach leaves. Subsequently students test the effects of light levels on photosynthetic redox reactions. They also test the effects of commercial herbicides on chloroplasts. Student use a color change by dichloroindophenol (DCIP) as an indicator of electron transfer activity within Photosystem II. DCIP is an aromatic molecule that accepts electrons from reduced plastiquinone; there is some evidence that it may also accept electrons from molecules in Photosystem I too, but this remains controversial. In near-neutral pH solutions, DCIP is blue when oxidized but colorless when reduced. This color change allows students to monitor electron transfer visually as well as with a spectrophotometer.

For the second week of this unit, students design and execute experiments of their own. Students are encouraged to look further at the effects of other external environmental conditions on activity of Photosystem II such as temperature, light intensity, or light wavelength. Alternatively, they can try to extract and test chloroplasts from other vegetables or ornamental plants. Yet another option is to test the effects of pollutants, herbicides, or other toxic environmental agents on photosynthetic rate.

Most students understand how cells can use covalent bond energy and electrochemical gradients to perform work. However, they often struggle to understand how and why electrons are transferred between molecules. As a result, they cannot see how such transfers can be harnessed to perform work within a cell. In practice, students also get bogged down frequently in the details of the redox reactions of the two photosystems. The author has chosen, therefore, to eliminate most of the details of the photosystems from the lab discussion. Instead, attention focuses on how oxidation and reduction cause molecular rearrangements. Start with the chemical changes occurring in DCIP. Students can see that DCIP changes shape and decolorizes when it gains electrons, which helps them to understand how electron transfers change the fundamental shape and character of molecules in general. The summary questions require students to explain what is happening at the molecular level as molecules are oxidized or reduced. When this concept is firmly in place, the mechanism for driving the proton pumps becomes logically straightforward. The proton pumps simply are larger molecules that require the energy released from redox transfers of several electrons in order to change shape.

It is important to ensure that students understand that DCIP is an artificial electron acceptor that replaces the normal electron acceptor in Photosystem II. Normally, electrons on reduced plastiquinone should be transferred down the redox gradient toward plastocyanin and ultimately to Photosystem I. Reduced DCIP removes these usable electrons from the pathway entirely, in essence short-circuiting one-half of the chloroplast's energy collection system. Loss of electrons would block photosynthesis and potentially could kill an intact plant. This helps students understand the part of Exercise 2 in which they determine whether commercial herbicides have any effects on photosynthesis.

Recommended Prelab Skills

Students should be able to use serological pipets and micropipets accurately. Prior experience using a spectrophotometer and a spreadsheet program to graph data, are both helpful but not essential.

Assessment

If pre- and postlab quizzes are used as part of formal assessment, the prelab quiz should focus on the general concepts: What is a redox reaction? What in general happens in the Hill reaction? Avoid asking detailed questions about differential centrifugation until students have actually performed it once or twice. For postlab quizzes, it is reasonable to expect students to be able to explain the basic principles of differential centrifugation and to explain the purpose and outcome of each step in the procedure described in Exercise 1. Students in a more advanced course might be asked to design a differential centrifugation procedure that enriches another organelle, particularly if they have had some background in calculating relative centripetal forces and know (or are given) relative densities of organelles. All students should be able to explain the general process of a redox reaction and how the addition or removal of electrons leads to molecular shape changes. It is unreasonable to ask for specific donors and acceptors within either pathway, unless that material has been discussed in detail in the lecture part of the course.

Safety and Housekeeping

Both the DCIP and herbicide waste solutions are toxic. Students should clean up thoroughly and wash their hands after handling them and dispose of the solutions only in the labeled waste containers. Paraquat is a potentially hazardous waste, but it is hydrolyzed by strong bases. Destroy any leftover stock solution by mixing it with an equal volume of 2 M sodium hydroxide. Allow the solution to react 20 minutes, then flush it down the drain followed by copious amounts of water.

All DCIP solutions are EPA-regulated hazardous waste. They can be collected and pooled for destruction by photocatalysis. Dilute the pooled wastes to less than 1% (wt./vol.) DCIP in a clear glass container. Add 1 g/L titanium dioxide (TiO_2) and expose solution to sunlight or a strong UV light source. When the solution is decolorized (~24 hours), it can be flushed down the drain, followed by copious amounts of water.

Dried DCIP solutions stain cuvets and test tubes. To remove the stains, scrub tubes with 70% or 95% ethanol.

Other Tips

- In addition to participating in redox reactions, DCIP is a pH sensitive dye. DCIP solutions turn pink at pH 4.0, but return to blue if the pH is raised once again. This is a problem when students want to test the effects of acid conditions on photosynthesis. To compensate, students can run the reactions normally, then neutralize the acid by adding 1 mL of 1 M sodium carbonate or pH 8 Tris buffer. The color returns to blue if there is any remaining DCIP that has not been reduced; fully reduced solutions of DCIP remain colorless.

- Some students mistakenly think that redox reactions proceed slowly, which is why DCIP decolorizes slowly in Exercise 2. In fact, the color change is instantaneous for a single molecule; in bulk solution the process is slow because the redox reaction is ongoing. Instructors can dispel this misconception by conducting the following demonstration, which shows how fast DCIP can be reduced and decolorized.

  Prepare three solutions:

  20 mL 200 mg/mL DCIP in water
  20 mL of 1 mg/mL ascorbic acid
  2 mL of 0.1% solution of m-phosphoric acid in water

  Just beforehand, mix the ascorbic acid and phosphoric acid solutions. To demonstrate the reaction, add the DCIP solution. Decolorization should occur within 10 seconds or so as DCIP molecules are reduced by electrons donated by ascorbate. In contrast, chloroplasts requires several minutes to reduce the DCIP.

PREPARATORY NOTES

Quantities listed are for a lab section of 20 students working in groups of three or four.

Weeks 1 and 2: Shared Materials

- Fresh spinach, 1 bunch
- 200 mL Tricine grinding buffer
- 250 mL Hill reaction buffer
- 250 mL DCIP solution
- 250 mL distilled water
- 10 mL of each herbicide solution; store wrapped in foil
- Tabletop centrifuge with 50 mL carriers in place, chilled to 4°C in advance
- Hemacytometers OR cell counting grid, microscope slides, and #1 coverslips
- Aluminum foil
- ~20 disposable plastic transfer pipets, 5 mL
- ~ 1 meter square piece of cheesecloth
- Scissors
- 5 mL and 10 mL serological pipets
- Pipet pumps
- Extra 50 mL disposable centrifuge tubes

Weeks 1 and 2: Materials at Each Work Station

- Ice bucket
- Paper towels or tissues
- Test tube rack with 20 large (15 × 150 mm) test tubes
- Test tube rack with 12 cuvets for spectrophotometer
- P-1000 and P-200 micropipets
- Boxes of blue and yellow tips, autoclaved
- Mortar and pestle
- 2 beakers (50 mL)
- 4 centrifuge tubes (50 mL)
- Spectrophotometer
- 4–5 microtubes (1.5 mL) (do not need to be sterile)
- Microtube rack
- Table lamp with an incandescent 60-watt bulb

Optional Materials for Experiments (make only if requested)

- Other leafy green vegetables
- Experimental test solutions (as requested by students)

Sources of Materials

High-quality spinach is vital to the success of this unit. The best results are obtained from fresh spinach grown in pots specifically for the laboratory exercises. Alternatively, use bundles of spinach (stems included) purchased fresh daily. Bundled spinach usually is stored in high-humidity coolers, so remains viable for several days; only use spinach that is still on the stems and has not wilted. *Do not* use precut spinach that has been sealed in bags, as it rarely works.

Solutions, Reagents, Equipment

Chloroplast Grinding Solution

Mix:

900 mL distilled water
37 g sucrose
0.5 g NaCl
8.96 g tricine

Adjust pH to 8.0 with 1 M NaOH. Bring to 1 L. Store at 4°C for up to three weeks. Discard if cloudy (indicates bacteria). Final buffer is 400 mM sucrose, 10 mM sodium chloride, and 50 mM tricine.

130 mM DCIP Solution
Dissolve 0.038 g DCIP (dichloroindophenol) in 1 L distilled water.

Hill Reaction Buffer (100 mM Phosphate Buffer, pH 6.5)
Mix:

950 mL distilled water
4.3 g Na_2HPO_4
9.5 g KH_2PO_4

Adjust pH to 6.5 with NaOH (1 or 10 M is fine). Bring to 1 L and store at 4°C.

Herbicide Inhibitor Solutions
Prepare stock herbicides as listed in Table 6.4. The final concentration of all three compounds is ~10 mM. Store refrigerated (wrapped in foil) and dispose of according to local regulations after each semester. If students want to test additional concentrations, all three herbicides can be diluted further in water.

Table 6.4

Preparing herbicide solutions

Herbicide	Prepare Solution of:	Mode of Action
2, 4-D	2 mg/mL in ethanol	Stimulates ethylene; plant overgrowth. No effect on electron transport.
Paraquat (methyl viologen)	2 mg/mL in water	Bipyridilium molecule diverts electrons from Photosystem I.
Diuron	2 mg/mL in ethanol	Blocks transfer of electrons through Photosystem II.

Notes:

- Do not use Roundup or other colloid-forming chemicals; the control mixtures are too cloudy to obtain a stable baseline absorbance level.
- Many other herbicides target photosystems. The major chemical families include
  - ◊ triazines, which block Photosystem II, such as atrazine;
  - ◊ uracils, which block Photosystem II, like bromacil and terbacil;
  - ◊ amides and ureas, which block Photosystem II, like dimefuron and siduron/tupersan; and
  - ◊ dinitrophenols, which uncouple the ATPase, including dinoseb and DNOC.

Cell Counting Grid

Many instructors hesitate to use hemacytometers in teaching labs because they are expensive and break easily. There is a simple disposable substitute that is less accurate than a clinical hemacytometer, but is sufficiently accurate for most student exercises. A standard hemacytometer chamber is 1 mm × 1 mm square and 0.1 mm deep and contains 1/10,000th of a milliliter. A similarly sized uniform space can be created using coverslips and a printed acetate grid.

1. Using a vector-based drawing program (e.g., Adobe Illustrator), draw fifty-one 5 cm vertical lines and fifty-one 5 cm horizontal lines. Space the lines equally so they form a grid of squares, 5 cm × 5 cm overall. When done correctly, the lines will be spaced exactly 1 mm × 1 mm on-center.
2. Select all lines, and rescale their stroke weight as small as the program allows (typically, 0.05 or 0.1 point).
3. Copy and paste the entire grid to fill a standard printed page.
4. Using a laser printer (*not* an inkjet printer or photocopier), print the file in black directly onto an acetate film sheet. For convenience, print several simultaneously.
5. Figure 6.6, on page 262, shows the procedure for counting a sample with the preprinted grid. First cut one piece of grid from the sheet that is slightly smaller than a standard microscope slide.

Figure 6.6

Setting up a disposable hemacytometer

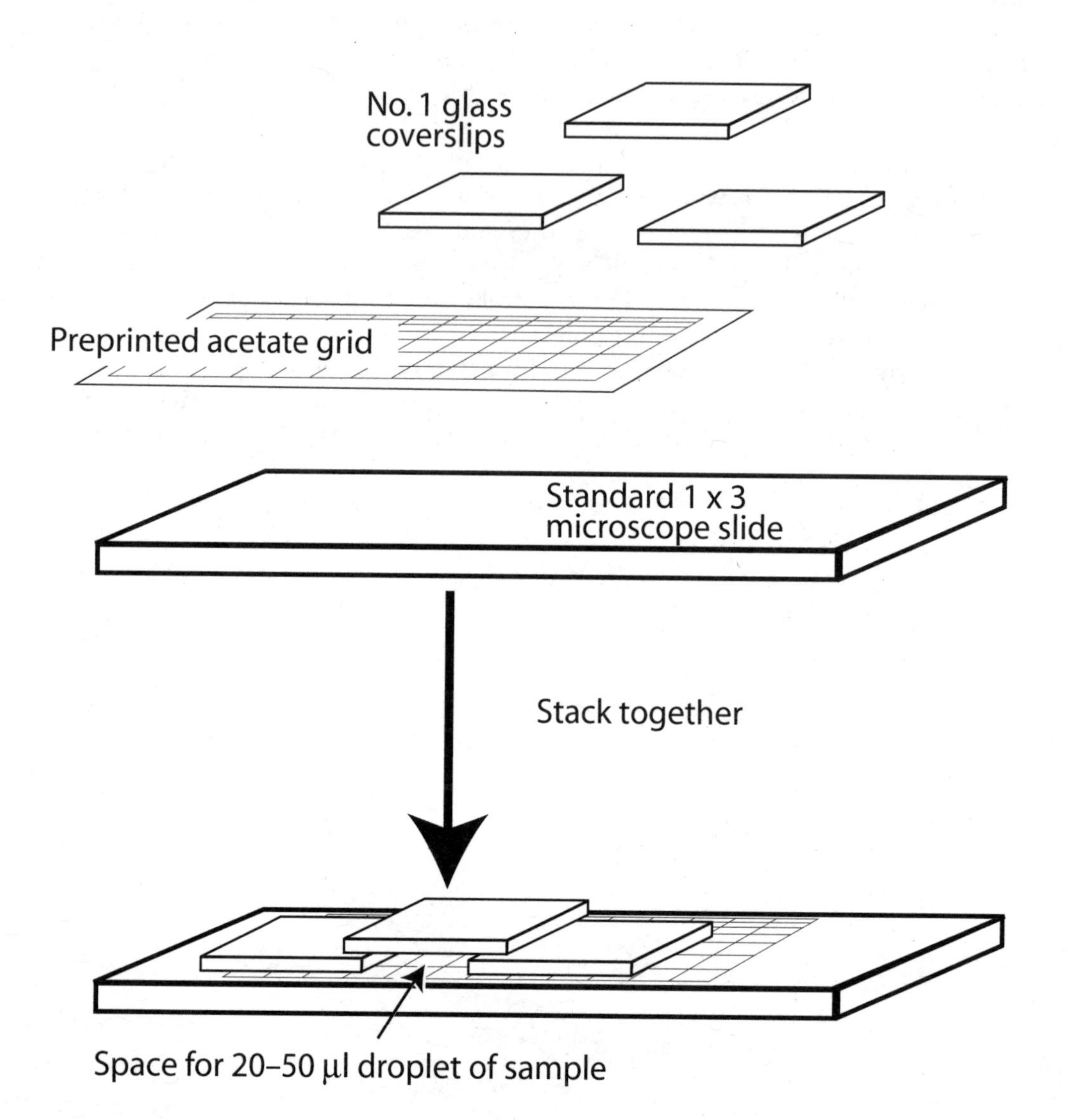

6. Lay the acetate grid on an unfrosted glass slide.
7. Place two #1 coverslips on top of the acetate, approximately 1 cm apart.
8 Place a 20–50 μL drop of sample between the side coverslips, then place a third #1 coverslip directly on top of the first two coverslips. The two coverslips on either side create an ~0.1 mm gap between the printed grid and the top coverslip.
9. To count, place the slide with the sample on a microscope stage. On low power, focus on the grid, and count every particle or cell within a single 1 mm square.

   - Depending on the quality of the printer, the lines on a grid may appear jagged. However, it is easy enough for students to estimate whether cells would fall inside or outside the square.
   - Static charge of the acetate usually immobilizes most cells, making counting much easier.
   - For greater accuracy, have students count three to four replicate squares and average them.

10. To determine the number of particles per milliliter of sample, multiply the number of particles in a 1 mm square by 10,000.

UNIT 7

Signal Transduction

Student Pages*

BACKGROUND

In this unit, you will be looking at the signal transduction pathways that let cells sense and respond to their external and internal environments. The model organism is the free-living unicellular algae, *Chlamydomonas reinhardtii* (Figure 7.1, p. 266). Chlamydomonas is extremely common in freshwater worldwide. It can also live in soil, either in a nonmotile form or swimming in the water between soil grains.

*Teacher Pages begin on page 281.

Figure 7.1

Schematic diagram of the cellular structures within the unicellular algae, *Chlamydomonas reinhardtii*

The chloroplast fills most of cytoplasm; G = Golgi; P = pyrenoid body; M = mitochondria. In normal cells the two flagella are slightly longer than the long axis of the main cell body. (Harris, E. 1989. *The chlamydomonas sourcebook*. Amsterdam, The Netherlands: Elsevier Press, p. 2. Revised and redrawn with permission.)

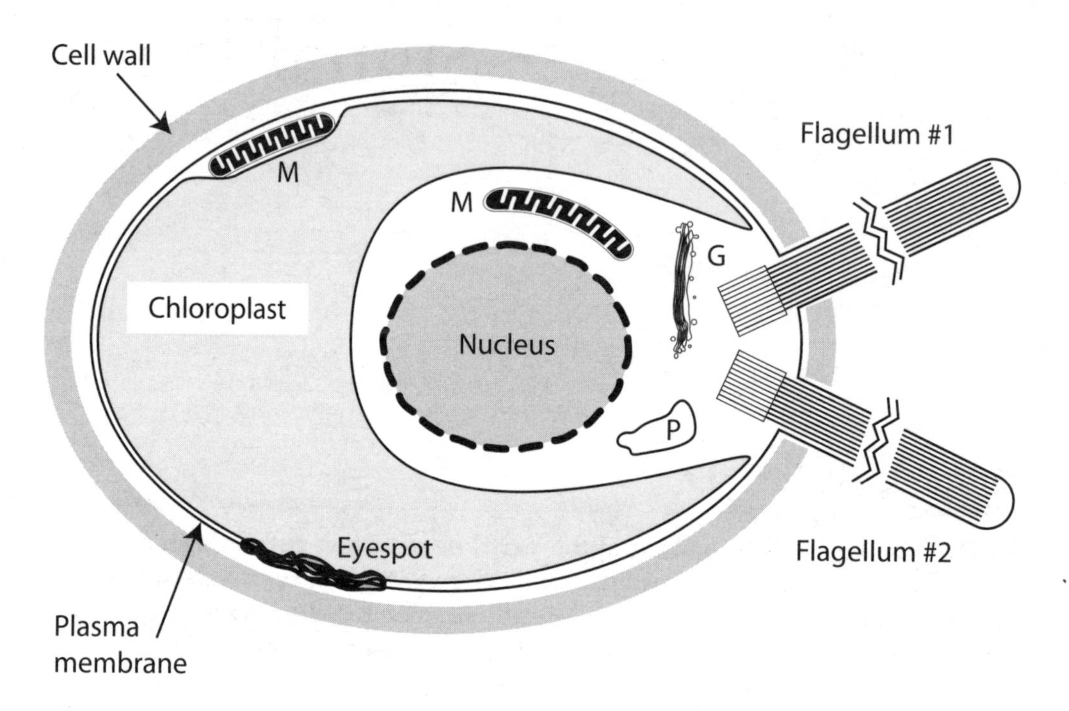

You may wonder why anyone *cares* how this bit of pond scum senses and responds to its environment. There are several reasons:

- Chlamydomonas use many of the same signal transduction pathways as human cells. Most eukaryotic cells are more alike than they are different; it is a consequence of our shared evolutionary ancestry. What is discovered about signaling cascades with simpler model organisms often is true in our own cells.
- Algae are much easier to grow and handle than human cells, allowing you to perform experiments that are not feasible otherwise.
- As you will soon see for yourself, Chlamydomonas respond to their environments in ways that can be easily observed and quantified.

Goals for This Unit

During the first week of this lab unit, you will learn how to

- quantify the number of algae in a suspension culture,
- perform a phototaxis assay, and
- induce flagellar excision and measure regeneration.

At the end of the first week your instructor will assign you one or two review articles to read for homework. They summarize our general knowledge of the signaling pathways that control the behavior of Chlamydomonas. Using the review articles as your starting point, you and your lab partners will develop a hypothesis, then design and execute an experiment to test it during the second week of this unit.

General Safety Precautions

The algae you will use are harmless, but the reagents you add to them to alter signaling paths range from irritating to extremely toxic. Your instructor will inform you if a compound is particularly hazardous. Always wear latex or nitrile gloves when you are working with a toxic chemical. Also, wash your hands thoroughly before you leave the lab each day or if you spill any chemicals on yourself.

A Brief Life History of Chlamydomonas

Nutritional Requirements

When sufficient light and carbon dioxide are available, Chlamydomonas generate carbon-containing molecules via photosynthesis. These are subsequently used for energy and as a source of fixed carbon for metabolic reactions to synthesize other macromolecules. If light levels are too low for photosynthesis, the cells can absorb and metabolize pre-fixed carbon directly from their environment. They can use many different carbon sources, but can most easily assimilate two carbon acetate ions.

Algal cells also need fixed nitrogen to synthesize amino acids and nucleotides. Unlike some soil bacteria, algae cannot directly fix nitrogen gas (N_2) from the atmosphere. They must obtain prefixed nitrogen ions from the local environment. Chlamydomonas preferentially use ammonium (NH_4^+) first, but can adapt to other sources like nitrate and nitrite (NO_3^- and NO_2^-). The free-swimming algae are **chemotactic** (chemo = chemical; taxis = moving toward) for nitrogen sources, that is, they will swim toward a dilute solution of usable nitrogen ions.

Motility and Phototaxis

Chlamydomonas can take on several different forms. When growing on solid surfaces, each cell retracts its two flagella, rounds up, and becomes nonmotile. As the cells continue to divide, they form dark green colonial masses containing millions of cells, surrounded by thick, mucuslike material. This is the **palmelloid** form of growth.

If a palmelloid colony starts to dry out, or there is insufficient nitrogen to support growth, the individual cells change again. Each cell secretes a thick protective wall of glycoproteins and forms a **vegetative spore**. These spores can remain dormant for months to years, particularly in soil. Once sufficient water and nitrogen are available again, the spores rupture and quickly resume growing.

Alternatively, if a palmelloid colony comes in contact with water, it disperses into individual free-swimming cells. These cells do not move about randomly, but rather swim toward favorable environmental conditions (such as the nitrogen sources described above) and away from noxious ones (e.g., high salinity, or dangerous temperatures). **Phototaxis** (photo = light) demonstrates just how complex these movements can be. Free-swimming cells will move toward a light (positive phototaxis) until they reach an area of a particular brightness; if the light is any brighter, they swim away from it (negative phototaxis). Presumably, this allows the algae to congregate where the light is at the optimum intensity for photosynthesis, while avoiding solar heating.

In Chlamydomonas, phototaxis is controlled by an **eyespot** located to one side of the cell. This region contains a version of the same photosensitive **rhodopsin** pigment used by our own eyes. However, algal rhodopsin triggers an entirely different signal transduction cascade than does mammalian or insect rhodopsin.

Flagellar Excision and Regeneration

As they swim, Chlamydomonas will sometimes move into an area where the pH is less than ~5, which is too acidic for growth. This triggers **flagellar excision**, in which the cell sheds its two flagella. If the pH subsequently returns to normal, the cells will synthesize tubulin and other motor proteins, regenerate both flagella, and begin swimming again.

Excision is not a random injury caused by low pH. It is a coordinated process that is regulated by a specific signaling pathway. It is not known what the adaptive advantages of shedding flagella might be. Some have suggested that excision reduces the total surface area exposed during acidic conditions. However, there is little evidence (yet) that supports this hypothesis.

Reproduction

As long as nitrogen and carbon sources are available, Chlamydomonas reproduce vegetatively, by simple mitosis. If nitrogen is depleted for a short period (minutes or hours), vegetative cells will stop dividing until fixed nitrogen is available again. However, if nitrogen deprivation continues for more than a day or so, vegetative cells differentiate into reproductive cells called **gametes**. Gametes can be one of two mating types, mt+ or mt-; the ability of an individual cell to form either an mt+ or mt- gamete is an inherited autosomal trait (see Figure 7.2). If it helps, think of the two mating types as different sexes.

Figure 7.2

Life cycle of Chlamydomonas

Under normal growth conditions, both palmelloid and free-swimming vegetative cells reproduce by simple mitosis. If fixed nitrogen ions are unavailable for more than 24 hours, vegetative cells differentiate into gametes. Gametes look identical to vegetative cells, but have distinctive surface proteins on their flagellar membranes.

If gametes of opposite mating types are mixed, aggregation and mating begins in one minute or less. Quadraflagellate cells (cells with four flagella) appear within 30 minutes, and secreted gamete autolysin can be detected in the surrounding medium. Mating is essentially complete within 90 minutes, and thick-walled zygotes form within 24 hours.

Normally, zygotes do not germinate immediately, but require a rest period. Zygotes can remain in this resting state for months until environmental conditions favor vegetative growth again. (Redrawn with permission from original created by Bill Snell, University of Texas Health Science Center, Dallas.)

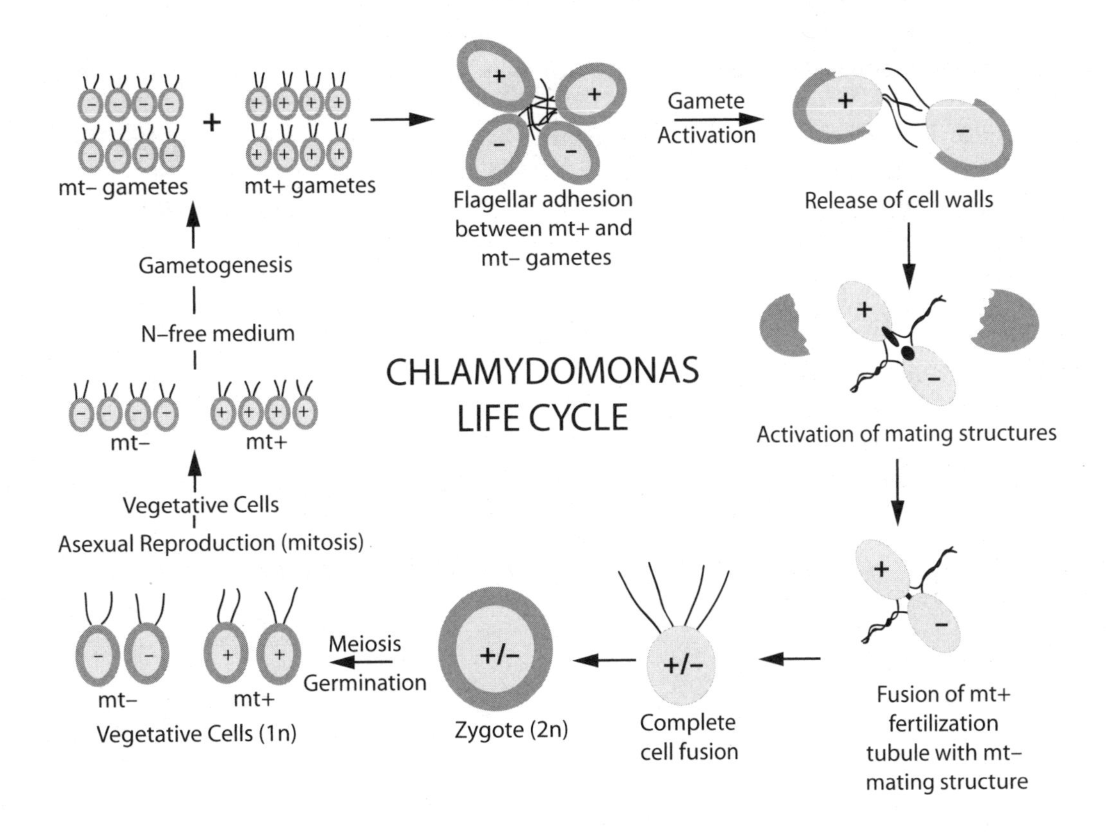

If several gametes of opposite mating types come together, they form a large mating aggregate of 3 to 20 cells. Pairs of mt+ and mt- cells will leave the aggregate and begin to fuse together, starting at the tips of their flagella. Within 30 minutes, the two cells will shed their individual cell walls and fuse into a single **zygote** with two nuclei and four flagella. Over the next 24 hours, this zygote will reabsorb its four flagella, produce a tough coat of glycoproteins, and form a **zygospore**.

Like vegetative spores, zygospores can withstand months of drought or poor conditions. When water and a fixed nitrogen source become available again, the zygospore ruptures to release two new mt+ vegetative cells, and two new mt- vegetative cells.

UNIT EXERCISES

Exercise 1: Baseline Observations

Your instructor will provide flasks of Chlamydomonas grown on **TAP** (Tris-acetate-phosphate) media solidified with agar.

1. Use a plastic transfer pipet to scrape out a small amount of Chlamydomonas culture; you only need enough to darken the tip of the pipet.
2. Smear the green material onto a clean glass slide. Do not add any liquid. Place a coverslip over the smear, and observe it using phase optics.

   Are the cells moving? What other characteristics could you use to describe them to someone else?

3. Prepare a second slide, but this time add ~25 μL of TAP liquid media to the smear. Use the transfer pipet to mix the algae thoroughly with the TAP. Add the coverslip, then immediately put the slide on the microscope so you can observe what happens as the cells rehydrate. Make sure everyone in your group has a chance to see what happens.

   What is happening to the cells? What specific changes do you see occurring? How long do they take to occur?

Exercise 2: Phototaxis

(*Note*: In Step 17, you must allow the cells to equilibrate for 60 minutes. During that time you can start Exercise 3. If the Exercise 3 takes more than 60 minutes to complete, do not panic. Letting the cells equilibrate for longer will not affect the outcome of this exercise.)

1. Your instructor will provide you with 100 mL of a liquid culture of Chlamydomonas. Before you can proceed, you must know the number of cells per milliliter. It is impossible to count the cells while they are actively swimming in a liquid culture, so you must kill and fix a small aliquot of them.
2. Gently but thoroughly mix your culture, and before the cells settle, remove a 500 μL aliquot. Place it in a 1.5 mL microfuge tube.
3. Add 500 μL of Lugol's solution to the algae in the microfuge tube. Quickly close the tube and invert it several times to mix the fixative and sample. The Lugol's solution both fixes (kills) the cells and stains the flagella and internal structures so they are easier to see.
4. Remove a 10 μL aliquot of fixed cells, and load it into one of the chambers of a hemacytometer. Load the other chamber with a second 10 μL aliquot of fixed cells.
5. Using phase optics, focus on the central grid of 25 squares. Count the cells in all 25 squares. Move the stage until you can see the grid of the other chamber, and count the number of cells in that set of 25 squares. Finally, add the two numbers together.

   How many cells did you count in the two central grids?

6. The central grid of a hemacytometer contains exactly 1/10,000th of a milliliter. However when you fixed the cells, you diluted them by one-half. So if you multiply the number of cells in both chambers by 10,000, you know how many cells are in 1 mL of your culture.

   How many cells per milliliter are in your starting culture?

7. If necessary, dilute your culture with additional fresh TAP media to a final cell density of 2×10^6 cells/mL. If the density is already below that, simply use the undiluted cells for the phototaxis assays.
8. Assemble three phototaxis tubes. Each tube consists of three pieces of a hard plastic graduated pipet connected by two short pieces of flexible Tygon tubing (see Figure 7.3, p. 272). Your instructor will show you how to assemble one.

Figure 7.3

Setting up and disassembling the phototaxis chambers

Panel A. One fully assembled phototaxis assay chamber. The pieces of plastic pipet and Tygon tubing have already been cut and will be provided for you. You will need to assemble them, then disassemble and clean them after completing the assay.

Panel B. Setting up for the assay. After you fill the chambers with either live or methanol-fixed cells (Step 1), they are laid flat, then covered with either aluminum foil (full dark) or a Kimwipe (diffuse low light) or are left fully exposed to the light (Step 2). After the cells have had 60 minutes to equilibrate, the three parts of the assay chamber are separated using pinch clamps and drained (Step 3).

A. An assembled phototaxis chamber

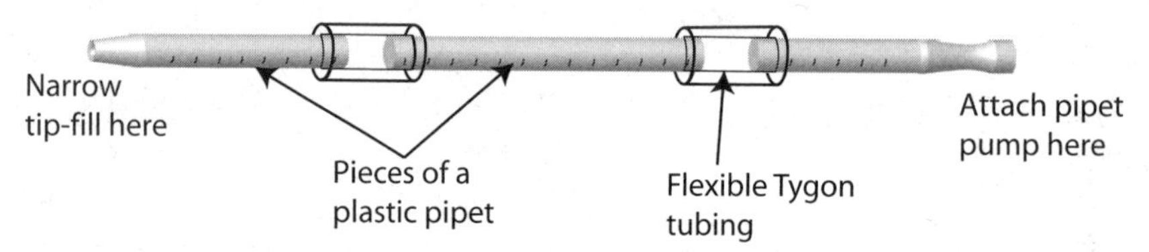

B. Steps in the assay

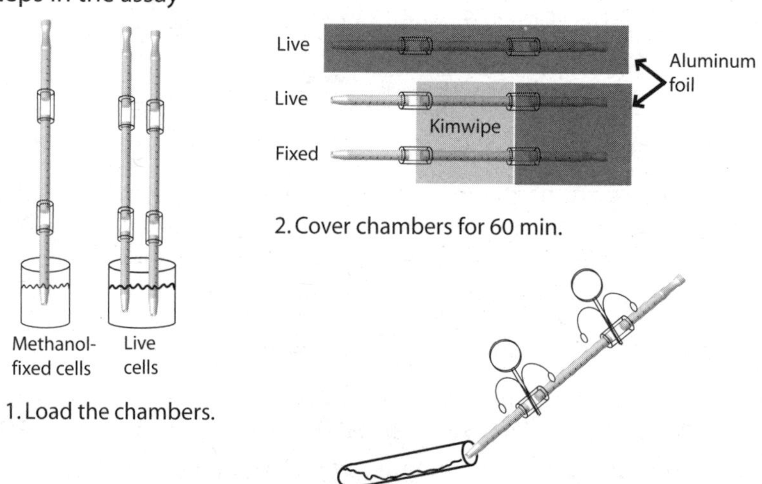

9. Rinse the phototaxis tubes with distilled water, and rest them on end so they can drain thoroughly.
10. While the three assay tubes drain, prepare some fixed, unstained cells. Put 20 mL of your diluted culture into a beaker or conical tube and add 5 mL of 100% methanol. Quickly mix. The methanol kills the cells and fixes them, without rupturing their membranes.

Once you begin filling your phototaxis tubes, you must work quickly. You may want to practice Steps 11 to 15 once or twice with plain water before you actually set up your assay.

11. Attach a pipet pump to the first phototaxis tube. Mix your diluted culture gently, then fill the phototaxis tube with cells. Leave only 1 cm unfilled at the neck of the tube. Do not remove the pipet pump yet.
12. Have a partner stretch and wrap a piece of Parafilm over the narrow end of the assay tube so it does not leak.
13. Gently remove the pipet pump. Seal the upper end of the tube with another piece of Parafilm. Immediately lay the filled tube flat on a sheet of white paper.
14. Fill and seal a second phototaxis tube with live cells. Fill the third phototaxis tube with the cells that were fixed in methanol. Then lay the two tubes flat on the white paper.
15. Gently cover one of the tubes containing live cells completely with aluminum foil. These are your live positive control cells.
16. Use another piece of aluminum foil to cover one end (~1/3 of the tube) of the remaining two phototaxis tubes. Cover the middle segment of each of the tubes with a single layer of Kimwipe. Leave the other ends of the two tubes uncovered.

    You already have a control tube of live cells in complete darkness. Why is it necessary to also have a control tube containing dead, methanol-fixed cells?

17. Place an incandescent reading lamp or other bright light directly over the uncovered ends of the two tubes, and turn on the light. Let the cells equilibrate 60 minutes before proceeding.

After one hour, you must separate and drain the three segments of each chamber, and determine the number of cells per milliliter in each of the segments.

18. Label nine glass test tubes so that you can collect the culture from the lower, middle, or upper segment of each of the three phototaxis assay tubes.
19. Gently remove the foil covering the tube of live cells that was completely covered. Keep the tube level. Working carefully but quickly, place two pinch clamps on the Tygon tubing so that cells cannot move between the three sections of the phototaxis chamber.
20. Hold the smaller end of the phototaxis chamber over the correct test tube. Remove the Parafilm, then carefully ease the plastic tube out of the Tygon tubing. Let all the cells and buffer drain into the collection tube. If there is any remaining cell suspension in the Tygon tubing near the clamp, remove it with a disposable pipet, then add it to the collection tube.
21. Remove the clamp and drain the middle part of the chamber into a second test tube.

22. Remove the second clamp and drain the last part of the chamber into a third test tube.
23. Gently uncover the tube of fixed cells that were exposed to varying levels of light. Again, use the two clamps to separate the three sections. Drain each section into separate, labeled collection tubes.
24. Finally, uncover the live cells that were exposed to three different light levels. Once more, clamp off the three sections and drain them into their correct collection tubes.
25. Working with one collection tube at a time, mix the cells gently but thoroughly, then transfer a 500 μL aliquot to a clean microfuge tube. Add 500 μL of Lugol's solution, cap the tube, and invert it three to four times.
26. Once you have fixed aliquots from all nine collection tubes, use a hemacytometer to determine the number of cells in each chamber. Record your results in Table 7.1.

Results for Exercise 2

Table 7.1

Cell counts from phototaxis assay

Treatment	Segment of Phototaxis Tube	# Cells in BOTH Hemacytometer Chambers	# Cells/mL (= # in chambers × 10,000)
Live cells, in complete darkness	Bottom		
	Middle		
	Top		
Fixed cells, in three light levels	Bottom (bright light)		
	Middle (diffuse light)		
	Top (low light)		
Live cells, in three light levels	Bottom (bright light)		
	Middle (diffuse light)		
	Top (low light)		

Summary Questions for Exercise 2

1. For the live cells that were completely covered, were there more cells in one area versus another of the tube? If so, what external stimulus might they be responding to?

2. For the fixed cells exposed to light, were there more cells in the completely exposed area, the area covered with white paper (diffuse light), or under the foil (low light)? What do you think this means?
3. For the live cells exposed to light, were there more cells in the completely exposed area, the area covered with white paper (diffuse light), or under the foil (low light)? What do you think this means?

Exercise 3: Flagellar Excision and Regrowth

As was explained before, Chlamydomonas shed their flagella in acidic conditions. In this exercise, to simulate swimming into an acidic environment, free-swimming, vegetative Chlamydomonas will be exposed to a sudden pH shock.

Your instructor will prepare the cells you will use. The cells were grown to one-half saturation in TAP, then collected by gentle centrifugation. The pelleted cells were resuspended in fresh TAP media at pH ~7.0 and diluted to approximately 5×10^6 cells/mL. You will be given a sample of these "unshocked control cells" and cells shocked as described in Step 3, page 276.

The algal cells will be allowed to recover from the pH shock in either TAP medium alone or in TAP plus a metabolic inhibitor (your instructor will tell you more about the effects of the inhibitor used). You will monitor flagellar regrowth by collecting samples at several times after pH shock and estimating the relative rates of flagellae under each condition.

1. Label 12 microtubes as shown in Table 7.2.

Table 7.2

Setup for flagellar excision assay

Tube #	Label
1	No shock, control
2	Plus shock, no recovery
3	Shock only, + 5 min. recovery
4	Shock only, +10 min. recovery
5	Shock only, + 30 min. recovery
6	Shock only, + 45 min. recovery
7	Shock only, + 60 min. recovery
8	Shock + inhibitor, + 5 min. recovery
9	Shock + inhibitor, +10 min. recovery
10	Shock + inhibitor, + 30 min. recovery
11	Shock + inhibitor, + 45 min. recovery
12	Shock + inhibitor, + 60 min. recovery

2. Place 0.5 mL of Lugol's solution in each microfuge tube.
3. The instructor will start the pH shock by adding 10 mL of 0.5 M acetic acid to 250 mL of cell culture in a beaker. This will reduce the pH to ~4.5. After one minute, the instructor will add 10 mL of 0.5 M potassium hydroxide (KOH) to the culture to return the pH to between 6.5 and 7.0.
4. When the pH returns to 6.5–7 again, the instructor will give you two 10 mL aliquots of shocked cells in conical tubes. Label the tubes as "Shock only" and "Shock + inhibitor."
5. Add 1 mL of TAP to the tube labeled "Shock Only." To the second tube, add 1 mL of the 10x inhibitor solution provided by your instructor.
6. Place both tubes flat on a sheet of white paper, about 12 inches away from the reading lamp on your bench.

(*Note*: Carefully read Steps 7 and 8 before you continue the procedure. Once you begin, you *must* stay on task and on schedule.)

7. After five minutes, mix each tube by gently inverting it two to three times. Then remove 0.5 mL of each culture, and transfer it to the corresponding, labeled microfuge tube to fix the cells. Close the conical tubes and return them to the light.
8. Continue collecting samples as described in Step 7. You will need to collect and fix samples of both cultures 10, 30, 45, and 60 minutes after the pH shock.

   If at all possible, collect an additional 0.5 mL sample of both cultures at each time point. Have one member of your workgroup immediately place a 10 μL droplet on a clean glass slide (not your hemacytometer), add a coverslip, and look at the live cells under phase optics. Group members should make notes about the behavior of the live cells at each time point.

Once you have collected samples at 5 and 10 minutes, you will have enough time between collections to score flagellar regeneration. Do not make one person do this alone; work as a team to score your samples.

9. Mix the sample of fixed but unshocked control cells by inverting the microfuge tube several times. Remove a 10 μL aliquot of cells and load it onto a hemacytometer.
10. Using the lowest magnification objective, focus on the grid in the center of the chamber. After the grid is in focus, switch to the 40x objective, and change the lower phase ring to "40."

    Once you switch to higher magnification, adjust the view with the fine focus only; do not try to use the coarse focus knob!
11. Look carefully at the cells; >90% of unshocked cells should have one to two thin flagella. If you cannot see any in the unshocked control cells, check

to make sure that your microscope is set for phase optics and that the condenser has not been moved away from beneath the slide.

Once you see flagella for the first time, and know what to look for, you should see them on the majority of the cells in the sample. If you cannot see them on a particular cell do not be alarmed; some get sheared off by pipeting or fold back beneath the cell when it is loaded into the hemacytometer.

12. For this assay, it is not necessary to measure the exact length of the flagella as they regenerate. Instead, you only need to estimate their length, relative to overall length of the main cell body. Compare the length of the flagella to the long axis of the cell (i.e., from the base of the flagella to the opposite end of the cell). Use the following scale:

    0 = flagella are not visible at all
    1 = flagellar stumps are barely visible
    2 = flagella are less than 1/2x the length of the cell
    3 = flagella are between 1/2x and 1x the length of the cell
    4 = flagella are 1–1.5x longer than the long axis of cell
    5 = flagella are two or more times longer than long axis of cell

    Using this scale, the normal, live vegetative cell would be considered a "4," because the flagella are longer than the rest of the cell, but not two times longer. It is unusual to see a cell with flagella that rate a "5" on this scale.
13. Record your results in Table 7.3.
14. Rinse off the sample of unshocked cells from the hemacytometer.
15. Mix the first sample of fixed, pH shocked cells, and load a 10 μL sample into the hemacytometer. Focus on the cells at low power again, then switch to 40x magnification.
16. Scan several fields of the sample to get a sense of how many cells have flagella and how long they are. Next, find a field that contains cells that appear about the same as in your other fields of view. Now use the 0–5 scale given above to score at least 10 different cells in that field of view. Record your results for each sample in Table 7.3.

    (*Hint:* You can count 10 cells very quickly if one person scores the cells' flagella and calls the scores out to a partner who records them in the table.)
17. Score at least 10 cells in each of the remaining samples. Make sure you do not mix up the results for cells with versus without inhibitor.

Results for Exercise 3

Table 7.3

Results of flagellar excision assay

		No. Cells w/ Flagella That Have:					
Tube #	**Treatment**	**(0) None**	**(1) Stubs barely visible**	**(2) <1/2x cell length**	**(3) 1/2x–1x cell length**	**(4) >1x cell length**	**(5) >2x cell length**
1	No shock, control						
2	Plus shock alone, control						
3	Shock only, 5 min.						
4	Shock only, 10 min.						
5	Shock only, 30 min.						
6	Shock only, 45 min.						
7	Shock only, 60 min.						
8	Shock + inhibitor, 5 min.						
9	Shock + inhibitor, 10 min.						
10	Shock + inhibitor, 30 min.						
11	Shock + inhibitor, 45 min.						
12	Shock + inhibitor, 60 min.						

Summary Questions for Exercise 3

1. How long did it take for flagella to begin regenerating in the "shock only" cells? In the cells that were mixed with inhibitor? What conclusions can you draw by comparing these two times?
2. Suppose you hypothesized that a particular signaling pathway was required for acidity to trigger flagellar excision. In general, how might you test your hypothesis? What controls would you need to use?

Your Own Experiment

This week you saw firsthand how Chlamydomonas respond when their environmental conditions change. So what signaling pathways are needed in order for algal cells to sense those changes, and respond appropriately? Your goal is to try and answer this question.

Your instructor will provide you with review articles to read for homework. Using them as a starting point, you and your lab partners must develop a hypothesis and design an experiment of your own that is related in some way to the signal transduction mechanisms that control algal behaviors. As you are reading the articles, keep these suggestions in mind:

- Chlamydomonas are very complex organisms. Focus on just one behavior that you are interested in exploring further. Once you have read the reviews you have been given, search for other articles on signaling pathways linked to that behavior.
- Scan the abstracts of any papers you find to see if they are relevant to your current questions. If a paper looks relevant, either download a PDF copy of the article or go to the library to pull it from the stacks.
- Do not rely on abstracts alone. Read the articles you find. As you read them, look at how the authors make their arguments. What is the evidence that Signal Path A is controlling Behavior B? Is there another possible explanation? Is more than one signal transduction pathway involved? What is still not known? How could you find out?
- When you come to lab next week, bring copies of any relevant articles you found to share with your partners.

Things to do or remember when you discuss possible experiments with your partners:

- You have a limited amount of time and cells with which to work. Every minute that you spend planning your experiment beforehand is going to save you 10 minutes later.
- Write out a formal procedure for what you plan to do. Walk through it on paper with your group. Do you foresee any problems? How might you solve such problems?
- Double-check your logic. Are you sure you have thought through your experiment? Is it going to give you a clear answer to your question?
- Look at the list of available materials and compounds. Is there something you need that is not on the list? If so, can you get it?
- You will not be allowed to just "test an interesting chemical." Your experiment must be driven by a hypothesis. Proper controls will need to be in place too. Have you thought about them as well?

Available Materials

Table 7.4 lists compounds that are available as 20x stock solutions. If you need a lower concentration than what is shown in the table, you will need to dilute the stock further. (*Note:* Gray boxes indicate that compounds with that particular effect are either unavailable or are too toxic to use in an open lab environment. Compounds whose names are in bold are highly toxic or have specialized disposal requirements.)

Table 7.4

Available Agonists and Antagonists

Other materials available include 20 mM solutions of K^+, Na^+, H^+, Ca^{+2}, lights of varying brightness, shade cloth, and colored acetate film. If there is something else you would like to use, but you are not sure if it is available, ask your instructor.

Target in Cell	**Activated by:**	**Final Conc.**	**Inhibited by:**	**Final Conc.**
Protein kinase A	Dibutryl-cAMP	10 mM	Curcumin	10 µM
Phospholipase C; IP_3	A23187	5 µM	Neomycin	1 mg/mL
Protein kinase C	Phorbol ester (Hazardous!)	1 µM	Tamoxifen citrate	100 µM
Surface Ca^{+2} channel	High levels of Ca+2 in media	1 µM to 1 mM	EDTA + Strontium chloride	1 mM each
IP_3–activated intracellular Ca^{+2} channels	A23187	5 µM	Lanthanum trichloride	50 µM
Ca^{+2}–activated intracellular Ca^{+2} channels	A23187	5 µM	Sodium butyrate	100 µM
cAMP phospho-diesterase			Caffeine + IBMX	Caffeine=20 mM; IBMX= 1 mM
Na+ channels	Monensin (Hazardous!)	10 µM	Lidocaine	1.25 mM
K+ channels	Nigericin	5 µg/mL	Tetraethylammonium chloride	5 mM
H+ channels	Dinitrophenol (Hazardous!)	50 µM		
Calmodulin			Trifluoperazine	50 µM

UNIT 7

Signal Transduction

Teacher Pages

INSTRUCTORS' NOTES

Background and Key Concepts

All cells must have mechanisms for sensing their environments, then responding to them. Some sensing requires an external stimulus to enter the cell and affect it directly. More often, though, an external event stimulates a receptor or sensor that in turn activates a signal transduction pathway. This lets cells react to environmental conditions without external stimuli actually having to enter the cell and allows for complex integration and regulation. Most cells have multiple signaling pathways in operation simultaneously. Very often, two or more different pathways will work together to cause a cell to respond to certain stimuli.

In eukaryotes, the first signaling pathways to be mapped out in detail were those linked to ion channel coupled receptors and to G-protein coupled receptors. Ion channel coupled receptors stimulate cellular responses directly—by altering membrane voltage potential, changing intracellular pH, or allowing in divalent calcium, which acts as a second messenger to activate intracellular, Ca-regulated proteins. In G-protein coupled signaling paths, the receptor often is a seven-transmembrane spanning protein that activates a heterotrimeric G-protein complex. Activated G-protein subunits move within the membrane to activate effector enzymes such as adenylate cyclase or phospholipase, which in turn generate second messengers. The second messengers move into the cell and activate various kinases that cause a cellular response. More recently, a variety of enzyme-linked receptors and signaling cascades have been discovered in which kinases are part of the sensory receptor. Novel, monomeric G-proteins may or may not be part of these enzyme-linked signaling paths. Other major signaling pathway mechanisms have been identified as well.

To determine which signaling mechanism a cell uses to sense and respond to a particular stimulus, that cell can be exposed to compounds that are known to activate or block specific steps in candidate pathways. Many cellular processes have been linked to specific signaling paths by systematically activating or blocking individual steps within one or more paths, then observing cellular responses. In this unit students get the opportunity to apply these same classical methods to discovering how cell behaviors are controlled.

The model organism for this unit is *Chlamydomonas reinhardtii,* a free-swimming algae. Chlamydomonas is a robust and highly engaging model organism for students. Both the vegetative stage and gametes are motile, with a repertoire of unambiguous behaviors that can be observed directly without specialized equipment. They have been called the "green yeast" because, like true yeast, they are easy to grow and manipulate in the lab. Many nutritional, motility, and other metabolic mutants are available, along with the chloroplast and nuclear genomic sequences. Relevant to this unit, Chlamydomonas use many of the fundamental ion channel-linked and G-protein coupled signaling pathways found in all eukaryotic cells. However, they also use some signaling mechanisms that resemble bacterial pathways, and still others found mainly in higher plants. Fundamental questions remain to be answered about how these pathways are used and interact, yet at the same time the basic research methods needed are well within the abilities of motivated undergraduates.

In preparing students for this unit, emphasize these concepts:

- Signal transduction pathways allow cells to respond to a range of chemical and other stimuli in the external environment, without the external factors having to enter the cell.
- Cells usually have multiple signaling pathways working simultaneously. These may work together to respond to certain stimuli.
- By methodically blocking single steps within a particular signaling pathway, it is possible to determine if and how a signaling pathway works in response to a certain stimulus.

General Teaching Strategy and Common Problems

This two-week unit was designed originally for a sophomore-level advanced cell biology course. In the first week students observe three stereotyped behaviors of Chlamydomonas. All three are controlled (at least in part) by the classical cAMP/PKA-, PLCγ/PKC-, and the IP_3/Ca^{+2}-dependent pathways. It is helpful if students have some background on these paths before they begin reading the articles for homework and designing experiments. It is not necessary for them to know about kinase-dependent pathways or about plant or bacterial signaling systems.

For the second week students design and execute experiments of their own. The typical question a student might ask is, Does Chlamydomonas use Pathway

A to trigger Behavior B, in response to Stimulus C? However, other kinds of questions can be asked and tested very easily.

If time is short, the second week can be eliminated, and the first week's exercises modified to make key points about signaling more directly. First, eliminate Exercise 1 entirely. Second, add a fourth phototaxis chamber in which students place live cells treated with lanthanum trichloride (which slows phototaxis by blocking IP_3-induced calcium release). Alternatively, have each group of students test different pathway inhibitors, then pool their data and interpret the data as a class. Third, have students perform the flagellar regeneration assay using A23187 or caffeine+IBMX, not cycloheximide. That lets students see signaling effects, rather than the effect of protein synthesis inhibition. Again, groups within a section could test different substances and share their data.

The mechanics of this lab are straightforward and students learn quickly how to count cells and estimate flagellar length. More often, they struggle to understand how an agonist or antagonist would be used. It is essential for the instructor to help them think through the fundamental logic. Stress that, if the agonist activates a path when the normal stimulus is missing, the agonist should trigger a behavior in the *absence* of its normal stimulus. Conversely, if an antagonist blocks a pathway, they should not see the correct behavioral response to a stimulus. If the agonist or antagonist do not target the pathway that elicits a particular behavior, then neither compound will have any effect on it.

Students frequently try to account for and integrate all potentially interacting signal paths simultaneously; this is impossible for them to do. Tell them to focus on a single clearly defined question and pathway, then design their experiments accordingly. For example, they might hypothesize that release of calcium from internal stores is needed for deflagellation. An appropriate question would be, Can an influx of external calcium trigger it too? It is unrealistic to expect them to account for potential contributions of extracellular, sarcoplasmic, and plastid pools of calcium simultaneously.

Students tend to over-interpret their data and make grand, sweeping conclusions. Encourage them to make very focused conclusions and to have solid evidence to back up any statements they make.

At the end of the first week students should be assigned one or two review articles to read for homework. Two particularly good articles are listed in the Supplemental References (p. 284). They are older articles but are written so that students will understand them with some effort. The instructor should provide electronic or photocopies if necessary, as many libraries may not carry these journals. Alternatively, search the literature for more recently published summaries and assign those instead.

Recommended Prelab Skills

Ideally students already know how to estimate cell density in suspension using a hemacytometer. It is helpful (but not essential) if they have previously learned how to estimate the relative size of one microscopic structure based on the size of another. Basic knowledge of graphing with a spreadsheet program will speed up data analysis.

Assessment

If pre- and postlab quizzes are used as part of formal assessment, the prelab quiz should focus on general concepts such as basic algal life history or general principles and purposes of signaling pathways. Do not expect students to know details about specific behaviors until they have actually observed them; asking prematurely suggests the answer is already known and must be memorized rather than discovered.

On a postlab quiz, it is reasonable to ask students to describe the behaviors and time needed for the palmelloid cells to become motile (Exercise 1). Students in a more advanced course might be asked to extrapolate from their observations and to predict what physical changes are occurring at the microscopic level. All students should be able to describe the general results of the phototaxis and flagellar regeneration assays.

Safety and Housekeeping

Although the algae are harmless, the chemicals used in student experiments range from irritants to potent carcinogens and mutagens. Have students wear gloves whenever they use inhibitors.

Supplemental References

Harris, E. H. 1989. *The Chlamydomonas sourcebook.* San Diego, CA: Academic Press. This is the single best one-stop source available for protocols and procedures, life history, genetics, algae culture, and terminology.

Quarmby, L. M., and H. C. Hartzell. 1994. Dissection of eukaryotic transmembrane signaling using Chlamydomonas. *Trends in Pharmacological Science* 15: 343–349. Although this paper is 15 years old, it remains a sound review of the general signaling pathways involved in phototaxis and flagellar regeneration. More recent reviews that emphasize mutational analysis may be too technically complex for introductory-level students

Quarmby, L. M. 1994. Signal transduction in the sexual life cycle of Chlamydomonas. *Plant Molecular Biology* 26: 1271–1287.
Similar to above, but focused on gametogenesis and mating responses.

PREPARATORY NOTES

Quantities listed are for a lab section of 20 students working in groups of three or four.

Week 1: Shared Materials

- 5 T-25 flasks of algal cells grown for five to seven days in continuous light on TAP/agar, labeled "Palmelloid Cells"
- 500 mL of algae, grown to ~5 × 10^6 cells per mL (~1/2 saturation), in TAP
- 2 bottles (500 mL) of TAP/HT-free liquid media, autoclaved (one bottle is for students and one is for the instructor to use when diluting cultures)
- 2–3 dozen plastic transfer pipets
- ~1/3 box of glass microscope slides
- Cover glasses for microscope slides
- 100 mL bottle of methanol
- Parafilm
- Aluminum foil
- 50 mL conical tubes, nonsterile
- 5 aliquots (1 mL) of 10x cycloheximide (50 μM)

Week 1: Materials at Each Work Station

- Yellow and blue micropipet tips, boxed
- 1.5 mL microfuge tubes, ~50
- 10 mL aliquot of Lugol's fixative (in tubes wrapped with foil)
- Hemacytometer with cover glass (or disposable equivalent; see Unit 6)
- 3 phototaxis chambers, disassembled
- 3 pinch clamps
- Distilled water in a wash bottle
- 1 manual pipet pump
- Box of Kimwipes
- Incandescent desk lamp with 60-watt bulb
- Tube rack with 10–12 glass test tubes (15 mL capacity)
- 1 beaker (250 mL)

Week 1: Materials for Instructor/Prep Area

- 100 mL 0.5 M acetic acid
- 100 mL 0.5 M potassium hydroxide
- 50 mL conical tubes, nonsterile
- 15 mL conical tubes, nonsterile
- 500 mL beaker or Erlenmeyer flask

Week 2: Shared Materials

- 2 cultures (500 mL) of algae, at ~5×10^6 cells per mL in TAP
- 2 bottles (500 mL) of TAP/HT-free liquid media, autoclaved

Week 2: Materials at Each Work Station

- Yellow and blue micropipet tips, boxed
- 1.5 mL microfuge tubes, ~50
- 10 mL aliquot of Lugol's fixative (in tubes wrapped with foil)
- Hemacytometer with cover glass (or disposable equivalent)
- 3 phototaxis chambers, disassembled
- 3 pinch clamps
- Distilled water in a wash bottle
- 1 manual pipet pump
- Box of Kimwipes
- Incandescent desk lamp with 60-watt bulb
- Tube rack with 10–12 glass test tubes (500 mL)
- 1 beaker (250 mL)

Week 2: Materials for Instructor/Prep Area

- 100 mL 0.5 M acetic acid
- 100 mL 0.5 M potassium hydroxide
- 50 mL conical tubes, nonsterile
- 15 mL conical tubes, nonsterile
- 500 mL beaker or Erlenmeyer flask

Optional Materials for Experiments (make only if requested)

- 1 mL aliquots of 20x stocks of signaling agonists, antagonists; store frozen
- 20 mM solutions of KCl, NaCl, HCl, $CaCl_2$
- Lamps of varying brightness
- Shade cloths of different densities
- Colored acetate film

Sources of Materials

The best results are obtained from stock cultures purchased from a research collection such as the Chlamydomonas Genetics Center. All three exercises (palmelloid transition, phototaxis, and flagellar regeneration) can be performed with either of the following strains (many others work equally well):

CC-125: wild type mt+ 137c
CC-620: wild type mt+ 137c R3 NM subclone (high mating efficiency)

Preparing Cultures of Chlamydomonas Vegetative Cells

Between semesters, maintain stocks on slants of TAP/agar that are kept in low light at room temperature. Slants can be viable for six to eight months, but should be passed through one cycle of liquid TAP every three months. Starting two weeks beforehand, inoculate algae from slants back into liquid TAP, and start the amplification process.

To Start a Liquid Stock Culture From an Agar Slant

1. Use a sterile pipet to scrape a pellet of palmelloid algae from an agar slant, about the size of a paper match-head. Transfer the materials into a 125 mL flask containing 50 mL of sterile TAP supplemented with 50 µg/mL of ampicillin.
2. Place the culture under continuous light. Let the algae grow to 50% saturation (5×10^6 cells/mL), agitating at least once daily. To quantify cells, fix an aliquot by mixing 1:1 with Lugol's fixative. Count using a hemacytometer.
3. Once the new liquid stock culture has reached 50% saturation for the first time, the population growth rate will stabilize at about one to two population doublings per 24 hours.
4. Routinely check *every* liquid stock culture for bacterial growth within 48 hours of inoculating them. Signs of contamination are

    - culture changes from green to gray-green, brown, or yellow,
    - bacteria are visible under phase optics,
    - algal cells are no longer motile,
    - large number of dead or dying cells are present, and
    - culture has a rancid or foul odor.

To Amplify an Existing Liquid Stock Culture

1. Prepare 250 mL of TAP medium in a 500 mL Erlenmeyer flask. Autoclave and add ampicillin.
2. Check the current stock culture to make sure it is free of bacteria.
3. Transfer 25 mL from the stock culture to 250 mL of fresh TAP (1:10 dilution). Place the new culture in continuous light. Mix at least daily.
4. After ~72 hours, the new culture will approach the same cell density as the originating stock culture.

To Make an Agar Slant From a Liquid Stock

1. Prepare TAP/agar and supplement with ampicillin. Aliquot into sterile tubes or tissue culture flasks (vented cap flasks work well). Fill the container at least one-third full of TAP/agar.

2. Overlay the agar with 50% saturated algal culture grown in TAP. Tilt the container to coat the agar surface with algae. Let sit five minutes so algae will settle.
3. Turn the flask upright, and use a sterile pipet to remove the excess liquid. Place the flask in continuous light until there is a dense lawn of algal growth.

Solutions, Reagents, Equipment

TAP Nitrogen/Salts Stock Solution

15 g NH_4Cl
4.0 g $MgSO_4 \cdot 7H_20$
2.0 g $CaCl_2 \cdot 2H_20$

Dissolve in distilled water then bring to 1 L.

TAP Phosphate Stock Solution

28.8 g K_2HPO_4
14.4 g KH_2PO_4

Dissolve in 60 mL distilled water, then bring to 100 mL.

Hutner's Trace Minerals Solution

This protocol (from *The Chlamydomonas Sourcebook* by Elizabeth Harris) requires two to three weeks to complete. Aliquots can be stored frozen for at least five years.

Prepare all of the solutions listed in Table 7.5.

Table 7.5

Stocks for Hutner's Solution

Component	Quantity (g)	Volume of water (mL)
$ZnSO_4 \cdot 7H_2O$	22	100
H_3BO_3	11.4	200
$MnCl_2 \cdot 4\ H_2O$	5.06	50
$CoCl_2 \cdot 6\ H_2O$	1.61	50
$CuSO_4 \cdot 5\ H_2O$	1.57	50
$(NH_4)_6Mo_7O_{24} \cdot 4\ H_2O$	1.1	50
$FeSO_4 \cdot 7\ H_2O$	5	50
EDTA, disodium salt	50	250

1. Dissolve the EDTA in boiling water. Prepare the ferrous sulfate immediately before adding it, to avoid oxidation.
2. Mix all of the solutions except the EDTA in a 2 L flask. Bring it to a boil, then add the boiling EDTA solution. The mixture will turn bright green. Once all material dissolves, allow the solution to cool to 70°C. Place the flask into a water bath set to 70°C to maintain temperature.
3. Use hot 20% KOH solution to slowly adjust the pH to 6.7; remember to standardize the pH meter at the same temperature. DO NOT use NaOH. Once the pH reaches 6.7, check the volume. Bring the final volume to 1 liter with water, and place a cotton plug in the neck of the flask. The solution will be clear green at this point.
4. Allow the solution to stand at room temperature for one to two weeks, shaking it vigorously once a day to aerate and mix. The solution will turn purple and develop a rust-brown precipitate. Filter out the precipitate using two layers of Whatman #1 filter paper, repeating until the solution is clear. Break into 10 mL aliquots and freeze.

TAP Medium

Mix:

2.4 g Tris base
950 mL distilled water

When the Tris base dissolves completely, add these components *in order*, mixing between each addition:

25 mL nitrogen/salts stock solution
0.375 mL phosphate stock solution
1.0 mL Hutner's trace minerals solution
1.0 mL glacial acetic acid

Bring to 1 L with distilled water. Check pH; it should be between 6.5 and 7.5. Autoclave for 20 minutes on liquid cycle. Liquid media is stable for two to three weeks at room temperature. If a precipitate appears or the TAP starts looking cloudy, discard it.

To make TAP/agar, add 15 g/L minimal agar immediately before autoclaving. To prepare TAP/HT-free, omit the Hutner's trace minerals.

Lugol's Iodine Solution

Mix:

120 mL distilled water
20 g potassium iodide

When the KI dissolves, add:

10 g iodine crystals

Stir continuously to dissolve as much iodine as possible. After two to three hours, remove undissolved iodine crystals by filtering the solution through two coffee filters. Aliquot into tubes. Store wrapped in aluminum foil at room temperature. (*Caution*: Elemental iodine is hazardous! If you prefer not to work with elemental iodine, tincture of iodine works as well.)

0.5 M Acetic Acid

Mix:

130 mL glacial acetic acid (17 M)
970 mL distilled water

0.5 M Potassium Hydroxide (Neutralization Solution)

Mix:

28 g potassium hydroxide
950 mL distilled water

Dissolve KOH in water, then bring to 1 L.

50 mM Cycloheximide

Mix:

1.5 mg cycloheximide
100 mL ethanol

Mix to dissolve. Store at -20°C.

Signaling Agonists and Antagonists (see Table 7.6)

Table 7.6

Preparing stocks of agonists and antagonists

Compound, With Conc. of the 20x Stock	Quantity	Volume of Solvent
Dibutryl cAMP 200 mM	0.5 g	10 mL water
Curcumin 200 μM	1.48 g	20 mL ethanol
A23187 100 μM	0.01 mg	200 mL ethanol
Neomycin 20 mg/ml	0.4 g	20 mL water
Phorbol ester 20 μM (Hazardous)	0.012 g	Dissolve in 1 mL DMSO; dilute 1:1000 in water
Tamoxifen citrate 2 mM	0.1 g	100 mL water

(cont. on p. 291)

Table 7.6 *(cont. from p. 290)*

Calcium chloride dihydrate 20 mM	0.3 g	100 mL water
EDTA 20 mM	10 mL of 500 mM stock	240 mL water
Strontium chloride 20 mM	0.5 g	100 mL water
Lanthanum chloride 1 mM	0.0123 g	100 mL water
Sodium butyrate 2 mM	0.02 g	100 mL water
Caffeine 200 mM	1.9 g	100 mL water
Isobutylmethyl xanthine 20 mM	0.045 g	10 mL ethanol
Monensin 200 μM (Hazardous)	0.014 g	100 mL water
Lidocaine 25 mM	0.1345 g	20 mL water
Nigericin 25 μg/mL	1.0 mg	40 mL ethanol
Tetraethylammonium Cl 100 mM	0.17 g	10 mL water
Dinitrophenol 1 mM (Hazardous)	0.018 g	100 mL water
Trifluoperazine 1 mM	0.048 g	100 mL water

(*Note:* Phorbol ester is a potent mutagen and carcinogen. Monensin and dinitrophenol are both acute toxins. Additionally, DNP is an EPA-regulated hazardous waste. If you choose to provide these compounds to students, set up separate waste collection containers for each compound. Consult an MSDS for proper disposal or destruction methods.)

Making Phototaxis Chambers

Prepare these from 10 mL polystyrene tissue culture pipets. Remove the cotton plug with forceps, and rinse the pipet with water to remove any residue. Use regular scissors to score the plastic at the preprinted 1 mL and 5.5 mL volume marks. Bend the pipet; it will break cleanly at the score lines. The resulting three pieces hold approximately the same volume of liquid.

For connectors, cut two 3 cm lengths of Tygon tubing (5/16″ ID, 1/16″ wall thickness). Shorter connectors create less dead space but are more difficult to clamp without creating leaks. Do not try to use thicker walled tubing, as the pinch clamps may not be strong enough to close it properly.

The author's students have also used a single, continuous piece of Tygo tubing for the entire phototaxis chamber. Clamping continuous tubing is easier, but draining it for counting can be difficult. Readers should experiment to decide which method they prefer.

Animal Hormones

Student Pages*

BACKGROUND

One advantage that multicellular organisms gain over unicellular ones is specialization. Individual cells no longer have to perform every process necessary to maintain life. However, specialization creates a new problem: how to control and coordinate different tissues and organs. Two interrelated systems serve this function: hormonal systems and nervous systems. Plants do not have a nervous system, and rely on hormones alone, while animals use hormonal and nervous systems together.

In this unit you will be exploring hormonal regulation of insect development. Often the endocrine (hormonal) and nervous systems are described separately, as if they were two independently operating systems. In reality the two systems are intimately connected. There are many different examples of links between nervous systems and endocrine systems, in both vertebrates and invertebrates. A good example of this interrelationship is the human pituitary gland.

You probably know that the pituitary gland releases several hormones that stimulate or inhibit release of hormones by other endocrine glands throughout the body. In this way the pituitary gland controls a range of functions, including basal metabolic rate (via thyroxines), sexual development (estrogens, androgens, and progestins), and energy and ion balance (corticoid hormones). The pituitary gland also releases some hormones that control body functions directly, such as water balance (antidiuretic hormone), childbirth (oxytocin), and general growth (growth hormone). However, the nervous system controls the release of hormones from the pituitary gland. The pituitary gland gets its regulatory inputs from many sources, but the primary source is a region of the brain called the hypothalamus. This area integrates signals coming from sensory neurons throughout the body.

*Teacher Pages begin on page 307.

It also senses body temperature, ion balance, pH, and other factors directly. Finally, the hypothalamus can sense the levels of various hormones in blood and body fluids. Once all of this information is integrated, the hypothalamus responds by stimulating or inhibiting hormone secretion by the pituitary gland. Signals pass from the hypothalamus to the pituitary gland in two ways: as releasing factors (hormones that have the pituitary gland as their target organ) and as action potentials moving along neurons that connect the two regions of the brain.

Hormonal Control of Insect Development

In insects, two hormones control the timing of normal molting and the formation of the pupa: **juvenile hormone** and **ecdysone**. The relative levels of each of these hormones produced during various stages in an insect's life cycle are outlined in Figure 8.1.

Figure 8.1

Schematic diagram of the relative levels of ecdysone and juvenile hormone during growth and pupation in insects (Dubrovsky, E. B. 2005. *Trends in endocrinology and metabolism.* Vol.16. Amsterdam, The Netherlands: Elsevier Press, pages 6–11. Redrawn from original with permission.)

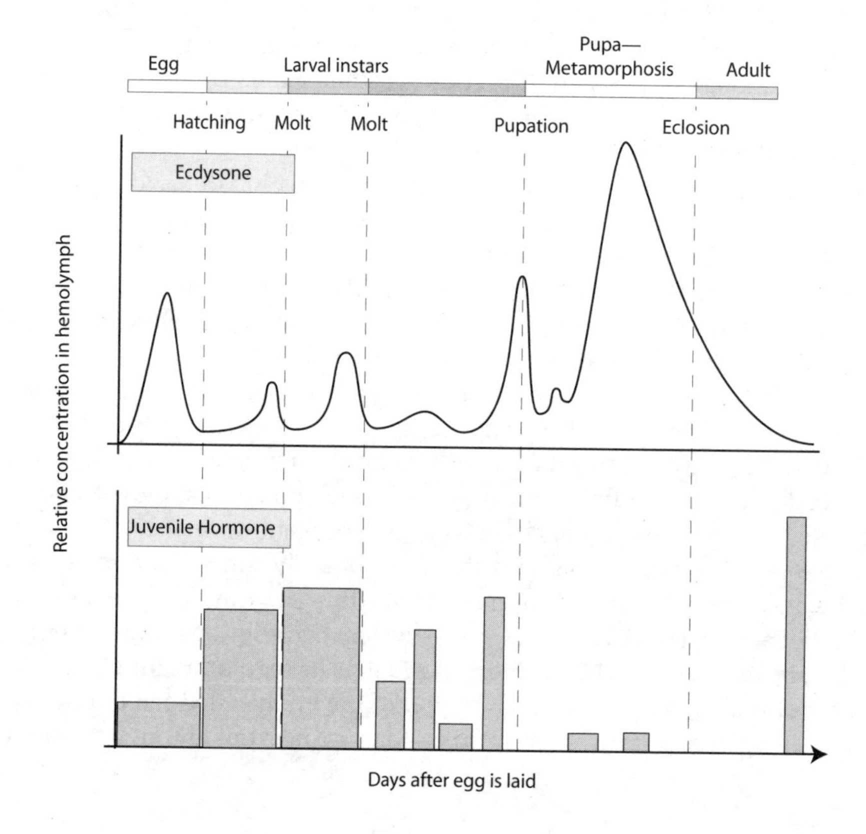

While the nervous system helps control the release of both juvenile hormone and ecdysone, it is the hormones that actually stimulate the cells in the larva to rearrange and form a pupa, which then metamorphoses into an adult. During development, brief bursts of ecdysone production stimulate molting. Conversely, a larva grows toward maturity as the level of juvenile hormone declines. When juvenile hormone disappears, the larva forms a pupa and metamorphoses. The next spike in ecdysone levels triggers the emergence of the adult.

Goals for This Unit

So what will happen to the normal development process in an insect if its hormone levels are disturbed? Answering that question is your goal in this unit.

This unit requires three weeks to complete. During Week 1, you will look at fruit flies (*Drosophila melanogaster*) in various stages in their life cycles. You also will expose larval flies to a range of concentrations of juvenile hormone. Finally, you will begin preparing for your own experiment by starting additional culture vials of fruit flies. Between Week 1 and Week 2, you will need to read more about other insect hormones and about the pathways and processes they control. This background material will help you design a more robust experiment. During Week 2, you will measure and score the flies you treated with juvenile hormone the previous week. Based on what you know to be the normal pattern of development, you will describe how increasing juvenile hormone changes the pattern. Finally, you and your lab partners will discuss your outside reading, then design and start your own experiment testing the effects of other hormones or hormone blockers on fruit fly growth and development. In Week 3, your group will collect data from the experiment that you designed and executed and then analyze the data.

General Safety Precautions

The growth medium and yeast you will use are both nontoxic. The flies do not carry any known pathogens. Wash your hands thoroughly when you finish each day, however.

Most of the hormones you will be using are not toxic or hazardous to humans except in large quantities, but some of the hormone antagonists may be toxic. Before you use any compound, ask your instructor to explain the relative hazards associated with it. If you want to know more about a compound's toxicological properties, ask the instructor to provide you with an MSDS describing it.

UNIT EXERCISES

Exercise 1: Anatomy and Development of *D. melanogaster*

Background

Fruit flies are sexually dimorphic, meaning males and females have distinctive external anatomic features. (To see photographs of males and females with their differences noted, go to Figure 2.2 on page 112.)

- Males have a dark brown or black sex comb on each of their front legs. This is the most reliable feature for separating males from females.
- Males have five ventral abdominal sternites (plates). Females have seven sternites.
- The male's abdomen tends to be more rounded and blunt with darker markings at the tip, especially on the ventral surface.
- Males are often slightly smaller (shorter and thinner) than females.

Fruit flies have a four-stage life cycle: egg, larval, pupal, and adult stages (see Figure 2.3 on p. 118). In optimal conditions the entire cycle takes 11 to 14 days. A female fruit fly can lay several hundred eggs during her reproductive period. Each egg is about 0.5 mm long and has two filaments projecting from the anterior end. After about a day at 25°C, a larva emerges from the egg. The larva is eyeless and legless but has black mouthparts at the anterior end and conspicuous spiracles (or air pores) through which it breathes. As it grows, the larva goes through three developmental stages, called **instars**. At the end of the first and second instars the larva molts and sheds its exoskeleton and mouthparts, allowing the organism to increase in size. The third instar lasts about four days, then the larva stops feeding and crawls out of the media onto the walls of the culture container or some other relatively dry surface. There, it forms a pupa.

Inside the pupal case, larval structures and tissues break down and are reabsorbed and adult tissues begin to grow. At the end of four days, an adult fly breaks out of the pupa. When the fly emerges, it is elongated and pale, and the wings are folded. During the next hour, the body becomes more rounded, the wings expand to their normal shape, and the body pigments darken. After eight to ten hours, the flies are sexually mature and ready to mate and produce another generation. Adult fruit flies live an average of 37 days after emerging from the pupal case, but do not grow any larger.

Procedure

1. Take one of the vials of flies labeled "Observation Vials" back to your desk. Begin by observing the flies while they are in the vial.

Where are the eggs and larvae found? Where are the brown pupae found? How could you quantify the number of larvae or pupae in a single vial?

2. To remove the winged adults in the vial, rap the container on the counter to force them to the bottom. Quickly remove the plug and invert the tube over the funnel of a fly "morgue." Tap the tube gently a few times to force the adults into the morgue.
3. Use a spatula to scoop out a match-head-sized chunk of the growth media from the vial. Try to obtain one to two obvious larvae. Place the material in a depression slide or watch glass, and add two drops of water to dissociate the culture media. Transfer the preparation to a dissecting microscope.

   Are any eggs present? If not, why might they be absent?

   Larvae preparing to pupate will stop moving, and can be mistaken for dead. Is there any way that you can distinguish between live versus dead larvae, BESIDES movement?

4. Later you will have to separate the two sexes. Obtain a second practice vial containing dead flies, and pour the flies onto a white index card. Working under a dissecting microscope, use a camel's hair brush to separate your flies into males and females. Have your lab partner and instructor check your sorted flies for accuracy. Once you are finished, return the flies to the practice vial so others may use them.

Exercise 2: Creating a Dose Response Curve for Juvenile Hormone

Background

Most hormones are active at extremely low concentrations. They also act in a dose-dependent fashion, that is, they have larger or different effects as their concentration increases. Physiologists frequently test the effects of hormones at different concentrations, then plot the organism's response(s) against concentration of hormone. This **dose response curve** allows us to estimate the minimum and maximum concentrations of hormones that would be needed in order to obtain physiologically relevant responses.

Dose response curves are used in a variety of situations. What are three situations in which, or reasons why, you might want to have a dose response curve?

For this exercise, different groups within the class will test the effects of different concentrations of juvenile hormone. The stock solutions of juvenile hormone range from 10^{-10} to 10^{-6} M. Your instructor will assign your lab group a particular concentration of hormone to test. Each work group will prepare and measure flies in three control vials and three treatment vials. Data from the entire class will be pooled for the final analysis.

Go back and look at the normal function of juvenile hormone during insect development. What do you predict will happen to fruit flies that are exposed to juvenile hormone while they are still larvae? What is your rationale?

Procedure for Week 1

1. Obtain six vials of flies from the box labeled "Assay Vials." Take the vials, a dropper bottle of the correct concentration of hormone solution, and a dropper bottle of water back to your desk.
2. Label all six vials with your group members' names and the date.
3. Label three vials as "Control #1," "Control #2," and "Control #3." Label the other three vials as "Plus Juvenile Hormone #1," "Plus Juvenile Hormone #2," and "Plus Juvenile Hormone #3." Also include the concentration of hormone you will be testing.
4. You will be applying the aqueous hormone and control solutions directly to the surface of the culture media. The juvenile hormone will diffuse into the media, and be taken in by larvae as they feed near the surface. Remove the foam plugs from the three control vials; if there are still adults in the vials, empty them into the fly morgue. Use the transfer pipet to add five drops of water to the surface of the medium of one vial. Be careful not to add it all in one spot, or so fast that it does not soak in. Add it in drops across the entire surface, and allow the solution to soak into the culture medium before moving or storing the vials.
5. Add five drops of water to each of the other two control vials. When all three vials have absorbed the applied liquid, replug them.
6. Unplug the three vials to be treated with juvenile hormone. Add five drops of your assigned concentration of juvenile hormone solution to each vial. Be sure to apply the solution across the entire surface of the vial, not all in one spot.
7. When the hormone solution has been absorbed, replug the vials. Double-check they are labeled properly, and place them in the storage box provided. Your instructor will return them to the incubator and retrieve them for you next week.

Procedure for Week 2

This week you will count, measure, or describe the following features for the flies that you treated last week with juvenile hormone or with water as a control.

- Total number of viable flies per vial
- Length of adults
- Sex ratio of adults
- Population fraction (relative number of larvae versus pupae versus adults)
- Morphological changes of adults

Each work group will report their summary data to the rest of the class. You will answer the Unit Summary Questions based on the data from the entire class, not just your own.

8. Lay your six vials on their sides in the refrigerator in crushed ice. (The cold immobilizes the adults so that you can transfer them without any flying off.) Once the adults are immobile (about 15 minutes in the refrigerator), transfer them from each vial into new, separately labeled, empty vials, and plug the vials. (*Hint:* If you have handled fruit flies previously, you probably know how to transfer adults from one vial to another while they are still active. If so, it is not necessary to chill the vials first.)
9. Place the six vials containing your adults in the freezer to kill them. Alternatively, dump them into a shallow dish containing 70% ethanol; your instructor will tell you which method to use.
10. While the adults are freezing, you can make the necessary measurements of larvae and pupae. Count the number of pupae on the sides of each vial, and record the data in Table 8.1, page 300. Do not count pale, empty pupal cases from which adults have already emerged; only count the denser, darker pupae that still contain flies.
11. Fill the vial labeled "Control #1" with tap water, and stir it to liquefy the culture medium. Before it settles, pour the contents of the vial onto a red plastic plate. Look carefully for larvae; usually they are easy to see moving against the red background.
12. Count all the larvae in the vial that are longer than 1 mm, and record the number in Table 8.1.
13. Use the same technique to count all the larvae longer than 1 mm in each of your remaining control and treatment vials. Record the counts for each tube in Table 8.1.
14. By now your adult flies should be dead. Retrieve them from the freezer.
15. Empty the adults from the vial labeled "Control #1" onto a white index card. Sort them into males and females. Record the number of each in Table 8.2.
16. Using either the micron scale in the eyepiece of your microscope, or a scale mounted on a glass slide, measure the length of five male flies and five female flies. Record their lengths in Table 8.2.
17. Once you have made all the measurements for the vial labeled "Control #1," pour the waste material into the collection container.
18. Repeat Steps 15–17 for the other two control vials and for the three vials treated with juvenile hormone.
19. As you work, make note of any morphological differences in Table 8.2. For example, do all the males treated with juvenile hormone seem to be missing their wings? Do all of the flies seem paler in the treatment versus control groups?

Results

Table 8.1

Numbers of larvae and pupae in vials

Vial	# Larvae	# Pupae
Control #1		
Control #2		
Control #3		
Plus JH, #1		
Plus JH, #2		
Plus JH, #3		

Table 8.2

Numbers, sizes, and sex ratios of adults

Vial		Total # Adults	Size of Five Adults (mm)					Mean Size (mm)
Control #1	Males							
	Females							
Control #2	Males							
	Females							
Control #3	Males							
	Females							
Plus JH, #1	Males							
	Females							
Plus JH, #2	Males							
	Females							
Plus JH, #3	Males							
	Females							
Unusual morphological differences:								

Summary Calculations

First you will calculate the population fraction, which is just the ratio of larvae to pupae to adults in each vial. To calculate fractions, take the number of larvae, pupae, and adults (both male and female) from Tables 8.1 and 8.2 and enter them in Table 8.3. Add up the three numbers for each vial, and enter the results in the column labeled "Sum of All Stages." Calculate percentages as:

(# Larvae) / (Sum of all stages) * 100%

Finally, determine the mean % of larvae, pupae, and adults in your control and hormone-treated animals.

Table 8.3

Percent of population at each life cycle stage

Vial	# Larvae	# Pupae	Total # Adults	Sum of All Stages	% Larvae	% Pupae	% Adults
Control #1							
Control #2							
Control #3							
Control, mean							
Plus JH, #1							
Plus JH, #2							
Plus JH, #3							
Plus JH, mean							

The numbers in darker boxes in Table 8.3 (mean number of flies in all stages and the population fractions) are the ones you will need to report to the rest of the class.

The next question is whether hormone treatment altered the normal 1:1 sex ratio. In Table 8.4, page 302, write in the raw numbers from Tables 8.1 and 8.2, and calculate the percent males and females in the population.

Table 8.4

Sex ratio of adults

Vial	# Males	# Females	Total # Adults	% Males	% Females
Control #1					
Control #2					
Control #3					
Control, mean					
Plus JH, #1					
Plus JH, #2					
Plus JH, #3					
Plus JH, mean					

Next, determine whether hormone treatment changed the size of the flies. In Table 8.5 and Table 8.6, write in the raw numbers from Tables 8.1 and 8.2. and calculate the percent males and females in the population.

Table 8.5

Mean sizes of adult flies

Vial	Males in vial (mm)	Females in vial (mm)
Control #1		
Control #2		
Control #3		
Control, mean		
Plus JH, #1		
Plus JH, #2		
Plus JH, #3		
Plus JH, mean		

Table 8.6

Summary data from entire class

					Sex Ratio (%)		Mean Size (mm)	
Treatment	Total #	% Larvae	% Pupae	% Adults	Male	Female	Males	Female
Control								
____ M JH								
____ M JH								
____ M JH								
____ M JH								
____ M JH								

Unit Summary Questions

Use the class data in Table 8.6 to answer these questions.

1. Is there any difference in the total number of flies (larvae, pupae, and adults combined) found in each vial with increasing dose of juvenile hormone? Would you have expected a difference in overall numbers? Why or why not?
2. Does treatment with juvenile hormone change the ratio of larvae to pupae or adults? If so, in what way? Is this a result you would have expected? Why or why not?
3. Is there a difference in the sex ratio after treatment with juvenile hormone? Would you have expected one? Why or why not?
4. Does hormone treatment affect the size of male or female flies? Is this an expected result? Why or why not?
5. Is there any effect of juvenile hormone that appears to be dose-dependent, that is, increases with the hormone concentration?
6. Look back at the original prediction you made last week. Was it close to the actual outcome? If not, what was different, and why might the difference have occurred?
7. In this exercise you treated flies only once with juvenile hormone. Based on the observed outcome, what would you predict would happen if larvae were treated every day with juvenile hormone, rather than just once? Why?

Your Own Experiment

Background

Insects use many hormones besides juvenile hormone to control development and growth. The most obvious example is ecdysone, but there are many others. Many are small peptides (short amino acid chains), while others are small amines or lipids. Often an insect hormone performs double duty as a neurotransmitter as well. There are several examples of this in vertebrates. One is epinephrine, which works as a neurotransmitter in the autonomic nervous system, but also circulates in blood as a hormone to trigger the "fight or flight" response. Biogenic amines that have dual functions in insects include dopamine, serotonin, tyramine, and octopamine. Insects also use some hormones found in vertebrate animals, like melatonin.

Between the first and second week of this unit, you should have done some background reading and research on insect hormones. Now it is time to get together with your workgroup and plan an experiment to test a question of your own.

Look back at Figure 8.2 and at the Available Materials list, below. Then think about the physical characteristics of flies you measured in Exercise 2. Formulate several if/then statements describing the effects of the available compounds on Drosophila development. For example, if cortisol were on the list, you might come up with the following:

- IF cortisol (a mammalian steroid) mimics ecdysone, THEN feeding cortisol to Drosophila larvae should prevent them from molting or pupating properly.
- IF larvae do not pupate properly when fed cortisol, THEN there should be significantly fewer pupae attached to the sides of the tube.
- IF cortisol acts like ecdysone in insects, THEN the effect on pupation should be dose-dependent.

Once you have generated several hypotheses, the *group* should decide on just one to test. Plan out the experiment, and decide how many vials of flies you will need. Make a list of all the materials you will need. Think about the data you will need to collect, and how you will collect it, in order to analyze your experiment next week.

Materials Available

The following hormones, hormone blockers, and other supplies are routinely available. If you need other materials, ask your instructor whether they are available.

- Hormone stock solutions (in water)
    - ◊ Ponasterone A (synthetic ecdysone); 10-6 M
    - ◊ Azadirachtin (ecdysone blocker); 10-6 M
    - ◊ Methoprene (synthetic juvenile hormone); 10-6 M
    - ◊ Precocenes (extracted from Ageratum sp.; a JH blocker); 10-6 M

- Biogenic amines (in water)
    - ◊ Serotonin; 100 mM
    - ◊ 4-chloro-DL- phenylalanine methyl ester (lowers serotonin levels by inhibiting tryptophan hydroxylase); 10 mM
    - ◊ L-DOPA (a stable dopamine agonist); 10 mM
    - ◊ Chlorpromazine HCl (a dopamine receptor antagonist); 100 mM
    - ◊ Melatonin; 10-6 M
    - ◊ Octopamine; 10-6 M
    - ◊ Tryptamine; 750 mM
- Micron scale rulers for measuring lengths of larvae or adults
- Ethyl alcohol morgues for destroying adults
- Red plates (for sorting, counting larvae)
- Spare vials, plugs, brushes, other handling supplies

Preparing Vials of Flies

To have larvae present when hormones are added, vials of flies must be started seven days before starting your experiments. The instructions are for preparing 12 vials (10 for the experiment and two extra in case flies in some vials do not survive).

1. Label 12 plastic vials with your group name, lab section, and the date. Take them to the side bench, and add one level scoop of dehydrated instant fly medium to each vial.
2. Add one scoop of spring water or distilled water (not tap water) to each vial. Immediately swirl each vial gently to mix the media and water. The media will solidify in one to two minutes.
3. Add one or two grains of commercial bread yeast to each vial. *Warning:* If you add more than three grains of yeast to a vial, your flies could die of carbon dioxide poisoning.
4. Put a plug in the top of each tube, and wipe up any media or yeast you have spilled. Return to your bench with your vials.
5. Your instructor will provide you with pre-anesthetized flies for you to use to set up your cultures. Use a soft brush to sort the flies into males and females. Have a lab partner check your flies to make sure you have sorted them correctly.
6. Put the females into individual piles of six or seven flies each. Make sure that there are the same number of females in each pile. Once you have 12 piles you can give any unused flies that are still unconscious to another group. If you do not have enough females to make 12 piles, ask someone to share.
7. Add three to four males to each pile.

8. Use a soft brush to gently sweep each pile of sorted flies into a separate vial. Insert the foam stoppers, and lay the vials on their sides until the flies regain consciousness. Do not set the vials upright until the flies are awake again.
9. Your instructor will place the vials in an incubator until next week, when you will remove the adults and add hormones for your experiment.

Animal Hormones

Teacher Pages

INSTRUCTORS' NOTES

Background and Key Concepts

Most organismal physiology courses explore processes and mechanisms by which hormones regulate growth, homeostasis, and adaptation. In lab, students may treat pea or bean plants with auxin or gibberellic acid, then measure their change in height. While a reliable demonstration, this exercise is not particularly engaging, and it does not challenge students to develop a deep understanding of hormonal mechanisms. This lab was designed so that students still see the effects of hormones firsthand, but also get to design and execute experiments of their own utilizing hormones.

Prepare students for this unit by posing these questions:

What are hormones, and why are they important? What are the major features shared by all hormones? What are the advantages and disadvantages of using hormones rather than a nervous system to control homeostasis and adaptation?

What are the three main hormones that control development and maturation of insects? What are their functions?

For each of the hormones just described, what is likely to happen if an insect is exposed to higher than normal levels of that hormone? What if an insect is exposed to a compound that blocks the activity of that hormone? What is the rationale for these predictions?

General Teaching Strategy and Common Problems

This three-week unit is scheduled near the beginning of a freshman-level organismal biology course. The audience is a mix of nonmajors and majors. The first week, students complete Exercise 1 and set up the vials for Exercise 2. In Exercise 1, students make baseline observations of the stages in the life cycle of Drosophila. In Exercise 2, students conduct a structured inquiry to determine the effects of increasing concentrations of juvenile hormone (JH) on development of Drosophila larvae. Students should think through the steps of Exercise 2 and make predictions about outcomes before they start the physical procedure. If necessary, make students write down their predictions and hand them in.

In the second week, students take measurements and perform the data analysis for Exercise 2. Most students are very surprised to see that a low dose of JH does not prevent development (which is what they usually predict will occur.) Instead, adult flies are 20–50% larger than untreated, sex-matched controls. This outcome forces students to re-evaluate their initial assumptions. Ultimately, many students conclude (correctly) that low doses of JH delay maturation, so the larvae feed longer and grow larger. This leads to larger adult flies. What students may not notice initially is that there are significantly fewer adults in JH-treated vials. If a group explores this observation further in their own experiment, they are likely to discover that higher or repeated doses of JH disrupt development sufficiently to kill most of the larvae before they pupate. Only a few larvae survive to grow abnormally large.

In the second week of this unit, student groups also set up an experiment they have designed. In the third week, they collect the results of their own experiment and begin analysis. Ideally, students will not finalize their experiment until after they complete data collection and analysis of Exercise 2, and discuss those data as a class. The instructor should remember that most students in an introductory course have not had any prior opportunity to develop testable hypotheses or do experiments of their own, and they are likely to struggle. For this reason, the number of compounds available for testing is limited, and the unit focuses on sound experimental design. Remind students that their experiments *must* be based on a hypothesis; they cannot test something just because "it might be interesting." Give groups ample time to discuss their ideas for experiments. Students frequently make simple errors in logic at this stage. Ask groups to present a "draft" of their experiment informally to the rest of their class for critique, then help them revise their plan based on classmates' comments.

If a shorter unit is needed, Exercises 1 and 2 can be completed in two two-hour lab periods, and the students' experiments deleted.

Recommended Prelab Skills

Students should know how to use a dissecting microscope. Basic knowledge of graphing with a spreadsheet program will speed up data analysis.

Assessment

If pre- and postlab quizzes are used as part of formal assessment, the prelab quiz should focus on the general concepts, such as, What are hormones, and why are they important? What are the major features shared by all hormones? What are the advantages and disadvantages of using hormones rather than a nervous system to control homeostasis and adaptation? Wait until the postlab quiz to ask questions about the actions of the three major insect hormones or the life cycle and sexually dimorphic traits of Drosophila.

Safety and Housekeeping

Old vials of flies should be plugged tightly and frozen overnight to kill the flies before disposal in general trash. Escaped flies become a nuisance if not kept in check. Keeping wine traps around at all times while the lab is running will reduce their numbers greatly.

Other Tips

- Some students have trouble learning to sort flies by sex. As a guide, mount examples of three or four males and females on an index card with double-sided tape.
- For their experiments, many students will test whether a particular treatment changes the number of larvae, pupae, or adults. The outcome of statistical tests on the raw counts often are not significant, because there can be considerable variation in the total number of animals in the different vials. The results are more consistent if the students count adults, larvae, and pupae, then normalize the numbers of each by expressing them as percentages of the total population of flies at all life stages (i.e., the sum of larvae, pupae, and adults) in each vial.
- Often students want to know whether a particular treatment changes the size of the adult flies. They may forget there is a significant difference in the size of male and female flies normally. Both control (nonhormone-treated) and hormone-treated adult flies should be separated by sex before they are measured. Suggest that students compare average sizes of males only, females only, and both together, with and without hormone treatment.

Supplemental Reference

Colombani, J., L. Bianchini, S. Layalle, E. Pondeville, C. Dauphivillemant, C. Antoniewski, C. Carré, S. Noselli, and P. Léopold. 2005. Antagonistic actions of ecdysone and insulin determine final size in Drosophila. *Science* 310: 667–670.

PREPARATORY NOTES

Quantities listed are for a lab section of 20 students working in groups of three or four.

Week 1: Shared Materials

- 5 shell vials (15 days old) containing live larvae, pupae, and adult Drosophila, labeled "Observation Vials"
- 30 shell vials (7 days old) containing adult Drosophila, labeled "Assay Vials"
- Serial dilutions of 10^{-6} M to 10^{-10} M Methoprene (aka, Precor, a synthetic juvenile hormone) in water
- 1 L powdered Drosophila culture medium, in resealable food storage box
- 4 measuring scoops (8–10 mL volume)
- 75 polystyrene (not polypropylene) shell vials (2–3 cm diameter)
- 75 foam plugs for shell vials
- 2 microtubes with ~0.2 mL of granulated dry baker's yeast
- 2–3 wine traps, placed around lab room

Week 1: Materials at Each Work Station

- Soft paintbrush
- Dissecting microscope
- 3 × 5 index cards
- 4–5 plastic transfer pipets
- Fly morgue (jar one-quarter full of mineral oil, with a funnel taped in the top)
- Permanent marker

Week 1: Materials for Instructor/Prep Area

- Drosophila adults, ~900, <1 week post-emergence
- FlyNap or other anesthesia supplies

Week 2: Materials at Each Work Station

- Soft paintbrush
- Dissecting needle
- Dissecting microscope
- 3 × 5 index cards
- Vials of flies that students treated with JH in Week 1
- Vials of flies that students set up for their experiments in Week 1
- Dissecting microscope
- Fly morgue (jar one-quarter full of mineral oil, with a funnel taped in the top)
- Permanent marker

Week 2: Materials for Instructor/Prep Area

- 12–15 spare vials of Drosophila, prepared by instructor the previous week
- FlyNap or other anesthesia supplies

Optional Materials for Experiments (make only if requested)

- 10^{-6} M Muristerone A *or* Ponasterone A (synthetic ecdysone)
- 10^{-6} M Methoprene (aka, Precor, a synthetic juvenile hormone)
- 10^{-6} M Precocene I or II (aka, 7-methoxy- *or* six 7-dimethoxy-, two 2-dimethy, l-3-chromene, a juvenile hormone blocker)
- 10^{-6} M azadirachtin (ecdysone blocker)
- 100 mM serotonin
- 10 mM 4-chloro-DL- phenylalanine methyl ester (lowers serotonin levels by inhibiting tryptophan hydroxylase)
- 10 mM L-DOPA (a stable dopamine agonist)
- 100 mM chlorpromazine HCl (a dopamine receptor antagonist)
- 10^{-6} M melatonin
- 10^{-6} M octopamine
- 750 mM tryptamine

Week 3: Shared Materials

- 5–10 red plastic picnic plates, nondivided
- 3–4 stage micrometers, mounted on glass microscope slides
- 2–3 wine traps, placed around lab room

Week 3: Materials at Each Work Station

- Soft paint brush
- Dissecting needle
- Dissecting microscope
- 3 × 5 index cards
- Fly morgue (jar one-quarter full of mineral oil, with a funnel taped in the top)

Maintaining Organisms

In Unit 2: Mendelian Genetics, you will find detailed instructions for maintaining Drosophila, preparing morgues and wine traps, disposing of flies, and general care. Flies can be grown in shell vials for this unit, but for multiple lab sections it may be more convenient to raise larger cultures in glass milk bottles, jars, or disposable 200 mL plastic Erlenmeyer flasks (available from lab supply houses specializing in Drosophila.) For each group of 20 students working in groups of four, set five to seven shell vials (observation vials for Week 1) and six 200 mL culture bottles (flies for students' experiments). These stocks are set up 14–15 days before the first week of the unit, so that larvae and pupae are still present in the observation vials and the emerged adults are less than one week old. An additional 30 vials are set up for the juvenile hormone dose response assay seven days later (i.e., seven to eight days before the first lab).

Inevitably some students' vials will not have larvae by the second week of this unit. The instructor should prepare 12–15 extra vials of flies at the end of the first week, which can be used as replacements.

For this unit, flies are anesthetized using FlyNap or other triethylamine-based anesthetic. Instructors can knock out enough flies for an entire class to use at the beginning of lab, and students can observe them for longer periods. Detailed instructions for using FlyNap come with the kit. Be careful, as over-gassed flies will die. As soon as flies are unconscious, pour them from the anesthesia container onto an open plate so the gas can dissipate. The reagent loses potency over time, so replace the solution each semester.

An alternative is carbon dioxide plus cold, which is described in Unit 2. Flies wake up more quickly from carbon dioxide/cold, and may escape, but sterility and death occur less frequently. Ultimately the choice comes down to instructor preference.

Solutions, Reagents, Equipment

Hormone Solutions

Prepare 10^{-2} M stock solutions of the hormones and store them at -20°C between semesters. Stocks of sterones are prepared in 95% ethanol; all other hormones and compounds are prepared in distilled water. Working hormone stocks of 10^{-6} M and lower are prepared fresh each semester by serial dilution of the 10^{-2} M stock solutions with water. Working stocks are prepared no more than one week in advance of the lab, then stored refrigerated and wrapped in foil.

The list of hormones that students might use for experiments is long, and not all of them should be made available at one time. In freshman-level courses at the author's institution, instructors use just the hormones that directly control maturation and ecdysis. Typically, a single dose of methoprene delays emergence, so adult flies are larger than normal. Both synthetic ecdysone and juvenile hormone blocker trigger early molting, so at low doses the adult flies are smaller than normal. At high or repeated doses of any of the three hormones, maggots fail to pupate or they die in their cases.

Biogenic Amines

Serotonin, dopamine, and other biogenic amines also act as hormones to control fly development. Students who wish to test these compounds will need more extensive background knowledge than is provided in this unit; experiments with biogenic amines can be reserved for more advanced students or classes. All of the biogenic amines listed can be purchased from most research chemical suppliers.

Neuromuscular Control

Student Pages*

BACKGROUND

Animals use a combination of hormones and nerves to coordinate and control homeostasis and to respond to environmental stimuli. In this unit, you will learn more about the neuromuscular system. Like hormones, nerves and muscles allow organisms to adapt to the internal and external environments. However, the neuromuscular system is different in that it responds within milliseconds, rather than in seconds to days. It is also unique to the animal kingdom; plants and fungi have never evolved any analogous system of rapid intercellular communication.

Skeletal muscle in both vertebrates and invertebrates consists of hundreds to thousands of **myofibers**, which are cylindrical, multinucleated, fused aggregates of cells. The individual myofibers are held together by connective tissue to form functional skeletal muscles. The myofibers are innervated by motor neurons that have their cell bodies located in the gray matter of the spinal cord. The actual contact points where axons end and release neurotransmitters onto the myofibers are called **neuromuscular junctions** (shown in Figure 9.1, p. 314).

*Teacher Pages begin on page 329.

Figure 9.1

Fine structure of a skeletal neuromuscular junction

The left side shows how the nerve containing bundles of axons splits into individual axons, which form the synaptic termini on skeletal myofibers. The right side shows the structure of the skeletal myofiber beneath each synaptic terminal. The voltage-gated Ca^{+2} channels admit calcium when an action potential arrives, and cause the vesicles of neurotransmitter to fuse and release their contents onto the plasma membrane of the underlying skeletal myofiber.

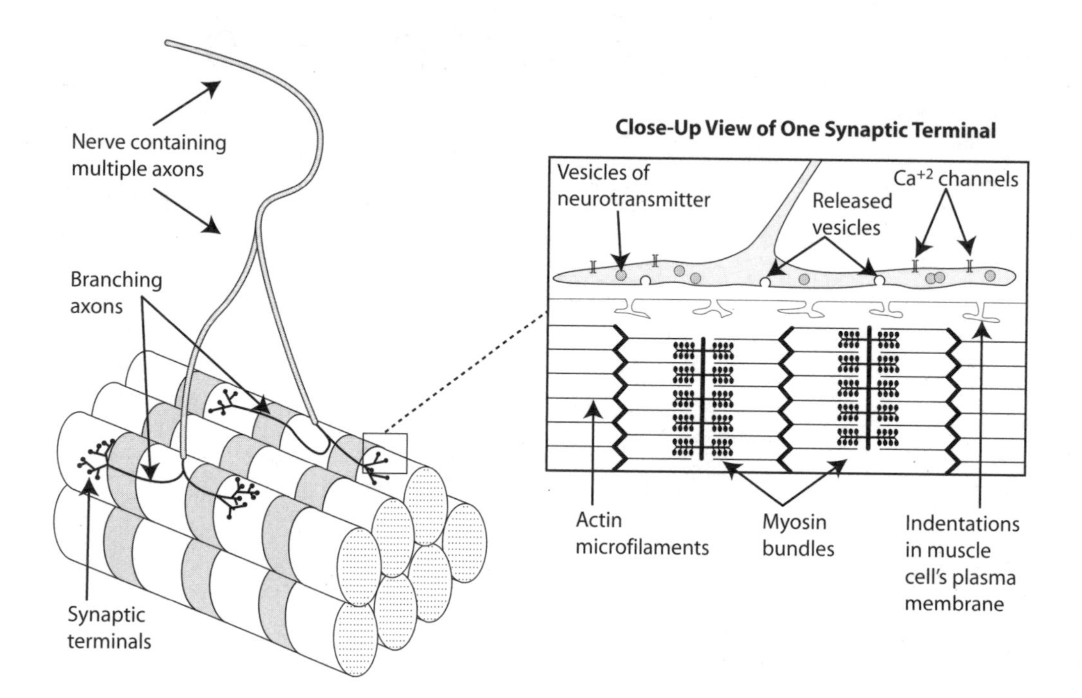

The basis of all muscle movements is the **twitch**. Nerve impulses arriving at the neuromuscular junction cause the release of the transmitter acetylcholine (ACh) from the presynaptic nerve terminal. Acetylcholine binds to its receptor on the membrane of the myofiber, opening sodium channels that generate an action potential in the myofiber. That action potential spreads along the plasma membrane of the entire myofiber, stimulating the release of Ca^{+2} from the sarcoplasmic reticulum (an intracellular calcium storage organelle). The release of Ca^{+2} activates the contractile proteins, causing the entire myofiber to shorten slightly (i.e., generating a twitch). A single twitch is very brief, usually lasting about 200 msec (0.2 sec).

The preceding model of muscle contraction raises several questions:

If a single twitch only lasts 0.2 seconds, how can a muscle generate and hold contractions that last for seconds or minutes?

How can one mass of myofibers generate different degrees of force? For example, humans can use the same finger and hand muscles to gently hold a flower for minutes without crushing it, to type rapidly, or to hold onto heavy furniture while moving it. How can the same set of myofibers apply such a wide range of force over different lengths of time, if all movements are based on a simple twitch of the myofibers?

In this unit, you will be looking for answers to these questions and to questions of your own.

Goals for This Unit

This unit requires two weeks to complete. During the first week, you will learn how to dissect the sciatic nerve and gastrocnemius muscle from the leg of a frog (*Rana* sp.), and connect them to a force transduction apparatus. Next you will learn to record myograms from the muscle preparation and make several baseline observations. Finally, you will try to demonstrate the processes of summation, tetany, and muscle fatigue in the leg muscle. At the end of the first week, you and your work partners will develop your own hypothesis about control of muscle contraction and design an experiment of your own to test that hypothesis. In the second week of this unit, you will actually perform your experiment.

General Safety Precautions

There are no toxic hazards during the first week. You can safely handle the frog legs without gloves, but they will be available if you prefer them. When you have completed the exercises, place the remains in the designated containers. DO NOT put animal remains in general trash. Wash your hands thoroughly before leaving the lab. The electrical output of the stimulator box is not enough to harm you. However, the electrode tips may be sharp. Be careful not to stab your fingers or hand. Some of the compounds available for your own experiment are moderately to extremely toxic. If you use a toxic compound, wear gloves and wash your hands thoroughly before leaving the lab. You will also be handling syringes in this lab. Your instructor will explain your school's policies for using and handling syringes. Dispose of them in the specified sharps containers only, NEVER in the general trash. If you get stuck with a syringe, even if it is empty and sterile, report your injury immediately to your instructor.

UNIT EXERCISES

Exercise 1: Measuring Nerve and Muscle Activity

As you collect data for this and the following exercise, record the relevant values in the Summary of Results section.

Dissecting the Gastrocnemius Muscle and Sciatic Nerve

1. Your instructor will give each work group a frog leg. Place the leg in a plastic dissecting pan. With a pair of forceps, grasp the skin at the top of the leg and pull the skin down toward the foot. Lay the skinned leg in the dissecting pan, and moisten it with frog Ringer's solution.
2. Gently separate the inner thigh muscles with a blunt probe, beginning at the upper end and extending to the knee. Lying near the bone you should see the white sciatic nerve and accompanying blood vessels (see Figure 9.2.) The nerve usually is the diameter of the lead used in a mechanical pencil, or larger. Do not mistake the thin glassy strips of fascia (connective tissue) for the nerve. Gently peel the nerve free from its connective tissue, using forceps; stop 2–4 mm from the knee joint. Be careful in the way you handle the nerve. The less it is traumatized or comes into contact with metal, the better. Do not tug or stretch it either. Keep the exposed tissues moist (but not drenched) with frog Ringer's solution.

Figure 9.2

Location of the sciatic nerve and gastrocnemius muscle

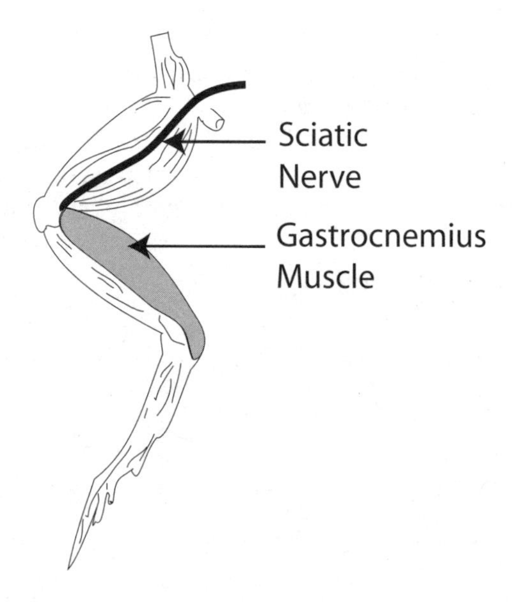

Using a Square Wave Stimulator to Determine Threshold

Before you set up of the computer recording system, you will stimulate the intact muscle and nerve to determine the minimum voltages that elicit a simple muscle twitch. The minimum applied voltage necessary to elicit a twitch is called the **threshold voltage**. The value of the threshold voltage depends on the characteristics of the nerve or muscle.

To determine the threshold voltages for direct muscle versus nervous stimulation you must use a square wave stimulator. This device allows you to precisely control the amplitude, duration, and frequency of electrical stimuli. Important features of the stimulator are as follows:

Output electrodes: Two pin electrodes are connected to wires and carry current from the stimulator to the nerve or muscle.

Stimulus mode controls: The "Mode" switch is used to select whether the box provides a single pulse or a continuous train of pulses. A red LED lights each time a pulse is produced.

Frequency, duration, and voltage controls: The "Events/sec" knob controls the number of pulses occurring in one second when the output control is in the continuous position. A selector switch multiplies the frequency by 0.1, 1, 10, or 100. Frequencies between 1 and 1000 pulses per second are possible. The "Duration" knob controls the time course of the pulse in milliseconds (0.04–50). Finally the "Volts" knob controls the amplitude of the stimulus (0.1 to 100 volts).

1. Set the stimulator to 0.1 volts and 15 msec duration. Place the leg in a flexed position. Hold the tines of the electrode against the belly (thickest region) of the gastrocnemius muscle and depress the mode switch to "Single." If the muscle twitches, reduce the duration to the lowest value that will produce a faint twitch of the toes. If the muscle does not twitch at 0.1 volts, 15 msec, increase the voltage in increments of ~0.1 volts until a slight twitch occurs. Record this voltage as the threshold stimulus for muscle.
2. Now stimulate the muscle with gradually increasing voltages, until the muscle contraction causes the foot to extend completely. Record this voltage as the muscle stimulus required for maximal contraction.
3. Next you will stimulate the nerve. Position the electrodes under the nerve and return the leg to a flexed position. Return the voltage control to 0.1 volts and stimulate the nerve by pressing the mode switch to "Single." If no muscle twitch is seen, proceed as described previously. Stimulate the nerve at increasing increments of 0.1 volts until a twitch is visible. This voltage can be recorded as the threshold stimulus for nerve. If a twitch or extension occurs with a stimulus of 0.1 volts, try reducing the duration of the stimulus.
4. Continue raising the voltage, and determine the minimum voltage that causes full extension of the foot by stimulating the nerve. Record this voltage as the nerve stimulus required for maximal contraction.

Recording a Basic Myogram

Now that you have determined the threshold stimuli that will induce a visible twitch on the muscle, you are ready to produce a **myogram** (Figure 9.3). A myogram is a visual recording of the different stages within a single muscle twitch, and is collected by connecting the muscle to a **force transducer**. The transducer works by changing mechanical energy (muscle movement) into an electrical signal. This signal is fed through an amplifier to a computer that records and displays the signal. For this lab, all the equipment is already connected properly for you, so you can focus on the biological concepts.

Figure 9.3

A basic myogram

The two traces shown were recorded simultaneously. The upper trace indicates the size and duration of the electrical stimulus applied to the nerve connected to the muscle. The lower trace shows the force generated by the muscle itself in response to the stimulus.

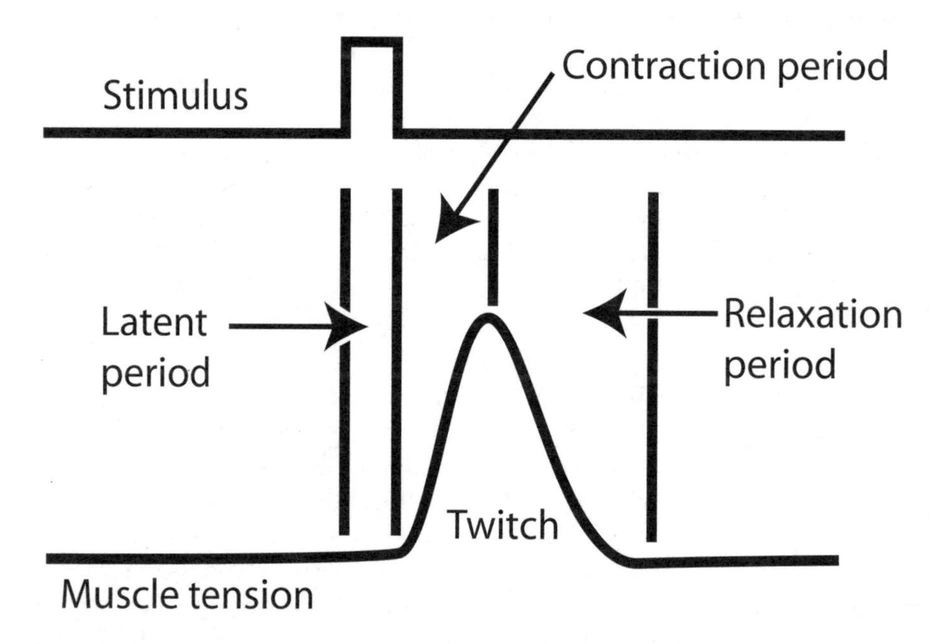

A typical myogram will have three distinct phases: a **latent period**, a **contraction phase**, and a **relaxation phase**. The latent period is a very short time lapse between the time of stimulation and the start of contraction. In most muscles of the body, it lasts less than 10 msec. In the contraction phase the muscle shortens due to the chemical changes that occur within the fibers. After the contraction phase has reached its maximum, the muscle returns to its former relaxed state (the relaxation phase). The duration of both the contraction and relaxation phases differs between various muscles in the body.

While you are attaching the frog leg to the transducer, keep the exposed tissues moist (not dripping) with frog Ringer's solution.

1. Use a blunt probe to free the gastrocnemius muscle from the underlying bone and adjacent muscles. Do not cut the large Achilles tendon that connects the gastrocnemius muscle to the foot yet.
2. Cut off a 1 m length of sewing thread, and double it. Insert a pair of forceps under the muscle near the Achilles tendon, and grasp one end of the thread. Pull the doubled thread through with the forceps and tie it tightly around the middle of the Achilles tendon.
3. With a pair of scissors, cut through the tendon distal (i.e., toward the toes) to where the thread is tied. Tie the other end of the thread to the flat spring of the transducer arm, which is mounted on a ring stand above the muscle.
4. To prevent the leg from moving during contraction, make sure the leg is firmly pinned to the dissecting tray. Position the electrode so that the nerve rests on the prongs, and pin the electrode holder so the electrode stays in place.
5. Adjust the tray and ring stand until the string is perpendicular to the tray and the gastrocnemius is vertical and held with moderate tension. Once the muscle and transducer are aligned, place pieces of clay under the four corners of the pan to prevent it from shifting. The final setup should look similar to Figure 9.4.

Figure 9.4

A prepared muscle, tied to the force transducer and ready to record myograms

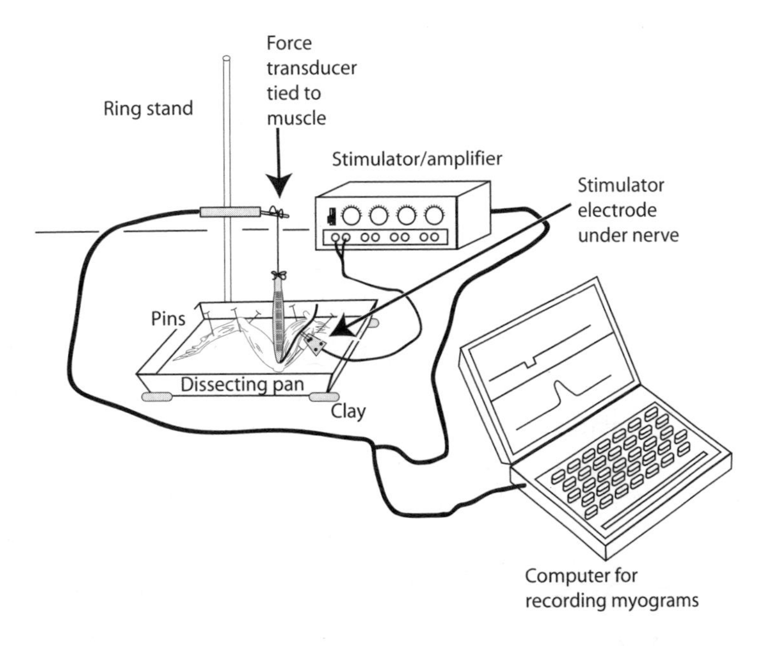

6. You will be using a data acquisition program called AcqKnowledge to record your myograms. This program should already be installed on the computer connected to stimulator box and force transducer. If the program is not already running, open it using the shortcut icon on the computer desktop.
7. Under the File menu, look for the command "Open," and use it to locate and start the empty template file that you will be using today. Your instructor can tell you exactly where it was installed. A screen should appear that looks like the one in Figure 9.5. The program is showing you two traces that correspond to two channels of data. The top trace is labeled "Stimulator" and shows the output voltage and duration of the electrical stimulus. The bottom trace is labeled "Transducer." It shows the relative force generated by the muscle in response to the stimulator output.

Figure 9.5

Diagram of the main AcqKnowledge interface

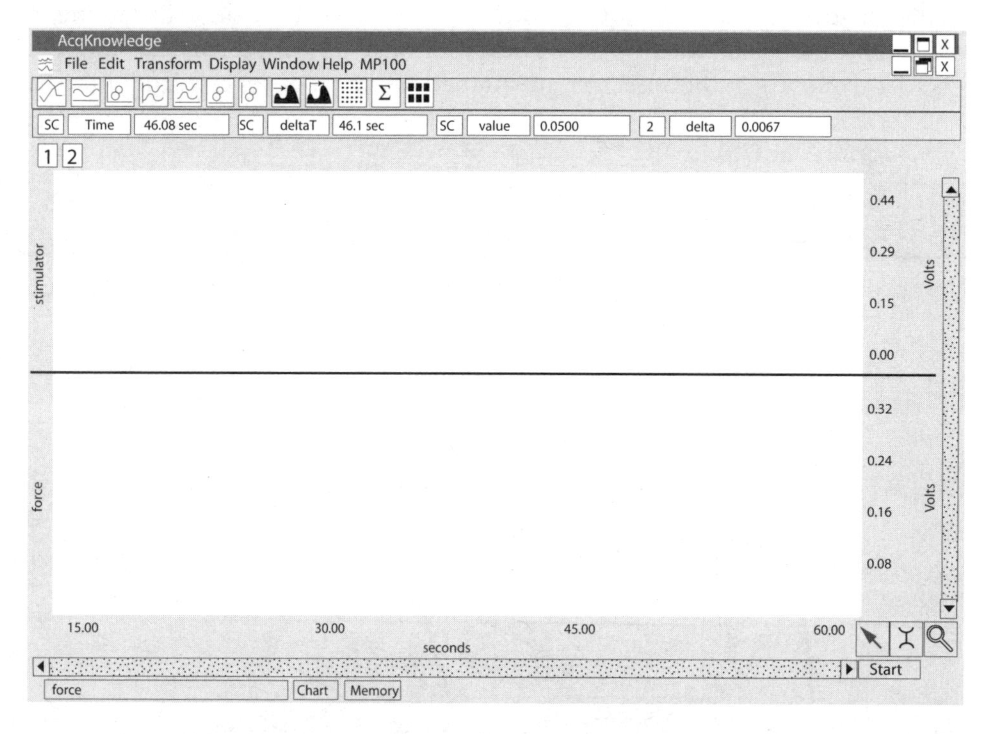

8. Each time you start a new recording, the preceding recording is overwritten and lost. To save recordings, select "Save As" under the File menu. You should select the appropriate drive and type in a descriptive file name. Once saved, the recording can be recalled later for analysis and printing. *Note:* Recording files can only be opened on a computer that has the AcqKnowledge software.

9. Set the stimulator to a duration of 15 msec and twice the minimum voltage required to produce a leg twitch by nerve stimulation. Set the mode switch to Off. You are now ready to record a simple muscle twitch myogram on the computer and determine the duration of each of the three phases in the contraction cycle.
10. From the menu, tell the program to begin recording. Stimulate the muscle by pressing the mode switch down once to Single. On the computer screen you should see a stimulator signal in the top panel and a myogram in the bottom panel. If either trace is off scale, click on the axes and make adjustments so the data are visible.
11. Once you have adjusted the scales so you can see the data being recorded, press the mode switch down to Single five more times, waiting about two seconds between stimuli.
12. Stop the data recording, and immediately save the file.
13. Your instructor will demonstrate how to use the AcqKnowledge software tools to measure the duration of the latent phase, contraction period, and relaxation period. He or she also will show you how to determine the amplitude of the contraction. Determine the mean values for each of these measurements for your frog leg muscle. Record them in the Results section.

Effects of Fluid Loading on Muscle Contraction

Anytime a muscle is stretched or deformed, its ability to generate force changes. You can demonstrate this effect in your muscle preparation by injecting it with frog Ringer's solution, then recording several myograms again.

1. Fill a tuberculin syringe with 0.05 mL of Ringer's solution.
2. To inject a muscle, turn the needle so that it parallels the long axis of the muscle. Gently insert the needle into the muscle, then slowly (over 10–15 seconds) inject the saline into the muscle.
3. Wait 2–3 minutes, then record three to five more myograms. Once again, measure the latent, contractile, and relaxation periods and amplitudes. Determine the averages of each one, then record the results.

Exercise 2: Temporal and Spatial Summation

Background

Up to this point, the nerve has been treated as single functional unit that conducts an action potential to a muscle, which in turn acts as a single functional unit to generate force. In reality, nerves and muscles consist of many hundreds of smaller functional elements. The sciatic nerve is a bundle of connective tissue in which are embedded myelinated axons of thousands of motor neurons. Within the muscle, each individual axon branches out and makes connections with more than one myofiber. These groups, or **motor units**, may consist of only a few myofibers, or hundreds of them, all under control of one axon. When an action potential travels down a single axon, every myofiber attached to it (i.e., the entire motor unit) will be stimulated to contract. The number of myofibers in the motor unit determines how much force it will generate with each twitch. If Motor Unit A has only three myofibers, it generates less force when activated than does Motor Unit B, which has 30 myofibers attached to one motor axon.

A single muscle can produce a wide range of contractile force because it contains multiple motor units of many different sizes. Consider your thumb and forefinger for example. If you are trying to remove a contact lens, or pick up a single crumb from the floor, your finger muscles need to generate very little force. In contrast, when you are trying to open a stubborn jar or bottle top, your finger muscles must exert tremendous force on the lid to prevent it from sliding through your fingers as your arm muscles work to break the seal. The same finger muscles are used in both cases, but motor units of different sizes are activated for each task.

Even when the motor units have the same number of myofibers, varying amounts of force still can be generated through a process known as **summation**. The term *summation* simply means that the force generated by multiple motor units is added together (summed) to produce muscle tension. There are two types of summation. In **spatial summation**, increasing the number of axons that are carrying action potentials causes an increase in the number of active motor units, which increases the force generated. To demonstrate spatial summation in the lab, the voltage used to stimulate the sciatic nerve is increased. In **temporal summation**, the axons of a given number of motor units are stimulated repeatedly. Assuming that the stimulus voltage does not change, each stimulus causes a twitch to occur in a fixed number of motor units. Yet if an axon is stimulated frequently enough, a new twitch begins in these motor units before the relaxation phase of the last twitch ends. As a result, the motor units never relax fully, and the individual twitches add up temporally (over time) to produce a single fairly continuous pull. Ultimately, temporal summation of twitches will cause a motor unit or muscle to generate the maximum force possible. This state of maximal sustained contraction is called **tetanus** or **tetany**.

Continuous Work Ultimately Produces Fatigue

A motor unit cannot exert maximal contractile force forever. If a muscle is stimulated for long enough, eventually it will run out of energy, and the actin and myosin cross-bridges will begin to separate. As a result, the muscle will begin to relax again. This process is called **fatigue**. Motor unit fatigue is different from the relaxation phase of a normal contraction because the muscle will not readily contract again if it is re-stimulated. See Figure 9.6 for an example of a myogram recording showing fatigue.

Figure 9.6

Myograms demonstrating spatial and temporal summation

Panel A. Spatial summation. The voltage applied to the nerve increases from left to right.

Panel B. Temporal summation, tetanus, and fatigue. The same voltage is applied each time, but frequency of stimulation increases from left to right.

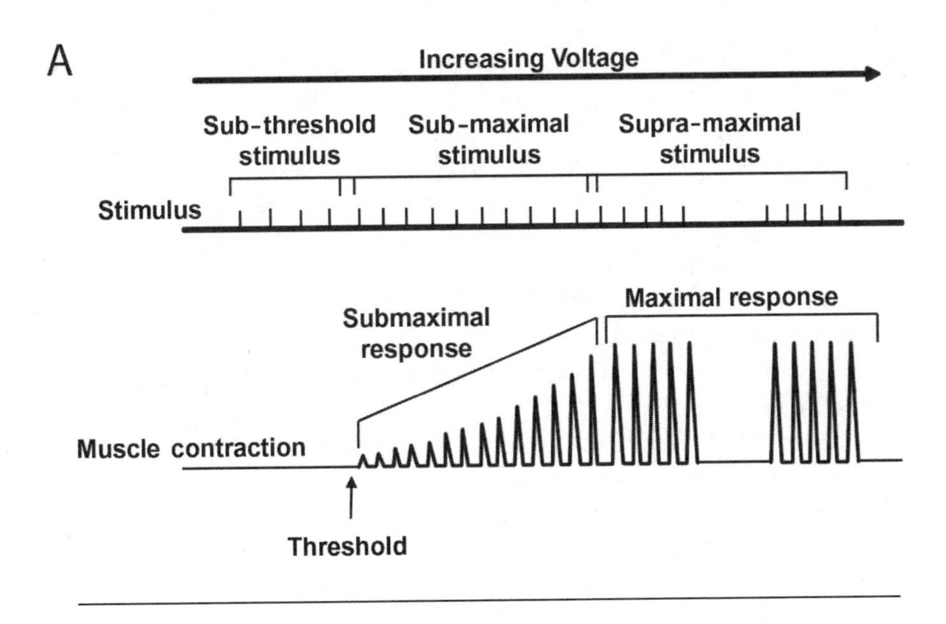

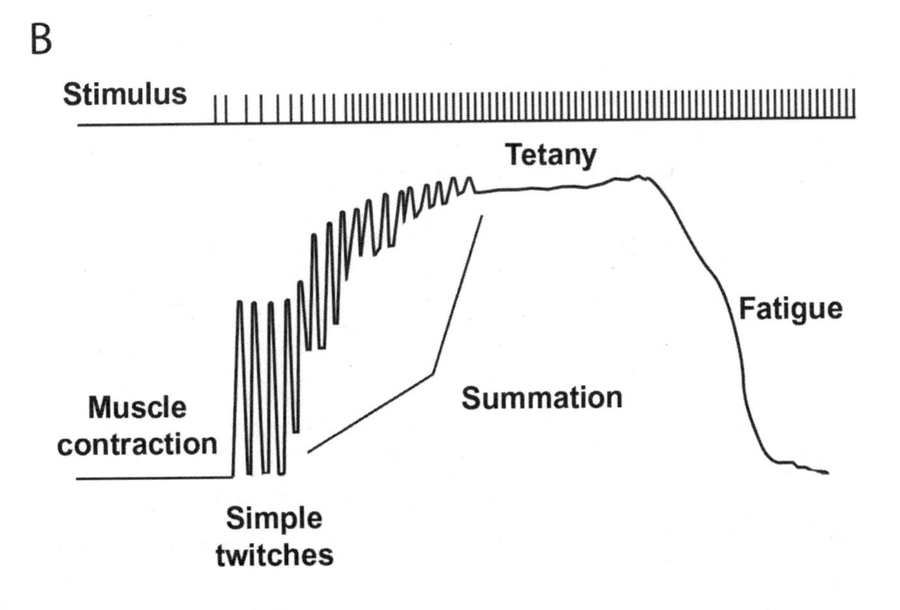

Procedure: Inducing Summation, Tetanus, and Fatigue

Continue using the same frog leg preparation from which you recorded earlier myograms.

1. Set the Stimulus Control to "Single Pulse." Reset the stimulator box output to 0.1 volts and 15 msec duration. Begin recording, and stimulate the muscle once at this voltage. If you do not see a response, continue recording and raise the voltage to ~0.15 volts.
2. Increase the voltage in ~0.05 volts increments and keep stimulating the muscle. At some point, you should start to see individual muscle twitches. As you raise the stimulus voltage, the peak contractile force should increase as well (see Figure 9.6A for an example). This happens because more and more motor units are being activated. In other words, spatial summation is occurring.
3. Keep raising the stimulus voltage and recording until an increase in voltage applied does not increase the contractile force generated. This is the voltage that induces maximal spatial summation across the entire gastrocnemius muscle. Record this voltage on your data sheet.
4. Reset the stimulus voltage to ~30% of the level required to induce a maximal contraction. Set the stimulus control box to deliver one pulse per second.
5. Begin recording a myogram. While recording, slowly increase the number of pulses per second until you begin to see the baseline disappear between individual twitches; this is the stimulus frequency at which temporal summation is starting to occur (see Figure 9.6 for an example). Record the number of pulses per second required to cause temporal summation.
6. Continue to slowly increase the number of pulses per second, until the peaks from individual twitches disappear, and you observe a single, smooth-appearing contraction. This is tetany. Record the stimulus frequency in your Summary of Results.
7. As soon as tetany occurs, note the exact time. Alternatively, start timing with a stopwatch. Now watch for the contractile force generated to start falling; this may take several minutes. When the force exerted has declined to 50% of the maximum value, note the time again. Calculate the time (in seconds) it took for fatigue to develop. Record this time in your Summary of Results.

Summary of Results

Initial Observations

Threshold stimulus when applied directly to muscle (V): _____
Muscle stimulus required for maximal contraction (V): _____
Threshold stimulus when applied through the sciatic nerve (V): _____
Nerve stimulus required for maximal contraction (V): _____

Baseline Myogram

Avg. latent phase (msec): _____

Avg. contractile phase (msec): _____

Avg. relaxation phase (msec): _____

Avg. amplitude of the contraction: _______

Fluid Loading

Avg. latent phase (msec): _____

Avg. contractile phase (msec): _____

Avg. relaxation phase (msec): _____

Avg. amplitude of the contraction: _______

Summation

Stimulation required to induce spatial summation across entire muscle (V): _____

Stimulation rate required to see temporal summation (pulses/sec): _____

Stimulation rate required to induce tetany (pulses/sec): _____

Stimulation time required to induce fatigue (sec): _____

Unit Summary Questions

When you turn in your answers to these questions, include a copy of your Summary of Results.

1. Which has the smaller threshold stimulus voltage, the muscle or the nerve? For which—the muscle or the nerve—is smaller voltage required for maximum contraction? What is the most likely reason for these differences?
2. Why is there a latent phase between electrical stimulus to a nerve, and the muscle contraction? In other words, why doesn't a muscle contract immediately when the nerve is stimulated? Given your answer, how could an organism reduce the latent phase for a contraction?
3. What was the effect of fluid loading on the gastrocnemius muscle? Was one particular stage in contraction affected more than the others? Why do you think fluid loading has the observed effects?
4. Different muscles perform different functions in an organism. As a consequence they have different biochemical composition, contractile properties, relaxation rates, or maximum force generating capacities. The gastrocnemius is primarily for locomotion, so it could be considered a "high intensity, short duration" type of muscle. In contrast, the long muscles in the back of vertebrates (on either side of the spine) maintain posture and support the body; they could be considered "low intensity, long duration" types of muscle.

   Suppose you strung one of the spinal postural muscles and measured the same properties as you did today. Based on its function, what differences (if any) do you think you would see in the following: latent phase, contractile period, voltage required to induce spatial summation, stimulation rate required to induce temporal summation, and stimulus needed to induce fatigue? What is your rationale for each one?

5. In the frog leg preparation, the gastrocnemius relaxes very quickly after a single twitch. In living organisms though, muscles can relax *much* more slowly. For example, stand on your tiptoes by contracting your gastrocnemius muscle. Now *slowly* return to standing normally, with your weight on your heels. It is very likely that you took longer than the average relaxation time you recorded above. So how does a muscle relax in a slow, controlled manner?

Your Own Experiment

Muscular contractions are the primary means by which most animals move. Even those that move by changing the pressure in a hydrostatic skeleton (e.g., earthworms) must contract muscles in order to generate the pressure required for movement. With the model system and data collection procedures you learned to use today, you can test many different factors, and determine how they would affect transmission of action potential along axons, neurotransmitter release, or contraction and relaxation of the muscle itself.

There are many different toxins and chemicals that can alter the ability of neurons to transmit action potentials and there are channel blockers that can alter the rate at which sodium, calcium, and potassium move across cell membranes. Many of these can be made available for experiments. Rather than just testing a chemical "to see what happens," as you design your experiment think about the reasons *why* it might be medically or economically important to be able to affect neurotransmission and muscle contractility. When would you want to stimulate them versus inhibit them? What chemical changes in membrane function would be needed?

There are many directions you can take. As you discuss what to do, keep these things in mind:

- Think about what must happen at each stage in a muscle contraction to produce the myogram you see. Focus on what would have to happen within the muscle to change one of the parameters you measured and *how* and *why* it would affect the overall myogram.
- Next, decide what you would have to do to the muscle to alter the parameter on which you focused. State it as a testable hypothesis.
- Plan out precisely what you will need to do and what data you will need to collect from your experiment.

- You will have more than one frog leg for your experiment. Plan to record pre- and post-treatment data on each leg, so each leg can serve as its own control.

Materials Available

If you wish to test something else that is not on this list, ask your instructor if it is available.

Modified Ringer's Solution

- Low calcium
- Calcium-free with EDTA (a calcium chelator)
- High calcium (50 mM)
- Potassium-free
- High potassium (20 mM)

Compounds (All in Ringer's Solution)

- Acetylcholine
- ACh receptor agonist
- Acetylcholinesterase inhibitors
- Sodium, potassium, or calcium channel blockers
- Ethanol (permeabilizes membranes)
- Lidocaine (blocks action potential propagation)

Other Equipment

Warm water baths, refrigerator, or ice baths to alter temperature of muscle

UNIT 9

Neuromuscular Control

Teacher Pages

INSTRUCTORS' NOTES

Background and Key Concepts

Most students in an introductory-level biology course will know that skeletal muscle generates movements by cleaving ATP and that such movements are initiated and controlled by motor nerves. However, very few can explain how a single muscle mass can develop a range of force lasting for varying lengths of time. This lab introduces students to the basic principles of muscle mechanics and neuromuscular control. It examines the processes of temporal and spatial summation, two fundamental mechanisms for generating complex movements.

Vertebrate skeletal muscle consists of hundreds of contractile fibers called myofibers. Individual myofibers are long, membrane-bound multinucleate syncitia containing actin, myosin, and related proteins. Myofibers are electrically insulated from each other, so they can be controlled independently. However, they rarely work in isolation; most are elements within a motor unit that consists of dozens to hundreds of myofibers. Each motor unit is controlled by a single skeletal motor neuron, which has a branched axon terminus that forms synapses on each myofiber in the motor unit. When an action potential nears the axon terminus of a skeletal motor neuron, the depolarization wave travels down all its branches simultaneously and stimulates all attached skeletal myofibers in the motor unit to contract. Electrical depolarization does not spread readily to myofibers outside of an activated motor unit, so the individual motor units within a muscle can be triggered to contract independently of one another, which in turn allows a single skeletal muscle to generate a wide range of force.

In the laboratory, students learn by direct observations that total force generated by a muscle depends on the number of motor units activated, a process known as spatial summation. They also see firsthand that the length of time a muscle maintains a contraction depends on the frequency at which action potentials arrive at the synapses of the motor units, a process called temporal summation.

To help students master this unit, emphasize these concepts:

- Cell membranes of both nerves and muscles can propagate action potentials.
- Action potentials can be triggered in both muscles and nerves with an external electrical stimulus. However, nerves can be stimulated to transmit action potentials by a much smaller electrical impulse than muscle.
- A single nerve consists of hundreds of separate axons that can transmit action potentials *independently* of one another. Low voltage stimulus activates only a few axons, while increasing the applied voltage increases the number of axons in a nerve that are transmitting action potentials.

As students progress through this unit, they learn how to record a basic myogram and how to measure the three major phases of a skeletal muscle contraction. They also learn the causes and effects on muscle contraction of spatial summation, temporal summation, tetany, and muscular fatigue.

General Teaching Strategy and Common Problems

Many instructors employ some variation of the exercises in this unit in an advanced undergraduate animal physiology lab course, but given students' responses to this model system, the author firmly believes it should not be restricted to advanced majors alone. It is included in this volume to demonstrate how even complex experimental models can be adapted to the needs and experience level of students in large survey courses.

As written, the two-week unit is appropriate for a freshman-level organismal biology course enrolling both nonmajors and majors. The first week, students complete the entire sequence of exercises, then plan experiments and submit an experimental proposal for their groups. In the second week, students conduct the experiments they designed. The first week of this unit can be completed in one three-hour lab period. Eliminating the final fatigue exercise ("Continuous Work Ultimately Produces Fatigue," p. 322) can shorten the unit, but this is not recommended. Students are fascinated by the process and give the unit significantly lower rating scores when it is eliminated. A better way to accommodate shorter lab periods is to space the first week's exercises across two weeks, then either eliminate the experiments or devote a third week to them.

Despite its popularity, some problems still arise regularly with this unit. Students at this level find it very hard to think simultaneously about the movements of actin and myosin proteins occurring inside the myofiber and about the basic physiological mechanisms controlling gross muscle movements. For this reason there is no discussion of the sliding filament model, and the unit focuses entirely on the mechanisms of gross muscle control. If an instructor wants to address sliding filament processes specifically, consider adding a follow-up microscopy exercise that employs glycerin-skinned rabbit psoas muscle for a model.

Students also can be put off by the underlying electronics. Some see the controller plus computer as a "magic, data-generating machine," while others become overly focused on the electronics. Explain precisely what the transducer is doing, so students understand what data are collected, but do not dwell on the data processing. Focus on data interpretation instead.

In the first exercise of this unit, students are asked to determine the threshold voltage that must be applied to the muscle and sciatic nerve to induce muscle contraction and also to determine the voltage needed for maximum contraction. This procedure establishes the dynamic range for subsequent exercises, but occasionally leads to confusion. In an intact animal, differences in muscle contraction strength are the result of different motor units being activated, that is, spatial summation. Different motor units are activated by stimulating more or fewer axons in the spinal cord, *not* by changing the voltage generated within axons. Once the leg is severed from the body, the ability to stimulate just one motor unit directly and not another is lost. Fortunately, a mechanical property of nerves compensates for this problem. Some axons are better insulated by myelin and are buried deeper in the neural sheath than others, so they need a larger voltage stimulus in order to generate action potentials. Stress to students that when they increase the voltage applied to the sciatic nerve, they are stimulating those deeper, more resistant axons to fire, and thereby stimulating additional motor units to contract.

For experiments, many students at this level want to see what Chemical X (usually some formulation of caffeine or alcohol) does, rather than develop a hypothesis. For nonmajor students, these kinds of experiments often are more interesting and instructive than rigorously designed ones. It is up to the instructor to decide whether to permit them. For majors, resist the urge to let them do so. Insist that students conduct hypothesis-driven experiments. If they cannot get started, ask, What physical or chemical change in muscle could modify a contraction? How? Why? Require them to put the rationales for their hypotheses into their experimental proposals, or take time to discuss their projects as a class. Also, do not put the common names for chemicals on the list of available agonists and antagonists; for example, use the term *sodium channel blocker* instead of lidocaine. Only tell students what chemicals they will be using once their experiment has been designed.

More advanced students may discover that the sciatic nerve can be split along the perineural fascial planes to create smaller bundles of axons. Each bundle can be electrically stimulated independently from the other. The procedure is not simple, but it does provide other experimental options for students with higher levels of technical skill.

Recommended Prelab Skills

Students will need basic computer skills (opening, closing, saving files). Knowledge of graphing with a spreadsheet program will speed up data analysis, but is not essential. Preparing the sciatic nerve and gastrocnemius muscle requires no prior dissection experience.

Assessment

If the instructor uses pre- and postlab quizzes routinely as part of formal assessment, he or she should be aware that there is relatively little general content in the Background section, and so there are relatively few options for prelab quiz questions. The first week, quiz questions should focus on general anatomy of the frog and on basic definitions. For the postlab quiz, students should be able to identify the three phases of a normal myogram, and to explain the biological and mechanical events that produce each phase. It is not useful for students to know the exact voltages required to stimulate the muscle and nerve, as these values differ among animals. They should be able to explain what the blade transducer does in general terms. Students should be able to differentiate between spatial and temporal summation. Tetany and fatigue are secondary concepts, so avoid asking questions about them on the post-lab quiz.

Safety and Housekeeping

Collect legs and other body parts and store them frozen prior to disposal. Carcasses that are contaminated with hazardous chemicals may have special disposal requirements. Consult your local ACUC (Animal Care and Use Committee) and your school's safety office to find out what your school's disposal procedures are.

Other Tips

There are three possible methods for administering test chemicals. The simplest is to directly inject solutions into the muscle using a tuberculin syringe. The gastrocnemius of an average adult frog can be injected with 0.1 mL of a chemical in Ringer's solution without causing significant fluid stretch overload. To make injections, insert the needle *parallel* to the muscle fibers, and infuse chemical slowly. The second method is to soak skinned legs for 10 minutes in a solution of Ringer's containing the compound being tested. The disadvantage of this method is that a pretreatment myogram cannot be recorded. The third administration method is to build a bath chamber around the muscle. Using scissors or a fine saw, remove the conical end of a 50 mL polypropylene tube. Pin a prepared frog leg in the dissecting pan, then tie off and cut the Achilles tendon as described. Pass the thread through the open 50 mL tube. Place a ring of modeling clay around the lower end of the tube, then press the end of the tube firmly onto the pan and leg to form a seal. Tie the thread from the muscle to the force transducer, check the tension, then fill the tube with room temperature Ringer's solution. Record a pretreatment myogram, then add the chemical to be tested to the chamber. Allow muscle to soak for 1 to 10 minutes, then record a posttreatment myogram.

To test the effects of temperature on muscle contraction, have students half fill two plastic bags with water and tie the bags shut. Heat or chill the bags to the temperature desired. Record a pretreatment myogram, then place the two bags on either side of the muscle for 10 minutes. Remove the bags, and record a posttreatment myogram.

PREPARATORY NOTES

Quantities listed are for a lab section of 20 students working in groups of three or four.

Week 1: Shared Materials

- 3 grass or leopard frogs, extra large, live
- Gloves, disposable
- 5–6 tuberculin syringes (1 mL)
- Sharps box (for used syringes)

Week 1: Materials at Each Work Station

- Dissecting scissors
- 2 pairs forceps
- Blunt probe
- Plastic dissecting pan filled with pin wax or silicone rubber
- Nondrying modeling clay (to stabilize pan and electrode)
- Dissecting pins
- Spool of sewing thread
- 100 mL squeeze bottle filled with frog Ringer's solution
- Ring stand
- Blade force transducer with ring stand clamp
- Square wave electrode stimulator
- Data acquisition unit with USB computer link

Week 2: Shared Materials

- 8 frogs, extra large, live
- Gloves, disposable
- Box of ~25, 1 mL tuberculin syringes
- Biohazard sharps box

Week 2: Materials at Each Work Station

- Dissecting scissors
- 2 pairs forceps
- Blunt probe
- Plastic dissecting pan
- Nondrying modeling clay (to stabilize pan)
- Dissecting pins
- Spool of sewing thread
- 100 mL squirt bottle filled with frog Ringer's solution
- Ring stand
- Blade force transducer with ring stand clamp
- Square wave electrode stimulator
- Data acquisition unit with USB computer link

Optional Materials for Experiments (make only if requested)

- Modified frog Ringer's solutions
  - ◊ Calcium-free
  - ◊ Plus 20 mM calcium
  - ◊ Plus 50 mM calcium
  - ◊ Potassium-free
- Compounds in Ringer's solutions
  - ◊ 0.01 mg/mL acetylcholine
  - ◊ 0.1 mg/mL nicotine (ACh agonist)
  - ◊ 1 mM tubocurarine (nicotinic ACh receptor antagonist)
  - ◊ 1% lidocaine (neuronal sodium channel blocker)
  - ◊ 5 µg/mL nigericin (potassium channel activator)
  - ◊ 5 mM tetraethylammonium chloride (potassium channel blocker)
  - ◊ 0.1 mg/mL caffeine (potentiates Ca^{+2} release due to signaling)
  - ◊ 5 µM A23187 (Ca^{+2} ionophore; triggers massive release)
  - ◊ Calcium-free Ringer's, with 10 mM EDTA plus 10 mM strontium chloride (blocks calcium channels)
  - ◊ 1% ethanol (raises membrane lipid fluidity)
  - ◊ 2% carbaryl or other insecticide (acetylcholinesterase inhibitor)

Sources of Materials

Each group of three to four students needs a workstation consisting of an electrode stimulator, force transducer, and data acquisition unit, which can be linked to a dedicated computer, or to the students' laptops (if available). In the author's course, a workstation consists of a Phipps and Bird 611 square wave stimulator with a two-wire needle electrode to stimulate the muscle and nerve. Muscle contractile force is recorded with an FT-2 blade force transducer connected to an ETH-200 bridge amplifier (both from CB Sciences, Inc.). The amplified signals are processed with a Biopac Systems data acquisition unit consisting of a UIM 100 interface module and an MP100A data acquisition module. Processed data are sent to a student's laptop computer via a USB1W I/O controller and displayed using Biopac's Acqknowledge PRO software. Both software and template files are installed from a CD or flash drive as students arrive in the lab. Any number of other combinations of equipment and vendors work equally well; check with members of your own department who have experience with electrophysiological recording for further assistance.

For most students the data acquisition software is not intuitive. They will require some type of general use guidelines until they become familiar with the interface. Most vendors make software guides available for students as free PDFs. These can be supplemented with premade blank template files, which let students begin collecting data more quickly than if they had to learn the entire interface first.

Maintaining Organisms

Housing and use of frogs for teaching is subject to federal regulation. Consult your institution's Animal Care and Use Committee for specific guidelines on documentation procedures and requirements. Typically, frogs can be housed in group tanks, but must be kept in a separate area from where students perform these exercises. It is best not to maintain a local colony, but rather to order frogs and schedule them to arrive just before the unit begins. This practice reduces animal losses due to "redleg," an infection caused by the motile gram negative rod bacterium, *Aeromonas hydrophila*. The bacteria release a neurotoxin that causes flaccid muscle paralysis and death. Frogs may arrive from commercial vendors with an obvious infection, or become symptomatic shortly after arrival. Inspect all animals when they arrive and daily thereafter. Segregate any that appear sluggish, resist moving if prodded repeatedly, or have spreading patches of dark pink skin on their bellies, feet, and ventral hind legs. If a suspect animal still can hop from the palm of a cupped hand, use it within a day. If a frog cannot hop out of a cupped hand, the neural damage is too extensive; euthanize the animal immediately according to your institution's procedures.

Harvesting the Legs

Again, local institutional regulations dictate allowed methods for euthanasia. Typically, frogs are anesthetized by packing them on ice for 30 minutes or by soaking them in 0.5% tricaine methane sulfonate (MS-222) in bicarbonate-buffered water just until the animal stops withdrawing its foot when the toe is pinched. Narcotized animals are then double-pithed with a dissecting needle. Using heavy scissors or pruning shears, cut the legs off at the point where they attach to the torso, then immediately place them on ice. Ideally legs are removed just prior to use, but if kept on ice, they can be held for two to three hours before muscular performance degrades.

Solutions, Reagents, Equipment

Frog Ringer's Solution (see Table 9.1)

Table 9.1

Basic frog Ringer's solution

Salt	Mol. Wt.	Final Conc.	For 1 Liter
HEPES (buffer)	238	5 mM	1.19 g
NaCl	58	125 mM	7.25 g
KCl	75	3 mM	0.225 g
$CaCl_2$	147	1.8 mM	0.265 g
Glucose	180	10 mM	1.80 g
$MgCl_2$	203	1.6 mM	0.324 g

Place 900 mL of water in a flask, add all salts listed in the table above, and stir to dissolve. Bring volume to ~990 mL with water. Use 1 M NaOH to adjust pH to 7.3, then bring to final volume of 1 L. Store refrigerated. Warm to room temperature prior to use.

For modified Ringer's, substitute the quantities as shown in Table 9.2.

Table 9.2

Modifications to Ringer's solution

Modification	For 1 Liter
Calcium-free	Omit calcium chloride
Plus 20 mM calcium	Use 2.9 g $CaCl_2$
Plus 50 mM calcium	Use 7.4 g $CaCl_2$
Potassium-free	Omit potassium chloride

Agonists and Antagonists

Table 9.3

Preparing neuromuscular agonists and antagonists

Solution	Action	Preparation
100 μM A23187	Raises intracellular Ca^{+2}	Dissolve 5.0 mg in 10 mL ethanol. Dilute to 100 mL in Ringer's
10 μg/mL acetylcholine	Natural ligand for junctional receptors	Dissolve 1.0 mg in 100 mL Ringer's
200 mM caffeine	Potentiates effects of Ca^{+2} release	Dissolve 1.9 g in 100 mL Ringer's w/ gentle heat
20 mM EDTA	Reduces available Ca^{+2}	Dilute 4 mL of 500 mM stock to 100 mL in Ringer's
1 mM lanthanum chloride	Blocks Ca^{+2} channels	Dissolve 12 mg in 100 mL Ringer's
25 mM lidocaine	Na+ channel antagonist	Dissolve 0.6 g in 100 mL Ringer's
1 mg/mL nicotine	ACh agonist	Dissolve 0.1 g in 100 mL Ringer's
25 μg/mL nigericin	K+ channel activator	Dissolve 25 mg in 10 mL ethanol. Dilute to 100 mL in Ringer's
20 mM strontium chloride	Blocks Ca^{+2} channels	Dissolve 0.5 g in 100 mL Ringer's
100 mM tetraethyl-ammonium chloride	K+ channel blocker	Dissolve 1.7 g in 100 mL Ringer's
1 mM tubocurarine	ACh receptor antagonist	Dissolve 68 mg in 100 mL Ringer's

Store in 2–5 L aliquots at -20°C until ready for use.

UNIT 10

Homeostasis and Heart Rate

Student Pages*

BACKGROUND

You probably have already learned before that **physiology** is the study of how an individual multicellular organism senses and adapts to changes in its external and internal environments. Various organs and tissues will modify their functions to maintain **homeostasis**, that is, the optimum energy stores, temperature, ion concentrations, water levels, and other conditions the body depends on for survival.

To maintain homeostasis, an organism must have predefined **set points** and at least one mechanism for changing the body to reach those set points. For example, your own body functions best when intracellular fluids have a salt concentration of 290–310 mOsm; this is your body's set point. As you sweat, your body water content falls and overall osmolarity of body fluids rises. Hyperosmolar blood can strain the heart, damage blood cells, and interfere with neurotransmission and muscular control (we experience this last effect as heat cramps). To return body fluids back to normal, the kidneys will excrete less water and more sodium into urine. At the same time, your brain sends signals that you are "thirsty" and tells skeletal muscles to carry you to a water supply and drink some of it. In short, several systems have cooperated to return the body to its optimum set point for body fluid osmolarity.

*Teacher Pages begin on page 349.

There are thousands of homeostatic set points for an organism. Some are quite stable, while others can be modified temporarily if external or internal environmental conditions change such that the existing set point is not appropriate anymore. For example, your body maintains a core temperature of ~37°C. If bacteria and viruses invade your body, the immune system usually destroys them quickly, and body temperature does not change. If routine defenses do not work fast enough and a bacterium or virus begins to replicate, your body can change its normal temperature set point, making your body as inhospitable as possible for the invader. When conditions are appropriate (the pathogen has been destroyed), the body returns to its original set point again.

Maintaining homeostasis is not the sole responsibility of the brain, the endocrine system, or any other single organ or organ system. It depends upon *coordinated* inputs and responses from every system in the body. Any external stimulus that affects one system will almost always affect others. A good visual analogy for the tight integration of different physiological systems is a spider's web—if you pull on just one strand, the entire web will move.

Heart Rate Will Be Our Model for Studying Homeostasis

Most animals above a certain size have hearts or some other pumping organs that circulate blood or body fluids. The absolute rate at which their hearts beat and the volume that they pump vary depending on size, body organization, and many other factors. However, all hearts share three features.

- They are autorhythmic. Unlike skeletal muscle, most cardiac muscle cells continue to pulse rhythmically even if they are removed from their parent organisms.
- There is a resting set point that is typical for that species. For example, normal human heart rate is 60–70 beats per minute at rest. That is true for every adult worldwide.
- Heart rate can change considerably from the resting set point as an organism adapts to changes in external or internal conditions.

Because hearts (and other systems) are so similar among organisms, a great deal can be learned about homeostasis by studying less structurally complex animal models.

For this unit, you will be using aquatic blackworms *(Lumbriculus variegatus)* (see Figure 10.1). Aquatic blackworms are members of the Phylum Annelida, which includes segmented animals like leeches and earthworms. As their common name implies, they live in slow-moving or stagnant freshwater. They are less than 5 cm long, and their skin is so thin you can see most of their organs with a light microscope.

Figure 10.1.

Internal anatomy of a blackworm

Sensory nerves on the surface and in the body are stimulated by external and internal conditions. They relay information to the brain as impulses called action potentials. Responses by the organs are controlled through peripheral nerves that branch out from the nerve cord. These peripheral nerves communicate with the organs using neurotransmitters such as acetylcholine.

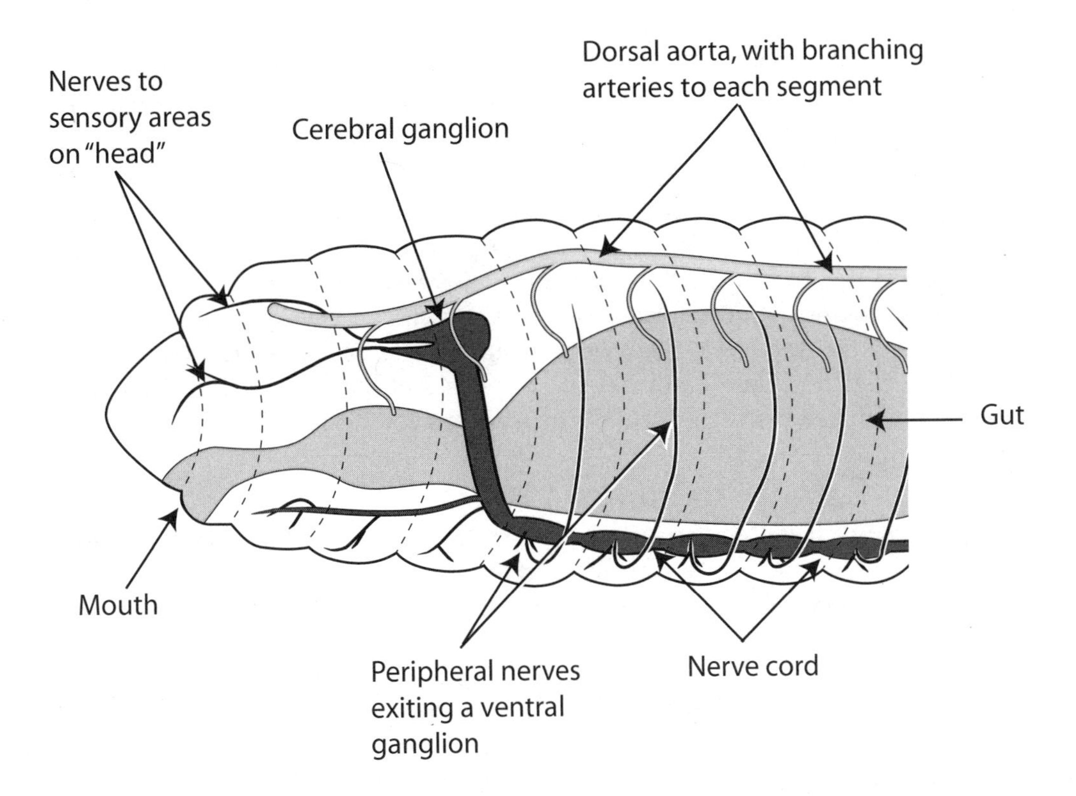

Despite their small size, they are physiologically complex organisms. A blackworm has a pulsating dorsal aorta that runs the entire length of its body and acts as a heart. It circulates both blood with nutrients from the digestive tract and oxygen that has been absorbed across the blackworm's skin (which doubles as a lung) to all the tissues of its body. In an intact worm, contractile waves pass from tail to head in a complex, ever-changing pattern. Some waves move the entire length of the body, while others stop partway toward the head, as some of the blood passes into lateral branch vessels.

Like humans, homeostasis in blackworms is controlled by a combination of signals from

- their brains and central nervous systems,
- circulating hormones from different sources in their bodies, and
- other chemical stimuli from inside their own bodies or from the outside environment.

If a blackworm's head and tail are removed, however, the nervous inputs that make the contractile waves so complex disappear. Instead, the dorsal aorta becomes autorhythmic, and the contractile waves move along smoothly and at a steady rate that can be counted easily.

Goals for This Unit

During the first week of this unit, your instructor will show you the dorsal aorta of a blackworm and how to prepare center segments of worms for observation. Once you have learned to identify the heart and prepare the worms, you will complete an exercise looking at how neurotransmitters and other chemicals affect heart rate. At the end of the first week, your work group will design an experiment of your own that tests one or more physiologic responses of blackworms. In the second week of the unit, you will perform the experiment you designed.

General Safety Precautions

Blackworms are harmless, but they are very fragile. Do not use forceps or narrow bore pipets to capture or transfer them; use only wide-bore pipets. Do not be squeamish about cutting the worms apart, since you are not killing them. Blackworms do not reproduce sexually, but instead asexually through fragmentation. The head and tail pieces you return to the stock tank will grow into new worms within a couple of weeks. The missing structures simply regrow from stem cells in their body walls. If you look closely at a smaller worm, it may be darkly pigmented on one end only; that is the original fragment from which the newer, lighter-colored portions grew.

You will be using a large number of glass slides and coverslips. Dispose of them in the designated glass waste containers, not in general trash. If you accidentally cut yourself on any broken glass, wash the cut thoroughly and tell your instructor.

Some of the chemical solutions you will use are toxic. Check each chemical container before you use it to determine the proper handling procedure. If you spill a toxic compound, inform everyone near you immediately. Do not ignore any chemical spills.

UNIT EXERCISE

Effects of Neurotransmitters and Hormones on Heart Rate

For this exercise, each work group will be working with a different neurotransmitter or neuroactive compound. Your group must determine the average heart rate for three blackworms at baseline, and again after treatment with your assigned neurotransmitter or compound. You will share your results with the rest of the lab and discuss them as a group. Compounds that you could be assigned might include

- acetylcholine (a neurotransmitter found in most organisms),
- pseudoephedrine (an analog of the hormone epinephrine), or
- "unknown" compounds and drugs.

As part of this exercise, you also will learn some basic data transformation methods used in physiological studies.

Procedure

1. Rinse three depression slides thoroughly with distilled water, dry them, then set them aside for a moment. Also wash three glass coverslips. Be careful, as the coverslips break easily.
2. Obtain a 60 mm petri dish and five 35 mm dishes. Label the five smaller dishes "A," "B," "C," "D," and "E." Finally, fill all six dishes with distilled water.
3. Take the 60 mm dish to the stock tank. Remove five blackworms from the stock tank, and place them in the dish. Try to get five worms that are a uniform overall body size. Avoid using any worm fragments that are in the process of regenerating. Blackworms are fragile animals, so you cannot pick them up from their culture container with forceps. Instead you must use a large-bore pipet to suck them up.
4. Back at your bench, remove one worm from your larger dish along with a little water. Place the worm on the clean glass slide. Working quickly, use the edge of a coverslip like a knife to cut off the anterior and posterior thirds of the worm. Place the front and rear portions of the animal back into the dish.
5. Use a pipet to gently rinse the middle third of the worm off the slide into one of the smaller (35 mm) petri dishes, "A."
6. Repeat the previous steps until you have prepared five middle segments.
7. Empty the front and back ends of the worms into the new stock tank at the front of the lab.

Next you will determine the baseline heart rate for three of the prepared segments (the other two are backups in case one of the first three animal segments ruptures).

8. Add one drop of water from the holding dish to the well of a glass depression slide. Use a pipet to gently transfer Animal "A" from its holding dish to the well in the slide.
9. Cover the well with a glass coverslip.
10. At low power, look for the reddish dorsal aorta. Watch one spot on the vessel; you will see contractions moving along the segment.
11. Concentrate on a single selected segment as close to the center of the segment as possible. While your partner keeps time, count the number of pulsations that move past a particular point in 20 seconds. Wait about a minute, then count for another 20 seconds. Wait again, then count the number of heart beats in 20 seconds for a third time. Record your results in Table 10.1.
12. Remove the slide from the microscope, but leave the worm on the slide. Set it aside so it will not be hit or knocked off the table.
13. Use the same method to determine basal heart rate for Animals "B" and "C." If a segment ruptures and dies during the procedure, replace it with one of the two backup segments. Be sure to obtain a basal heart rate for any replacement animal before you add your test substance.
14. Record your baseline results in Table 10.1. After you obtain baseline heart rates, you are ready to test your assigned neurotransmitter or unknown compound.
15. GENTLY lift the coverslip off of the first depression slide containing worm segment "A." Using a transfer pipet, remove as much plain water as possible from the well. Replace the plain water with water plus the drug you were assigned to test. Do the same for segments "B" and "C."
16. Allow the animals to soak for five minutes.
17. Starting with segment "A," count contractions of the dorsal aorta three times again. Repeat the process for segments "B" and "C" as well. Record your results in Table 10.1.
18. When you have completed Table 10.1, and are sure you no longer need the worm segments, empty them and the drug solutions into the disposal container provided. Thoroughly wash the depression slides you used and place them on the side benches to dry.

Results

Table 10.1

Blackworm heart rates before and after drug treatment

Animal	Trial #	Pulses in 20 sec. w/ no drug	Pulses in 20 sec., after adding _______
A	1		
	2		
	3		
B	1		
	2		
	3		
C	1		
	2		
	3		

Analysis: Normalizing Heart Rates by Calculating Percent Change

Under normal conditions, most animals of the same species will have a slightly different heart rate. This complicates comparisons among animals and treatment groups. To simplify the comparisons, your raw counts in Table 10.1 need to be expressed as the percent change from the normal basal rate for the *same* animal. This step is called **normalizing** the data. It is simply a mathematical way to account for any differences in basal heart rate among animals.

1. Once you have entered all the required data into Table 10.1, convert your observations into beats per minute by adding together the three 20-second baseline trials for each worm. Record the heart rates for each animal in Table 10.2. Do the same for each of the three worms after drug treatment.

Table 10.2

Normalized heart rates

Animal	Normal Heart Rate (bpm)	Rate After Treatment (bpm)	% Change in Heart Rate After Adding _______
#1			
#2			
#3			
Average Change in Heart Rate is:			

2. Use the equation in Figure 10.2 to calculate percent change in heart rate:

Figure 10.2

The equation for normalizing heart rate

$$\left[\frac{(\text{Rate with Drug} - \text{Basal Rate})}{\text{Basal Rate}}\right] * 100\%$$

For example, suppose one animal had a basal heart rate of 30 bpm (beats per minute) and a rate of 25 bpm after treatment with Drug X. The percent change in heart rate would be calculated as

$$[(25 \text{ bpm} - 30 \text{ bpm})/30 \text{ bpm}] \times 100\% = -16.7\%$$

(A negative sign indicates that heart rate dropped 16.7% after treatment with Drug X.)

3. Calculate the heart rates and percent change in heart rate for your drug treatment, then record them in Table 10.2. To get the average percent change in heart rate, add the results from the three replicate worms together, then divide by 3.
4. Report your final average changes in heart rate to the class using the table your instructor has placed on the board. Once everyone has recorded their results, copy the class data into Table 10.3.

Table 10.3

Summary of class data for heart rates

Drug Tested	Type of Drug (From Instructor)	% Change in Heart Rate

Unit Summary Questions

1. Did all three of your individual animals respond to the drug you tested in exactly the same way? If not, what might explain the differences between them? How might this be important to human biology?
2. In this exercise, you used various chemicals to alter heart rate artificially. How would you summarize the results obtained by the entire class? Did you see any general trends for particular drugs or neurotransmitters? If so, what were they?
3. This exercise looked at chemical effects on heart rate. What types of natural external environmental conditions might cause an organism's heart rate to increase? To decrease? How would you test these possibilities?
4. One example of homeostasis described in the Background section of this unit was of body temperature. The description mentioned something you did not explore today: how body functions return to their original homeostatic set points after a stimulus or stress has passed. There are two possibilities: (1) it is a passive process that happens just because the original stimulus is not present anymore, or (2) it is an active process, which requires the organism to make specific physiological changes to return to normal.

   Imagine that you conducted a further step in the exercise today, and rinsed away the drug you tested on the worm segments. Within 20 minutes, you observe that the heart rate returns to baseline. Based on this evidence, is the return to baseline a passive process or an active process? What is your reasoning for this conclusion? If you cannot make a conclusion either way, what additional experiments or evidence would you need to conduct in order to decide one way or the other?
5. When scientists measure a physiological response, they report their results as a change from baseline, just as you did. Why is it better to report percent change, rather than the absolute change (e.g., +6 bpm, or -13 bpm)? You also used multiple replicate samples. Why is it better to use an average of several replicates rather than just look at data from a single individual animal?

Your Own Experiment

Organisms use many internal chemical and electrical signals to control their organs. These signals do not change at random, though. Some kind of external or internal environmental stimulus triggers that change in chemical or electrical signals.

For example, a drop in our core body temperature is an internal stimulus, to which our body must respond. First, specialized thermosensory cells that are connected to neurons (the cells that make up nerves) send electrical signals along the neurons to the brain and spinal cord. These structures in turn send electrical signals to various target tissues (skin, muscle, and organs) telling the tissues how to respond to the temperature change (with goose bumps or shivering or by shunting blood to the colder body parts). At the same time, hormones are released from our pituitary gland that stimulate more heat producing metabolic reactions in our body. This continuous cycle of stimulus/response is what allows us to adapt very quickly to the smallest changes in our environment. Humans are not alone in this ability either. Every organism has the ability to adjust its body functions and thereby adapt to changes in its environment.

It is time for you to design an experiment that explores how internal or external stimuli affect the physiology of *Lumbriculus variegatus*. Think about how a particular environmental condition will affect the body of a blackworm. Would you expect the heart rate to change in order to adapt to that stimulus? In what way would it change? How quickly would heart rate change? How long would it take for heart rate to return to normal?

Although the exercises in this unit focused on heart rate, keep in mind that all of the body's systems must change and adapt together to maintain homeostasis. If you identify some other behavior or response you want to test, ask your instructor if you can test that in your experiment.

Remember to frame your hypothesis as an if/then statement. If you try to test too many things at once, you may end up with so many possible alternative explanations that your experiment becomes uninterpretable. It is always better to test several single, clear questions in a row than to try and answer every question you have in a single massive experiment.

You can work in small groups of two to four or the entire class can design and carry out a single larger experiment. As before, each student working in a group must participate in carrying out the assays and will be responsible for understanding what was done.

Most of the reagents in the list below are available routinely. However, if you'd like to try something different, ask your instructor if the necessary materials are available.

Materials Available

- Water at different salinities
- Water at low versus high osmolarity (clear stream water has low osmolarity, while stagnant water typically has higher osmolarity)
- Water with different dissolved ions: Na^+, K^+, Ca^{+2}, HCO_3^-. Water baths at different temperatures: ice bucket (0°C), refrigerator (4°C), room temperature (22°C), and warm water (30°C).
- Insecticides: Malathion, Sevin, Diazinon (they interfere with neurotransmission and are common contaminants in agricultural runoff)
- Waste water from agricultural (high nitrogen) or industrial (contaminated with organic solvent or heavy metals) sources.
- Pond water that has been bubbled with different gases (N_2, O_2, CO_2)
- Water at different pHs (to simulate acid, alkaline runoff)

UNIT 10

Homeostasis and Heart Rate

Teacher Pages

INSTRUCTORS' NOTES

Background and Key Concepts

The ultimate function of most physiological processes is to help organisms maintain overall homeostasis, while letting them adapt to changes in internal and external environments. For students to make sense of the complex relationships and feedback loops that connect various systems, they need to understand the concept of homeostatic set points. Most students can provide common examples of human homeostatic set points, like normal body temperature and normal resting heart rate. However, they usually do not realize that many other vital set points must be maintained. Some examples of equally important but less obvious set points they probably have not considered before are

- calcium concentration of interstitial fluids (affects overall muscle tone and heart function),
- baseline skeletal muscle tone (determines posture, venous return, and cardiac output), and
- blood osmolarity and protein content (controls rates of capillary export, import, organ function, and cellular metabolism).

Students usually come into this lab with many misconceptions about general processes of physiological regulation and homeostasis. Often they think of physiological processes as working in isolation from each other, with each individual set point being maintained relatively independently. Students also tend to focus on proximal regulatory processes and lose sight of the ultimate function of physiological adaptation, which is to keep the whole organism alive. Finally, few students can explain the general mechanisms by which physiological adaptation is controlled. This lab unit was designed to address all of these shortcomings in student understanding.

Three types of chemical signals control most physiological processes: neurotransmitters, circulating hormones, and small bioactive molecules (which can come from the external environment or from within the organism itself). Gas intake by the lungs is a good example of how all three types of signals are needed to regulate a physiological process. A human's baseline respiration rate (~16 breaths/minute) is set by a respiratory center located in the brainstem. The center transmits action potentials to the motor neurons in the phrenic nerve that innervate the muscles of the diaphragm. These neurons release pulses of neurotransmitter about 16 times a minute, triggering regular contractions that inflate the lungs. This baseline respiration rate rises when the hormone epinephrine is released from the adrenal gland during a fight-or-flight stress response. Epinephrine increases gas intake by dilating bronchial airways and by stimulating the respiratory muscles to contract more forcefully. The baseline respiratory rate also rises when excess carbon dioxide from the air or working tissues dissolves in the interstitial fluids and lowers their pH. Lowered pH around the brainstem stimulates the respiratory center to transmit action potentials more frequently. As the same time, acidification of the skeletal myocytes in the diaphragm can increase the contractile force they generate in response to each arriving set of action potentials.

To prepare students for this unit, stress that physiological processes do not happen in isolation from one another. One process may have large effects on many others. Also, the ultimate function of physiological processes and adaptation is not to maintain individual set points. Individual processes may be pushed far away from their normal homeostatic setpoints, if that helps keep the whole organism alive.

General Teaching Strategy and Common Problems

This lab was developed for a freshman nonmajors course, but may be tailored to fit a variety of audiences. Many students at the author's institution take this particular course as part of their premedical requirements, so in addition to the neurotransmitter acetylcholine and the hormone epinephrine, they test three pharmacologically active drugs: ethanol, lidocaine, and caffeine. Other instructors may choose to emphasize environmental contaminants, or physiologically relevant ions, as a way of engaging their particular students.

The lab procedures for this exercise are straightforward. Students may hesitate to cut their worms initially; inform them that partial cuts made with insufficient pressure are more likely to kill the animal than a firm, quick cut. If a worm is very large, dark pigment can obscure the students' views of the dorsal aorta. Encourage students to search carefully for a point where they can see the aortic pulsations clearly. If necessary, students can gently move the top coverslip, so the worm segment rolls over in the depression slide well and presents a different view.

Students struggle to understand why they must normalize the changes in contraction rates to percentage of baseline rates. Typically it does not help to explain the effects of natural variation in advance of starting the lab. Instead, allow students to collect baseline heart rates of their three test animals, then ask how much variation is present. Usually students can see for themselves that there is considerable variation that must be accounted for. This provides a natural segue into normalization of data.

In some trials, students may see relatively small (~5%) changes in heart rate. Often students dismiss these apparently small changes as meaningless. Remind them that in biological systems, small changes can have very large effects, particularly if they persist for a long time. For example, when a doctor measures a patient's blood pressure, a reading of 110/80 would be considered normal, while 110/90 would be cause for concern. The diastolic pressure has only increased by 12% (10 mm Hg/80 mm Hg), but over several years this change would be sufficient to stress the heart and increase the risk of heart disease.

Recommended Prelab Skills

Students should know how to operate a compound microscope. They can use a spreadsheet program to automate the routine calculations, but it is not required.

Assessment

If pre- and postlab quizzes are used as part of formal assessment, the prelab quiz should focus mainly on general principles of physiology such as the definition and purpose of a set point. Students must determine the pharmacological effects of the compounds being tested, so do not expect students to know how specific compounds will affect normal heart rate until after they have completed the exercise. For the postlab quiz, ask students to normalize a small data set, explain the purpose of normalization, and give their rationales for how they performed the normalization. Students also should be able to name or describe the three general factors or mechanisms that control most physiological processes. Students frequently see considerable variation in the drug effects between animals. On the postlab quiz ask students to give some possible explanations for these differences.

Safety and Housekeeping

Students will generate considerable glass waste in this unit. Make sure to have dedicated glass waste boxes available for disposal of slides and coverslips. Instructors are strongly discouraged from reusing coverslips and slides, for two reasons. First, students may not clean them thoroughly, leading to chemical cross-contamination. Second, students could be cut very seriously by a broken coverslip; it is more cost-effective to purchase a case of coverslips than send a student to the hospital for stitches. Remind students that the glass depression slides are *not* disposable.

Other Tips

- Blackworms can also be drawn up into large-bore capillary tubes for observation, rather than placing them onto depression slides. The primary advantage is that the worms are less able to move around, so are easier to observe. To provide suction, use a P-200 micropipet with a disposable tip attached or a tuberculin syringe with the needle removed. Flame polishing the tips of the capillary tubes reduces the likelihood of cutting an animal.
- Blackworms also can be viewed using a dissecting microscope. Place a covered petri dish filled with water between the depression slide and the substage light to protect the worms from the heat of the bulb.
- Many instructors conduct a similar exercise to this, but use water fleas (*Daphnia magna*) instead. One common limitation is that they have a high basal heart rate of 180–220 bpm, which is difficult for students to count. In contrast the blackworm's dorsal aorta pulses at a rate from 30 to 60 bpm under baseline conditions, making it much easier for students to count.

PREPARATORY NOTES

Quantities listed are for a lab section of 20 students working in groups of three or four.

Week 1: Shared Materials

- Stock culture of 50 to 60 blackworms
- Acetylcholine solution, 50 mL
- Pseudoephedrine solution, 50 mL
- Ethanol solution, 50 mL
- Lidocaine, 50 mL
- Caffeine, 50 mL

Week 1: Materials at Each Work Station

- 10 plastic petri dishes (60 mm)
- 50 plastic petri dishes (35 mm)
- Labeling tape or permanent marker
- Glass microscope depression slides, 3 per work group
- Box of 18 mm or 22 mm square #1.5 coverslips
- 6 to 8 plastic transfer pipets
- Hand tally counter (optional)
- Stopwatch (optional)

Week 2: Shared Materials

- Blackworm stock culture (50 to 60 worms)

Week 2: Materials at Each Work Station

- 10 plastic petri dishes (60 mm)
- 50 plastic petri dishes (35 mm)
- Labeling tape or permanent marker
- Glass microscope depression slides, 3 per work group
- Box of 18 mm or 22 mm square #1.5 coverslips
- 6–8 plastic transfer pipets
- Hand tally counter (optional)
- Stopwatch (optional)
- Test reagents requested by students (only make what is requested)

Maintaining Organisms

Blackworms and food pellets can be purchased directly from educational suppliers. Many aquarium supply stores also sell them as live food. Blackworms can be readily cultured in a 1 gal. aquarium or flat-bottomed pan. Add three to four inches of dechlorinated, aged tap water (spring water also works) and enough shredded brown paper towels to form a bottom substrate about 1 in. thick. Let the tank sit one day, then add a culture of blackworms and one commercial food pellet. When the tank water turns cloudy, or the worms begin floating with their anal ends parallel to the water's surface, exchange half the water for fresh, aged tap water or spring water. Aerating the container with an airstone reduces stagnation, but is not essential.

To expand a culture, prepare one or two new tanks. Stir the stock tank roughly to fragment some of the worms (they reproduce entirely by fragmentation and regeneration). Before the worms settle, divide the culture fairly evenly between the stock tank and new tanks. Top off the stock tank with water if necessary, and place one food pellet in each tank.

Solutions, Reagents, Equipment

Acetylcholine, 0.1 mg/ml

Dissolve 5 mg acetylcholine chloride in 50 mL of water. Store refrigerated. Typically slows heart rate.

Pseudoephedrine or Phenylephrine HCl, ~1 mM

Place one 30 mg tablet in 150 mL of water at room temperature, and allow the solids to disperse. Stir for 10 minutes to dissolve the compound. Filter suspension through a coffee filter or Whatman paper to remove undissolved cellulose and cornstarch binders. Store solution refrigerated. Accelerates heart rate. (*Note:* Do *not* substitute epinephrine for this experimental substance. It is an EPA-regulated, P-listed waste that must be disposed of as hazardous material.)

Lidocaine, 1 mM

Dissolve 0.01 g powder in 50 mL of water. Store refrigerated. Blocks sodium channels, slowing neurotransmission and heart rate.

Caffeine, 5 mM

Dissolve 0.05 g of powder in 50 mL prewarmed distilled water. Continue to warm until the caffeine goes into solution. Store refrigerated. Potentiates release of calcium from intracellular storage sites in response to action potentials, so usually raises heart rate.

Ethanol, 0.02%

Mix 10 μL of 95% ethanol with 45 mL distilled water. Store at room temperature. Changes membrane fluidity. Effects on heart rate are quite variable.

Test Solutions for Week 2

Only make these solutions if students request them for experiments.

- Different salinities: Prepare 0.6% NaCl (6 g/L) in distilled water. Dilute with water to obtain lower salinities. Worms should be able to survive in all solutions, although some expire when exposed to undiluted 0.6% NaCl.
- Low versus high osmolarity: Collect stagnant pond water locally and aerate for one day. Compare to aged distilled or spring water.
- Different dissolved ions: Prepare solutions containing 1, 10, or 100 mg/L of the selected ion. For cations, try to use chloride salts; for anions, try to use sodium salts.
- Insecticides: Purchase stocks from a garden center; check label for concentration of active ingredient. Dilute to *final* concentrations of 0.02% or 0.2% active ingredient in water. Most are lethal at 0.1% solution; encourage students to test lower concentrations.

- Agricultural wastewater: Prepare liquid plant food in distilled water according to package directions. Make 1:10 and 1:100 dilutions in distilled water. Generally, these have little effect on heart rate but will stimulate overall crawling and escape behavior.
- Industrial wastewater: To limit the potential risks to students, simulate industrial wastewater using a solution containing three metals: 2 mg/liter copper chloride, 0.5 mg/liter nickel chloride, and 0.2 mg/liter zinc chloride.
- Hydrocarbon-contaminated water: Shake together 1 mL xylene and 49 mL water. Allow to separate, then dilute aqueous phase 1:10 for use in the lab.
- High, low pH: Prepare 10 mM sodium phosphate monobasic and dibasic solutions. To make the acidic buffer, mix 9 parts monobasic with 1 part dibasic solution, then adjust pH to 5.0 by adding more mono- or dibasic solution. To make the basic buffer, mix 9 parts dibasic with 1 part monobasic solution, then adjust pH to 8.0 the same way as for acid buffer. Prepare acidic and basic water by diluting 1 part of buffer at appropriate pH into 9 parts distilled water (final concentration of 1 mM phosphate).

UNIT 11

Metabolism and Oxygen Consumption

Student Pages*

BACKGROUND

You know that all organisms must respond and adapt to environmental changes. But what is the source of the energy used for these functions? **Autotrophs** (plants and other organisms that manufacture food from inorganic precursors) capture sunlight or chemical energy in glucose and similar high-energy molecules via photosynthesis or other enzyme-catalyzed reactions. In contrast, **heterotrophs** obtain high-energy molecules by consuming autotrophs or other heterotrophs. Regardless of the origin of these molecules, eukaryotic autotrophs and heterotrophs use mitochondrial enzymes to break them down into lower-energy ones and to capture the energy released during the conversion in a common "energy currency": ATP. We often use the term **metabolism t**o describe the overall process of energy capture, conversion, and transfer within cells or an organism. The term **metabolic rate** refers to the actual amount of energy being expended by all physiologic processes at any given time in an organism.

One way to describe a metabolic process is to measure energy capture and release directly. We could measure ATP synthesis or consumption by biological reactions, but these assays can be extremely difficult to perform for a single reaction, let alone for whole cells or a complete organism. Fortunately, oxygen consumption and carbon dioxide production are excellent *indirect* indicators of energy production by an organism. To understand why, it is helpful to review the major reactions of metabolism (see Figure 11.1, p. 358).

*Teacher Pages begin on page 365.

Briefly, as an organism converts glucose to energy, molecules of carbon dioxide (CO_2) are released. These reactions also release high-energy electrons that are captured by two molecules, NAD+ and FAD, which transfer them to the **electron transport system** in mitochondria. As the electrons move between molecules in this system, they give up redox energy that is used to convert low-energy ADP into high-energy ATP. Electrons are removed by transferring them to molecular oxygen (O_2), which is obtained from the environment via gas exchange organs. Adding electrons converts the oxygen to water, which then joins the body fluids or is excreted as necessary. Since oxygen is needed for the mitochondria to produce the ATP that is consumed by metabolic processes, greater oxygen consumption indicates greater energy production and use. These processes are summarized in Figure 11.1.

Figure 11.1

Overview of the metabolic reactions in mitochondria

Glucose is hydrolyzed to pyruvate, then acetate, and finally carbon dioxide. Electrons that are released travel between molecules in the electron transport system, releasing energy used to generate ATP. The electrons ultimately are donated to oxygen, forming water. Note that there are multiple reactions in the citrate cycle that release NADH or CO_2; they have been listed together for simplicity.

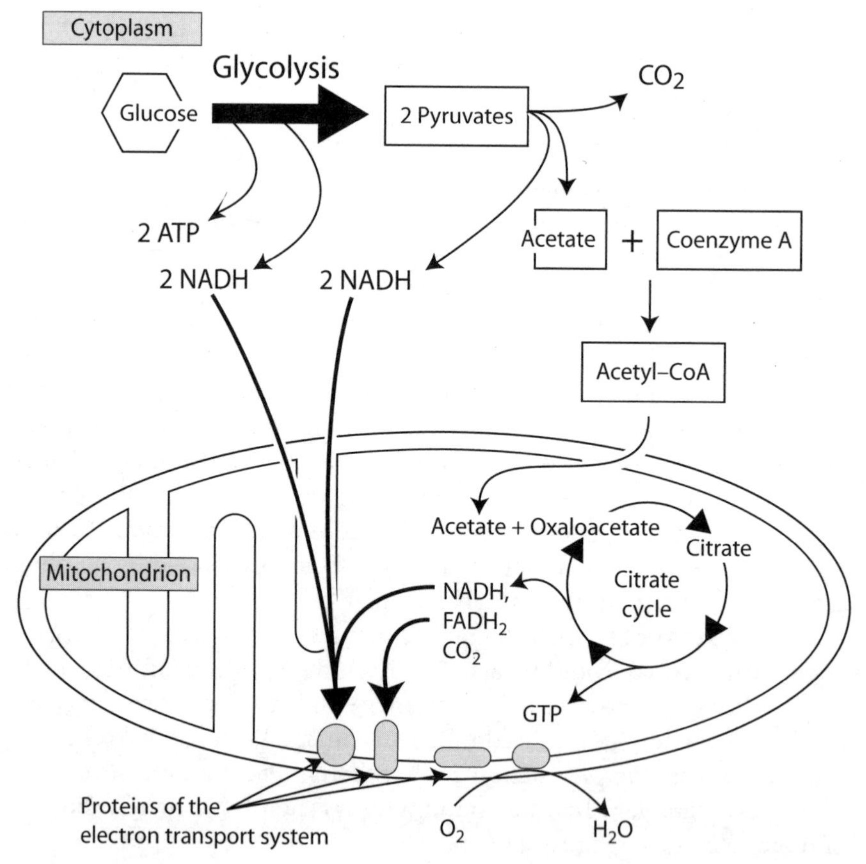

Goals for This Unit

First, you will conduct an experiment to determine if there is a relationship between environmental temperature and rate of metabolism in crayfish (*Orconectes* sp.). You will also look at whether there is any effect of body size on **specific metabolic rate**, which is the amount of oxygen consumed per gram of body mass. After completing the exercise, you will design an experiment to test the effects of environment, exercise, or other variables on metabolic rate.

General Safety Precautions

Crayfish might pinch you, but they are not dangerous to handle otherwise. No hazardous chemicals will be used in this unit. Wipe up all water that gets spilled onto the floor, so that no one slips and falls.

UNIT EXERCISE

Effects of Body Temperature on Metabolic Rate of Crayfish

Background

The rate of metabolism is not constant for an organism. Any internal process that requires that organism to expend energy is going to raise its overall metabolic activity. This in turn causes the animal's oxygen consumption to rise. Some processes obviously require energy, like movement or reproduction. Yet even changes in the external environment can affect metabolic activity. The question you must answer today is, What will those effects be?

Your model organism for today is a crayfish. Crayfish are ectotherms; unlike mammals, they do not use internal metabolic processes to maintain a constant body temperature. You will work in groups of three or four to measure oxygen consumption by crayfish of two different sizes at 22°C (room temperature) and 32°C, using a dissolved oxygen meter. You will use your observations to calculate the specific metabolic rate of crayfish, and determine whether external temperature affects it. Before you begin, answer these questions.

- At which temperature do you predict the crayfish will use more oxygen? Why?
- Which size of crayfish will use more oxygen per gram of body mass, the larger ones or the smaller ones? Why?
- Which crayfish will use the most *total* oxygen, a small crayfish or a large crayfish? Is this the same animal that you predict will use the most oxygen per gram of body mass? Why or why not?

(*Caution:* Crayfish can survive for several minutes out of water, but dehydration stress can alter their metabolic rates. Keep crayfish in water as much as possible, and limit their time out of water.)

Procedure

1. At your bench there should be a glass jar and a metal lid with a single hole in it. This is your test chamber. The hole in the lid allows you to insert the probe for the dissolved oxygen meter. There is also a pan of water that has been allowed to equilibrate to room temperature. *Do not* use this water to fill your test chamber; use only the aged tap water that has been bubbled with air.
2. To begin the room temperature (22°C) measurements, fill your test chamber with water from the tank labeled "Room Temperature Aerated Water" and return to your bench.
3. Set the test chamber into the plastic dishpan. The level of water in the pan should be ~2 cm from the top of the test chamber jar. If water spills out of the test chamber into the bath, that is no concern. If water overflows from the bath *into* the test chamber, empty the test chamber and refill it with fresh, aerated water.
4. Remove a crayfish from the tank labeled "Small Crayfish." Weigh it and record the weight in Table 11.1.
5. Gently place the crayfish into the test chamber and screw on the top. Again, do not be alarmed if water overflows into the bath.
6. Once the top is secure, use a beaker to completely fill the test chamber with aerated water, so that the top overflows slightly.
7. Remove the probe for the dissolved oxygen meter from its storage bottle, and insert the probe into the hole in the lid of the chamber (some water will overflow as the probe is inserted.) Make sure that the chamber is completely filled with water.
8. Turn on the meter itself, and set it to read oxygen in "0.01 mg/L."
9. Wait two to three minutes for the meter to stabilize.
10. The oxygen concentration reading will never stop changing completely, but it will stop fluctuating wildly and begin to decline steadily. At this point, take a baseline reading of the dissolved oxygen concentration. Record the value in Table 11.1 in the column with the head "Time Zero."
11. After 5, 10, 15, and 20 minutes, record the oxygen concentration again.
12. After you take the 20-minute reading, remove the probe and return it to the storage bottle.
13. You need to know the exact volume of water the crayfish was in during the experiment. Carefully pour the water from the test chamber through a funnel into a graduated cylinder. *Do not* remove the crayfish yet. Record the volume in Table 11.2.
14. For the 32°C measurements, refill the test chamber with water from the tank labeled "32°C Aerated Water."

15. Submerge the chamber in a water bath set to 32°C. Fill it to overflowing with warm, aerated water, and insert the probe for the oxygen meter.
16. Once again, wait for the meter to stabilize, then take a baseline reading of the dissolved oxygen.
17. Record the oxygen concentration in the water at 5, 10, 15, and 20 minutes.
18. When you are finished, unseal the chamber and measure the amount of water used. Finally, return the small crayfish to its original holding tank.
19. Repeat the entire process again, using a larger crayfish. Collect data at both room temperature and at 32°C.

Results

Table 11.1

Oxygen use over time by crayfish

					O_2 reading (mg/L) at:		
Size/Temp	**Wt. (g)**	**Starting Time**	**Time Zero**	**5 Min.**	**10 Min.**	**15 Min.**	**20 Min.**
Small, 22°C							
Small, 32°C							
Large, 22°C							
Large, 32°C							

Analysis: Calculating Oxygen Uptake

The readings you recorded are of oxygen concentration in the water, so you need to determine the actual amount of oxygen consumed by your crayfish. To calculate the amount of oxygen consumed by each animal per minute, multiply the change in O_2 concentration in the water by the total volume of water in your bottle (in liters), then divide that number by time, multiplied by the weight of the respective animal. This gives you specific metabolic rate, in mg O_2/g/min. Written out, this is the equation you must solve:

$$\text{Specific metabolic rate} = \frac{[\Delta O_2\ (\text{mg/L}) * \text{volume } H_2O\ (\text{L})]}{[\text{Time (min)} * \text{Weight (g)}]}$$

Table 11.2 will help you organize the numbers you need to calculate metabolic rate. Normally, you only need to calculate the final metabolic rate using the change in oxygen after 20 minutes. However, if for some reason you have an unusual data point that you cannot use, you can change the elapsed time from 20 minutes to 5, 10, or 15 minutes, and use the other data points you collected.

Table 11.2

Calculating specific metabolic rates

Size/ Temp	Wt. (g)	Elapsed Time (usually 20 min.)	Starting O_2 (mg/L)	Final O_2 (mg/L)	Change in O_2 (mg/L)	Volume H_20 (L)	Calc. Metabolic Rate
Small, 22°C							
Small, 32°C							
Large, 22°C							
Large, 32°C							

Once you have calculated oxygen consumption, plot your individual results using a spreadsheet program or graph paper. You will plot four separate values, one each for the large and small crayfish at 22°C and one each for the same animals at 32°C.

Unit Summary Questions

1. At which temperature did the crayfish use more oxygen? Was this expected or unexpected, and why?
2. Which size of crayfish used more oxygen per gram of body mass, the larger one or the smaller one? What did other work groups observe? Was this what you would have predicted would occur? Why or why not?
3. Go back for a moment and look at your *raw* numbers (i.e., those obtained before you calculated specific metabolic rate). Which animal used the most *total* oxygen, the small crayfish or the large crayfish? Is this the same animal that you predicted in advance would use the most total oxygen? If not, why do you think your original prediction was incorrect?
4. In mammals and birds, a graph of oxygen consumption versus temperature looks like an inverted bell or parabola (see Figure 11.2). Consumption usually is lowest when the external environmental temperature is nearly the same as that organism's normal body temperature. Oxygen consumption increases on either side of normal body temperature. This response is shown graphically in Figure 11.2.

Figure 11.2

Typical effects of environmental temperature on specific oxygen consumption in terrestrial mammals and birds

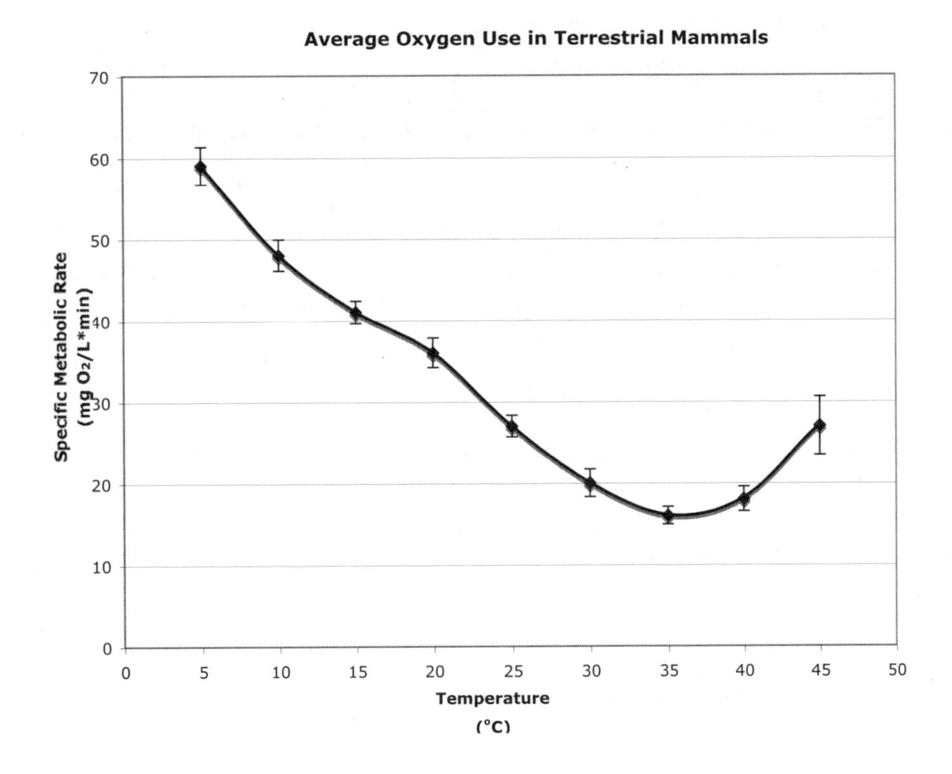

Looking at your own data, is the same relationship true for crayfish? If not, why might they be different? Why does oxygen consumption drop in the middle of the plot for mammals and birds when external environmental temperature approaches normal body temperature?

Your Own Experiment

Now it is time for you to design and execute your own experiment. Remember that the subject of this unit is metabolism; plan your experiment to test the effect of some physiologic or environmental stimulus on metabolic activity. If you are unable to think of an experiment, the following questions may give you ideas that you could explore.

1. Is the relationship between temperature and specific metabolic rate linear over a wide range of temperatures?
2. Part of the energy expended during normal metabolism is used to obtain oxygen and transport it to the tissues. If there is less oxygen available in water at a given temperature (i.e., it is *not* saturated), does the crayfish need to expend more energy to extract it?
3. Some basal metabolic reactions are not only temperature sensitive, but pH-sensitive as well. Does pH affect oxygen use? What about water salinity?
4. Often metabolic processes are not 100% efficient. In fact, many organisms will dramatically change the amount of energy they expend and how efficiently they complete basic metabolic reactions, depending on how much food they have available. Would nutrient restriction affect oxygen use in crayfish? How would you test that?
5. Do all ectothermic animals respond to temperature changes in the same way?

Materials Available

These materials are routinely available. If you wish to test something that is not on this list, ask your instructor whether it may be available as well.

- Water baths at different temperatures (4° to 35°C)
- Water at different salinities, up to 0.65% NaCl
- Water at different pHs
- Flow chamber to stimulate the crayfish to exercise
- Other species of aquatic arthropods
- Goldfish
- Tropical fish

UNIT 11

Metabolism and Oxygen Consumption

Teacher Pages

INSTRUCTORS' NOTES

Background and Key Concepts

Most eukaryotic cells synthesize the majority of their ATP in mitochondria, using energy extracted from high-energy electrons through redox reactions. This process requires oxygen, which is the final electron acceptor; as long as oxygen is available, ATP production continues. When a multicellular organism is in a state of energy balance, ATP concentrations in its tissues remain relatively constant and the rate of ATP production is approximately equal to the rate of ATP consumption. Moreover, the rate of oxygen consumption will be directly proportional to the rate of ATP synthesis. This relationship provides a convenient way to estimate metabolic activity of a single organism over time or under different conditions. It also provides a way to compare the metabolic activities of two different organisms. In this unit, students see this relationship between energy demand and oxygen consumption for themselves.

Students with more advanced knowledge of biochemistry or physiology may point out that oxygen uptake is not always directly proportional to the rate of energy use. During anaerobic conditions, or when the respiratory apparatus cannot supply sufficient oxygen, ATP can be synthesized by glycolysis, which does not require oxygen. Subsequently there will be a rebound period in which oxygen consumption is high even though no new ATP is being synthesized. In both situations, oxygen uptake is no longer directly proportional to the rate of ATP consumption. However, these are situations when an organism is *not* in a state of energy balance. The goal here is to look at an organism experiencing a mild stress that does not cause energetic imbalance.

To prepare students for this unit, ask them these questions:

What do the cells in organisms use oxygen for? How is that oxygen actually delivered from the external environment to the cells?

There are many different types of respiratory organs (e.g., various gills, different types of lungs, and different sizes of both). Why are there so many different types, if all cells are using oxygen for the same basic metabolic processes?

When an organism needs to produce more ATP, how does it increase the rate at which it takes up oxygen from its immediate environment? Do all organisms use the same methods to increase oxygen uptake?

General Teaching Strategy and Common Problems

This lab was designed for a first semester freshman course in organismal physiology. The audience consists of both freshman biology majors and nonmajors who are fulfilling general science requirements. The exercises for the first week can be completed in two hours if students work quickly, but it is better to plan for three hours.

The first week of this lab focuses on the effects of temperature on energy demand and oxygen consumption. Students usually already know that in endotherms oxygen usage changes with ambient temperature. Endotherms expend more energy when ambient temperature is significantly higher or lower than the core body temperature. This makes intuitive sense in that the organism must expend ATP to raise body temperature above the ambient conditions when it is cold and to cool the body when ambient temperature is above body temperature. Students are not likely to know that the pattern is different for ectotherms, which is one reason crayfish were chosen as the model organism for this unit rather than an endothermic vertebrate. Under controlled conditions in which the animal is restricted from moving to another location, metabolic activity of an ectotherm tends to be linearly related to ambient temperature. At first this relationship makes intuitive sense as well; cold ectothermic animals are sluggish and less likely to move or seek food than warmer ones. However, the link is not so tight when ectothermic animals are free to move about. Ectothermic organisms expend considerable energy on movement when the temperature gets above or below a certain point. These movements can generate additional metabolic heat and allow the organism to remain active. In other instances, the organism may be attempting to move toward a warmer location. In these cases, oxygen consumption will not decline linearly with temperature, but can actually go up as temperature goes down, much as it does in mammals and other endotherms.

To maximize students' learning, it is vital that the instructor not treat the first week as a demonstration exercise. Do not supply students with an "expected" answer, but rather require them to derive the relationships between temperature or size and oxygen demand in ectotherms for themselves. To reinforce the goal of interpretation of the observations, peer groups often are asked to share data, particularly when two groups have significantly different results.

There are certain problems with this unit that occur routinely. Conceptually, many students do not understand the connection between oxygen use at the cellular level and macroscopic respiration and gas exchange. Others may enter the lab thinking that oxygen is converted directly to CO_2. Remind them that oxygen is converted to water (not carbon dioxide) by the redox reactions that help generate ATP, while CO_2 is generated as a waste product of the catabolic chemical reactions of intermediary metabolism. The function of macroscopic respiratory structures is to move those gases in and out of the body.

Oxygen readings will rise and fall at random if the temperature of the water in the test jar is not kept stable. Students *must* keep the test jars immersed in a water bath at the appropriate temperature.

In the second week of this unit, student groups conduct an experiment that they design. Ideally, students should not finalize their experimental proposals until after they complete data collection and analyses the first week and discuss those results as a class. For a shorter unit, the second week can be eliminated without significantly compromising the learning goals.

Recommended Prelab Skills

A spreadsheet program can be used to speed up the routine calculations, but it is not essential.

Assessment

If pre- and postlab quizzes are used as part of formal assessment, the prelab quiz should not focus on the details of mitochondrial function. Instead ask students to make hypotheses, then predictions about what they expect to see when they measure oxygen use. Students also should be able to describe the differences between direct and indirect measurements. Typically the postlab quiz can draw heavily from the Unit Summary Questions at the end of the unit. Students should be able to summarize their observations from the first week and to interpret a graph showing the relationship between oxygen use and temperature. It is reasonable to ask specifically about how and why oxygen consumption was normalized to body mass.

Safety and Housekeeping

During this unit the students can be quite comical to watch, since most species of crayfish will rear up and display their claws. However, they pinch reluctantly. If a student does get pinched, there is almost no danger of serious injury. No hazardous chemicals are used in this unit. Remind students to wipe up all water that gets spilled onto the floor, so that no one slips and falls.

Other Tips

- Do not let students test temperatures outside the range of 10–35°C. Below 10°C, oxygen consumption is too low to measure reliably. Above 35°C, most crayfish will die within 15 minutes.
- If the reading on an oxygen monitor repeatedly jumps up and down without stabilizing, the membrane is probably damaged. Change it according to the manufacturer's instructions.
- Depending on the accuracy of their measurements, students may see an interesting paradoxical metabolic phenomenon. Theoretically, specific metabolic activity should be the same for two members of the same species experiencing the same metabolic demands. Yet when the specific metabolic activity of two organisms of the same species is compared, oxygen use per gram of body mass usually is slightly but significantly *lower* in the *larger* member of the species. Students may see this when they plot the *specific* metabolic rates of their smaller versus larger crayfish; the larger crayfish will use less oxygen per gram of body mass at a given temperature.

  When this phenomenon was first observed in mammals, it was assumed that smaller organisms had more surface relative to volume, so must expend more energy to maintain body temperature. However, the phenomenon has been seen repeatedly in plants and ectothermic animals, suggesting a mechanism beyond surface-to-volume effects. One hypothesis contends that structural supporting tissues make up a larger percentage of total body mass in larger organisms within a given species. A larger organism requires less energy per gram of body mass, because structural support tissues are less metabolically expensive to maintain. Do not make a special effort to emphasize this paradox; it is merely pointed out here in case students observe it.

PREPARATORY NOTES

Quantities listed are for a lab section of 20 students working in groups of three or four.

Week 1: Shared Materials

- 6 small crayfish (3 to 4 inches)
- 6 large crayfish (5 to 6 inches)
- 20 gal. aquarium filled with room temperature, aged tap water, with 2 bubblers

- 20 gal. aquarium filled with aged tap water, with 2 bubblers, and 2 aquarium heaters calibrated to 32°C
- 2 graduated cylinders (1000 mL)
- Large carboy of aged tap water at room temperature, with 2 bubblers
- Mop and wringer bucket for cleanup

Week 1: Materials at Each Work Station

- 1 glass canning jar (pint)
- 1 flat metal canning jar lid, with 1 in. hole drilled in center and rubber gasket inserted
- 1 screw ring for pint jar
- 1 beaker (1000 mL)
- Oxygen monitor or equivalent
- 1 medium dishpan filled with tap water, room temperature
- 1 small water bath, calibrated to 32°C (two student groups can share one bath)

Week 2: Shared Materials

- 18 to 24 crayfish (mixed sizes)
- 20 gal. aquarium filled with room temperature, aged tap water, with 2 bubblers
- 2 graduated cylinders (1000 mL)
- Large carboy of aged tap water at room temperature, with 2 bubblers
- Mop and wringer bucket for cleanup

Week 2: Materials at Each Work Station

- 1 glass canning jar (pint)
- 1 flat metal canning jar lid, with 1 in. hole and rubber gasket
- 1 screw ring for pint jar
- 1 beaker (1000 mL)
- Oxygen monitor or equivalent

Optional Materials for Experiments (provide only if requested)

- Water baths at different temperatures (4°–35°C)
- Water enriched with different gases
- Water at different salinities (upper limit is 0.6% NaCl)
- Water at different pH levels (4–9)
- Water contaminated with nitrogenous agricultural waste
- Water with different dissolved ions, up to 100 mg/L: Na^+, K^+, Ca^{+2}, HCO_3^-
- Water contaminated with industrial metallic waste

Maintaining Organisms

Small numbers of crayfish can be field collected or purchased from educational supply companies. Bulk quantities in a range of sizes can be purchased from commercial suppliers like the Louisiana Crawfish Company (*www.lacrawfish.com*), 140 Russell Cemetery Road, Natchitoches, Louisiana 71457.

Crayfish are cannibals, especially if animals of very different sizes are housed together, or they lack hiding places. Animals can be housed individually in quart jars filled halfway with aged tap water or springwater. Alternatively, up to a dozen animals of similar size can be housed together in one 20 gal. aquarium if they are provided with shards of broken flowerpots or other large debris under which to hide. Fill the tank with 6 in. of aged tap water, and add the potshards. Transfer crayfish to the tank carefully, as any injured animal is likely to be eaten. Whether the crayfish are housed individually or in groups, feed dry fish food daily, and change water weekly. If one animal dies, leave the tail from the carcass for the others to consume. The presence of one carcass reduces the likelihood of another being killed and eaten.

Solutions, Reagents, Equipment

Preparing Jar Lids

Many oxygen probes are designed to fit the neck of a standard 500 mL BOD bottle; to accommodate other sizes of probes, simply use a larger or smaller drill bit and gasket. Clamp the metal lid for a pint canning jar between two scraps of lumber. Use a 1 in. chisel bit to drill a hole through the wood and metal. Unclamp the lid and use a metal mill file to smooth the edge and remove metal burrs. For the gasket, insert a black rubber PCV valve grommet into the opening in the lid (available at auto parts stores).

Materials for Week 2

To enrich water with gases other than room air, place 1 L of aged tap water in a clean, empty 2 L soda bottle. Attach an airstone with Tygon tubing to the regulator outlet of a nitrogen, carbon dioxide, oxygen, or other compressed gas tank. Place the airstone in the water, then barely open the tank valve so that the airstone bubbles gently. Bubble the water with the selected gas for 10 minutes, then remove the airstone and cap the bottle tightly until ready to use. Just prior to an experiment, shake the soda bottle, dilute 100 mL from the bottle with 900 mL of normally aerated water.

Caution: Do not use undiluted water that has been saturated with nitrogen or carbon dioxide. Nitrogen displaces oxygen, while carbon dioxide acidifies the water and causes narcosis; either treatment will kill a crayfish within a few minutes. Water bubbled with pure oxygen can be used undiluted if necessary, but may be outside the range of the instruments being used by the students.

Specific preparatory instructions for the various test solutions are in Unit 10, Homeostasis and Heart Rate on blackworm homeostasis. It is not necessary to make them unless students request them for experiments. All solutions must be aerated for at least 20 minutes at room temperature prior to use.

If crayfish are exposed to nitrogenous agricultural waste, metallic ions, or extremes of pH, they should be allowed to recover for 30 minutes in plain aerated water before they are returned to the housing tank. Ideally, crayfish should not be used for experiments involving chemical additives in the water more than once per day.

Crayfish Exercise Chamber

Crayfish can be exercised in two ways. The simpler method is to remove them from their holding tank and prod them to move around the tabletop for 3 to 5 minutes. Oxygen demand rises dramatically for 15 to 20 minutes postexercise, but the animal is unharmed.

Alternatively, place 3 in. of water in the bottom of a 20 gal. aquarium. At one end, place a small submersible sump pump; using a short stretch of Tygon tubing, direct the pump's water outflow at the glass on the opposite end of the tank. When the pump is turned on, there will be a continuous current flowing across the bottom of the tank. Crayfish placed in the middle of the tank will have to expend energy to stay in place and away from the vibrating pump. Individual crayfish often will seek out an eddy point where the current is lower than in the rest of the tank; leave it for the students to figure out how to prevent the crayfish who seek this point from escaping exercise.

Other Organisms to Consider

Students may want to test oxygen utilization by other aquatic organisms. Goldfish work well for this exercise and can be used for the entire exercise if crayfish are unavailable. Tropical fish also work well. The lab preparer or instructor should monitor the oxygen probes carefully, because mucus from fish fouls the membrane of the probe very quickly. Probe membranes will need to be changed more frequently.

Tadpoles and axolotls (aquatic phase salamanders) also should work well. Large snails are another option for an invertebrate, although their rate of oxygen uptake is very low compared with that of crayfish.

UNIT 12

Transpirational Control

Student Pages*

BACKGROUND

To survive and function, every organism has both an optimum total volume of body fluids and specific concentrations of ions that it requires in those fluids. It also must have mechanisms for accumulating sufficient quantities of water and ions and for keeping them properly balanced.

Different environments and lifestyles pose very different osmoregulatory challenges for organisms. For example, marine vertebrates have no trouble obtaining enough Na+, Cl-, and other soluble minerals. However, seawater has approximately threefold higher osmolarity than the body fluids of most marine vertebrates. Marine vertebrates must expend considerable energy to prevent water loss due to osmosis, while simultaneously excreting the excess NaCl that inevitably diffuses into their body from surrounding seawater. In contrast, vertebrates living in freshwater streams have body fluids that are hyperosmolar relative to their surroundings. Their challenge is to extract enough Na+, Cl-, and other ions to function, while eliminating the excess water that enters their tissues via osmosis. Terrestrial vertebrates must successfully extract ions from their environment *and* obtain water.

*Teacher Pages begin on page 387.

There is no one solution to the problem of water and ion regulation, so different organisms typically use several strategies. In general, though, animals have one or more mechanisms that let them ingest unbalanced solutions, then actively excrete any excess water or ions they do not need. Plants take a different approach; they absorb a balanced solution of water, nutrients, and ions from the surrounding soil. Plants do not have water-specific pump proteins that they can bring in against the water potential gradient. Instead, they actively transport solutes into the root, and water follows passively.

Once this balanced solution has entered root xylem tubes, it is transported upward to the stems and leaves as part of the process of **transpiration**, without the need for excretion of excess water or solutes. During transpiration, water evaporating from the inner spaces within leaves escapes through small pores called **stomata**. Evaporation creates cohesive tension on the entire water column in the xylem and pulls water plus dissolved solutes up the xylem tubes from the roots.

Both Internal and External Factors Affect Transpiration Rate

Plants regulate their rate of transpiration primarily through **abscisic acid**, a hormone that stimulates guard cells to close stomata when conditions do not favor transpiration. However, the numbers of stomata that are open versus closed, and the size of those openings, are not the only factors controlling transpiration rates. External environmental factors, such as air temperature, relative humidity, and soil water availability, increase or decrease the rate of transpiration as well. These factors may reduce the overall rate of water evaporation from inside the leaves or prevent water from entering the roots at the start of the process.

At times, environmental conditions can be sufficiently extreme to stop transpiration entirely. Affected plants lose their ability to cool their leaves or to move water and dissolved nutrients from the roots to the leaves where they are needed. If adverse environmental conditions block transpiration long enough, the affected leaves or even the entire plant will die. A good example of this is a wilted houseplant. If water is withheld long enough, a plant reaches a state called its **permanent wilt point**. The plant cells still are alive at this point, but death is inevitable. The transpiration process has collapsed and cannot be restarted, even if the plant is watered thoroughly at this point.

Cellular processes begin shutting down long before a plant reaches permanent wilt point. Figure 12.1 shows the effects of the relative water content in plant tissues on photosynthesis, growth, and transpiration.

Figure 12.1

Effects of water content on physiological processes in plants

Relative water content of plant fluids is on the *x* axis and is stated as a water potential. The different activities are on the *y* axis; the size of the bars shows which activities still occur at a given water potential and the relative level of that particular activity. As water potential becomes more negative (toward right end of the *x* axis), the relative concentration of solutes becomes higher. In other words, tissue osmolarity rises as water potential becomes more negative. (Taiz, L., and E. Zeiger. 1998. *Plant physiology*. 2nd ed. Sunderland, MA: Sinauer Press, page 173. Redrawn from original with permission.)

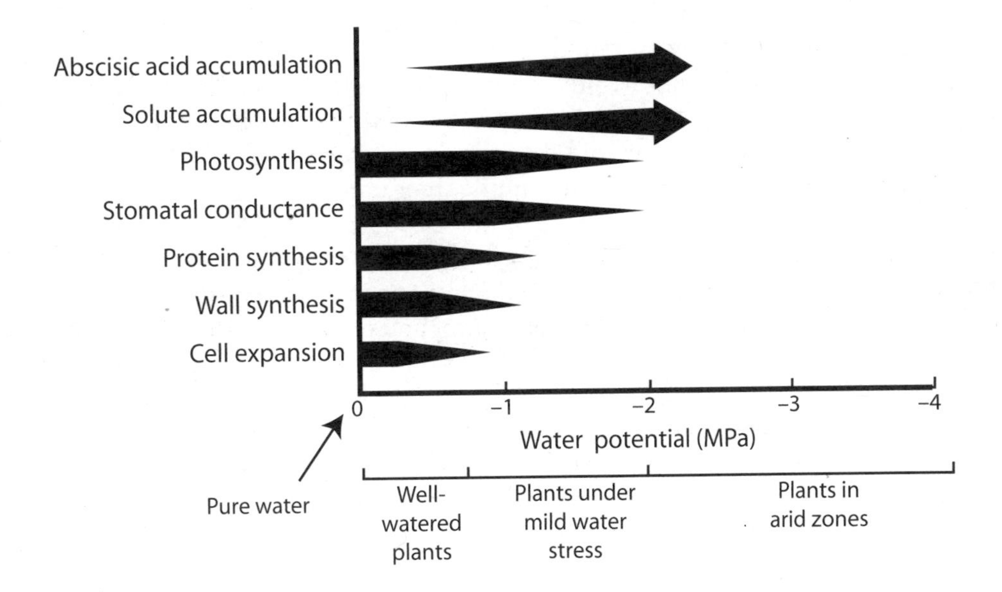

How Is Transpiration Rate Measured?

One standard method of measuring transpiration rate involves clamping a leaf in a **pressure chamber potometer**, so that just its cut stem protrudes. At normal atmospheric pressure, transpiration continues in the leaf for several minutes, so water in the xylem moves into the stem away from the cut end. Raising the air pressure in the potometer slows the rate of transpiration. When the pressure is high enough, xylem fluid stops moving away from the cut end and begins to flow back out. At this point, the positive gas pressure in the chamber is equal to the tension (negative pressure) on the water in the xylem created by evaporation.

Pressure potometers are very accurate, but the chamber can be very dangerous to operate. Fortunately, **weight potometry** provides a simpler way to estimate transpiration rates. A potted plant is wrapped in plastic so that only its shoots are exposed to air. All other routes of evaporation are cut off, so transpiration becomes the only route for water loss. If the wrapped plant is weighed before and after an experimental treatment, the difference is a reasonable estimate of the total quantity of water lost due to transpiration. Point-counting or another method is used to measure leaf area, then to estimate the amount of transpiration that occurs per unit of leaf area (i.e., per cm^2, per m^2) per unit time (minute, hour, day).

Goals for This Unit

In this unit you will use weight potometry to estimate the basal rate of transpiration for bean plants. For the exercise you will estimate the soil salinity necessary to stop transpiration by mung bean seedlings. Why this question is so important is explained in more detail later (see p. 377). Subsequently you will have the opportunity to devise and conduct your own experiment to see whether other environmental factors affect transpiration.

To conserve time during the first week of this unit, different lab sections will be conducting separate parts of the exercise. Some students will set up plants for assays, while others will collect and compile final data. Those who are collecting the data at the end of the assays will compile their data into a single large dataset and share it with the students who set up the plants initially. The instructions for the exercise are written as if you are performing the entire procedure; your instructor will tell you which part you will be doing. Even if you do not actually do the entire procedure, you must be able to explain it and answer questions about it. Be sure, therefore, that you understand what is occurring in each step of the entire procedure.

When you perform your own experiment for the second week of this unit, you will need to come back around 24 to 48 hours after setting up your experiment to collect your own data. So as you plan your experiment, make sure that at least one person in your group will be available to collect the data.

General Safety Precautions

None of the materials that you will use are toxic or harmful. Spilled water and vermiculite will make the floors slippery, so clean up any spilled materials immediately.

UNIT EXERCISE

Effects of Soil Salinity on Transpiration Rate of Mung Beans

Background

The Background section at the beginning of this unit explained how soil water availability can change the rate of transpiration. So what environmental factors alter soil water availability? The most obvious is lack of rain or irrigation; no water in the soil means no water for root transport and transpiration. However, other factors affect water availability. For example, soil with high clay content has an average particle size that is very small. Each clay particle has a charged surface that binds a thin boundary layer of water that is not removed easily. Thus, clay soil holds a great deal of water (and nutrients) once wetted, but it may not be readily available for plants to take in. Soils with a large amount of decaying organic matter in them hold considerable water as well, but more loosely. Soils rich in organic materials will release that water easily to roots. The trade-off is that nutrients quickly leach out of soils high in organic matter alone.

Agricultural practices also can affect soil water availability. For example, California's Central Valley is one of the largest vegetable producing regions in the nation. Most vegetables that are transported to the rest of the country from California are grown in this region. The Central Valley is semi-arid, and growers rely heavily on irrigation. After decades of this practice, thousands of acres of otherwise fertile land in the valley have been lost to **salinization**. All fresh water has a small amount of dissolved sodium, potassium, and other minerals. Irrigation deposits these mineral salts into the soil. In most areas, the salts are flushed from the soil by seasonal rainfall into streams and other runoff. In the Central Valley, though, most water leaves by evaporation, which does not remove the mineral salts. Irrigation followed by evaporative water loss has deposited enough salt to increase the water potential outside the root. Crop yields are reduced because plants must expend additional energy and resources to respond to the excess salt.

Theoretically, the water potential of salinized soil might become so high that crop plants cannot move water from the soil into xylem at all. As a result, transpiration should shut down. Your goal in this exercise is to test this hypothesis and to find out if high soil salinity can indeed stop transpiration.

Your model organisms for today will be mung bean seedlings. In the exercise you will determine the rate of transpiration for a set of untreated, negative control plants and for bean plants exposed to increasing soil salinity. At the end of the exercise you will correct for differences in plant sizes by estimating the total area of the leaves. Then the class will compile the data to create a larger dataset that everyone will use to determine the effects of soil salinization on transpiration.

Plants must expend energy and resources to keep salt out of the roots and to prevent osmotic water loss. What is the primary source of the energy being expended? Given this, why does salinization reduce crop yields?

What SPECIFIC processes are going on that are using this excess energy?

Can water be actively transported into root cells and the xylem? If so, why is a higher salt concentration outside the root a problem? If not, how is water brought into the root, and why does salt outside the root disrupt this process?

Most plants have some limited ability to adapt to salinization. What are two possible adaptations you might expect plants to have or use to reduce the effects of salinization? What is your rationale?

Procedure

Setting Up the Assay

1. Your instructor will assign each group in your section one salt solution to test.

   Solution that your group will test: __________________

2. Go to the side bench, collect the following items, and take them back to your bench:

   - 6 pots containing mung bean seedlings (Most pots will have 2 mung bean seedlings. If a pot has just one, do not be concerned; you will correct for this later. *Do not* use plants that are obviously wilted or dying. Select healthy-looking seedlings only.)
   - 50 mL conical tube for measuring solutions
   - 2 sandwich boxes or other containers for catching runoff from pots
   - 1 beaker with 500 mL of distilled water
   - 1 beaker with 500 mL of the salt solution you were assigned
   - 6 resealable plastic bags
   - 6 small pieces of paper (for labeling bags)
   - Labeling tape
   - A permanent marker

3. Label each of three pots with "Water Only." Number these pots #1, #2, and #3.
4. Label the other three pots with "Plus XX Salt." (Replace "XX" with the concentration of salt that you are testing.) Label individual pots in each group" #1," #2," and #3."
5. Place the three pots labeled "Water Only" in one sandwich box. Measure out 50 mL of distilled water, and slowly pour it through the soil mix in one pot. Do not let the excess run off the top of the soil mix. You want to flush the soil thoroughly with water. Flush the other two pots with 50 mL each of distilled water in the same way.
6. Empty the runoff from the sandwich box into the sink, then return the pots to the box to finish draining.
7. While the "Water Only" pots drain, place the three pots labeled "Plus XX Salt" in the second sandwich box. Measure out 50 mL of the correct salt solution, and slowly pour it through the soil mix in one pot. Again, do

not let the excess run off the top of the soil mix. You want to flush the soil thoroughly so that the salt solution takes the place of the water in the soil. Flush the other two pots with 50 mL each of salt solution.

8. Empty the runoff from the sandwich box, then return the salt-treated pots to their sandwich box to finish draining.
9. Go back to the first three pots you flushed, and dump out any water that has drained from the pots. Flush them a second time with another 50 mL of distilled water per pot. Place the pots in the sandwich box to drain well.
10. Flush the three pots labeled "Plus XX Salt" a second time with an additional 50 mL per pot of the correct salt solution. Again, leave the pots in an empty sandwich box to drain well.
11. Let all six pots drain for 10 minutes before you proceed.

The preceding steps changed the salinity of the soil in which the beans were planted. Now you must wrap the pots so that water can only be lost by transpiration from the leaves. *Caution*: Do not hold plants by the stems or leaves. Handle them by the pots only.

12. Open the six resealable bags, and place individual pots into them. Make sure that all of the leaves stay outside of the bag. Close the tops so that the mouth of each bag hugs the stem of the plants inside it. If necessary, use tape to make the opening around the stem as small as possible. Do not pinch the stems.
13. Once you have assembled your bags, place a small piece of labeling tape on one side of the bag, so you can write on it but not damage the plants inside.
14. Weigh each plant in its bag, and record these initial weights on the label on the bag *and* in Table 12.1, page 382. (The weights you record in Table 12.1 are a backup, in case the label from a bag is lost.)
15. Place all six of your bagged and weighed plants on the side bench under the grow lights or where your instructor tells you to. Students in another lab section later in the week will complete the analysis of the plants you prepared today.

Collecting the Data

16. Your instructor will assign each group in your section one set of samples to measure. These samples were set up by students in another lab section.

Plant treatment that your group is measuring: ____________________

17. Carry your assigned control and test plants back to your bench. Also bring back:

- 6 small weigh boats
- 1 clear acetate transparency with photocopied grid on it
- Scissors
- Clear tape

18. Each bagged plant should have a tape tag on it listing its initial weight. Keeping the pot in the bag, weigh it and record its final weight on the tag.
19. Once you have weighed all six plants, record both the initial weight and weight after 24 hours for each of the six pots in Table 12.1. For each plant, subtract the final weight from the starting weight to get the weight loss over 24 hours. Record the total water weight lost in Table 12.1.
20. Next, use scissors to cut off just the leaves of the mung bean plants in the pot labeled "Water Only, #1." Cut just the blades of the leaves; do not include the part of the stem that extends from the base of the leaf to the stalk of the plant. Place all of the leaves from the pot into a prelabeled weigh boat, and set them aside.
21. Remove the leaves from each of the remaining five pots. Be careful to keep the leaves for each pot in separate, labeled weigh boats.

Estimating Leaf Area

As one person removes the leaves from the bean plants, another should start estimating total leaf area for the plants in each pot.

1. Take the printed acetate grid and turn it over so the textured printed side is facing down.
2. Using clear tape, attach all of the leaves from the plants in the pot labeled "Water Only, #1" to the acetate sheet. Tape the leaves so they are as flat as possible on the sheet.
3. Turn the sheet face up. What you see should resemble the right side of Figure 12.2.
4. Looking straight down through the acetate sheet, count the number of squares that are completely filled by part of a leaf. Record that number in the first column (labeled "#F") of Table 12.2.
5. Next count the number of squares that are partially filled by the leaf. Most of these squares will be around the edges of the leaves. Record that number in the second column (labeled "#P") of Table 12.2.
6. Peel off the leaves from the first plant from the grid and dispose of them. Tape down the leaves from the pot labeled "Water Only, #2." Once again, count the number of squares that are completely covered, and partly covered, by leaves. Record the numbers in Table 12.2.

Once you have counted the number of squares overlying the leaves from each of your six pots, you must calculate the total area of the leaves from the plants in each pot.

7. Take the number of squares partly covered by leaves for each pot, and divide that value by 2; enter the result in the third column (labeled "#P/2") of Table 12.2.
8. To determine approximate leaf surface area, use the equation in Figure 12.3.

Figure 12.2

Schematic diagram showing how to mount bean leaves for point counting to estimate leaf area

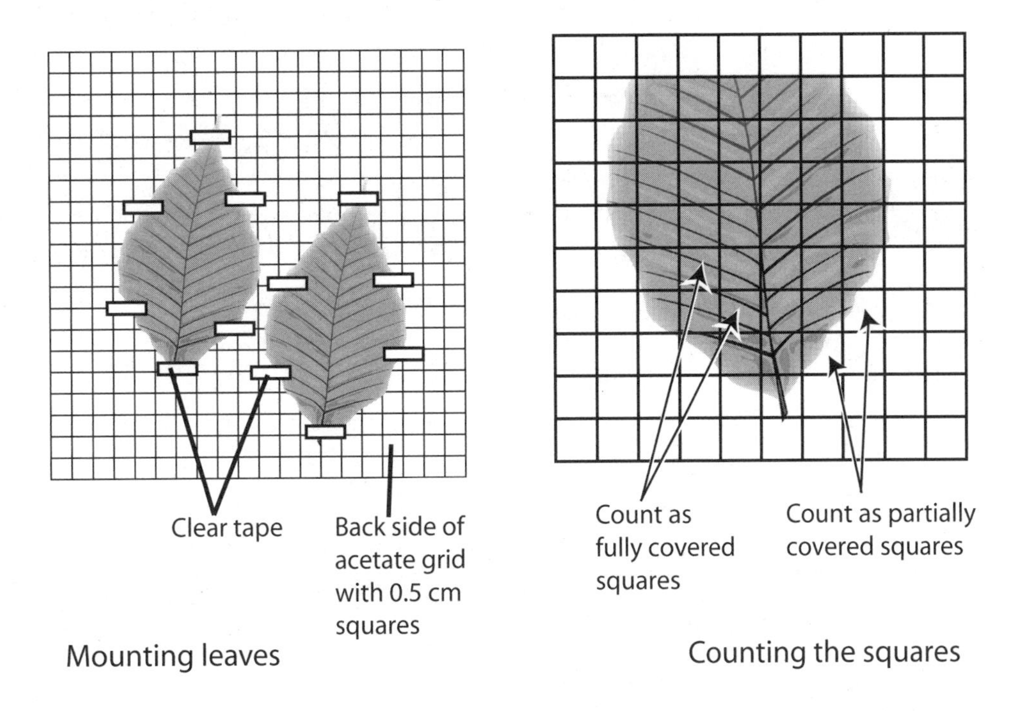

Figure 12.3

Equation for calculating leaf surface area from point-counting data

$$\text{Area (in cm}^2) = [(\#F) + (\#P/2)] * 0.25 \text{ cm}^2$$

Calculate the total leaf surface area for each pot, and enter the numbers in the last column of Table 12.2.

Results

Table 12.1

Pot weights of test plants

Plant # and Treatment	Initial Wt. (g)	Wt. After 24 Hr. (g)	Wt. Lost in 24 Hours (g)
Water only, #1			
Water only, #2			
Water only, #3			
Plus "XX" salt, #1			
Plus "XX" salt, #2			
Plus "XX" salt, #3			

Table 12.2

Grid counts and leaf area calculations

Plant # and Treatment	# Squares Fully Covered (#F)	# Squares Partially Covered (#P)	#P/2	Approx. Surface Area (cm^2) [#F+(#P/2)]*0.25 cm^2
Water only, #1				
Water only, #2				
Water only, #3				
Plus "XX" salt, #1				
Plus "XX" salt, #2				
Plus "XX" salt, #3				

1. The formula for calculating the transpiration rate per unit leaf area is shown in Figure 12.4. Using the equation and the values from Table 12.1 (the measured change in weight over time for each pot) and Table 12.2 (calculated leaf area for each pot), calculate the transpiration rate for each pot over the 24-hour test period.

Figure 12.4

Equation for calculating transpiration rate per unit of leaf area

$$\frac{[\text{Initial weight (g)} - \text{Final weight (g)}]}{\text{Time (24 hrs)} * \text{Area of leaves (cm}^2\text{)}} = \text{Rate of water loss [g /(hr} * \text{cm}^2\text{)}$$

2. Enter the transpiration rate for each pot in Table 12.3. Then calculate the mean and standard deviation for your three water-only pots and for the three pots you treated with salt solution, and enter those values in Table 12.3.

Table 12.3

Average transpiration rates of test plants

Plant # and Treatment	**Rate per Unit Leaf Area ($g/cm^2*hr.$)**
Control #1	
Control #2	
Control #3	
Mean ± st. dev.	
Plus "XX" salt, #1	
Plus "XX" salt, #2	
Plus "XX" salt, #3	
Mean ± st. dev.	

3. Your instructor will put a copy of Table 12.4 on the board, where each group will report their results to the class. Be sure to copy down the numbers given by other groups, so that you can answer the homework questions. The results will also be posted online, so the students who originally set up the plants you measured can also complete the questions.

Table 12.4

Summary of class data for transpiration rates

Plant Treatment	Mean Transpiration Rate per Unit Area	St. dev. of Rate per Unit Area
Control (distilled water)		
250 mOsm salt		
500 mOsm salt		
750 mOsm salt		
1000 mOsm salt		

Unit Summary Questions

When you turn in these questions, also hand in your answers to the questions on pages 377–378.

1. Graph the relationship between relative transpiration rate and soil salinity. Based on the class summary data and your graph, is there any evidence that rising soil salinity changes transpiration rate? What is your rationale for your conclusion (be specific)?
2. Did increasing the salinity of the potting mix ever stop transpiration completely? If not, do your data suggest that rising salinity could eventually stop transpiration entirely? What is your rationale?
3. Given the results you obtained, would you predict that a decline in transpiration rate is or is not a major contributor to plant death associated with soil salinization in the Central Valley of California? What is your rationale? If you cannot make a conclusion either way, what other data would you need in order to make such a prediction?
4. In this exercise you used pure solutions of sodium chloride. However, groundwater also contains potassium, trace metals, calcium, magnesium, and numerous other ionic salts. Do you think they all will have the same effect on transpiration? What is your rationale?
5. In the Central Valley, salinization occurs in different areas at different rates. What are some variables that might affect the rate at which salinization occurs at a particular location? Again, why did you choose those variables?

Your Own Experiment

Now that you are familiar with the basic method of weight potometry, your group must design an experiment to explore a specific element of water transport in plants in more detail. Spend some time discussing with your partners all the possible environmental variables that could affect transpiration rate. List them, and for each one, make specific, testable predictions of what will happen when you change that variable and why you think it will occur. From these predictions, choose one to test.

Next, plan out the experimental procedure you plan to use. Decide on the materials you will need, and complete an experimental outline. Make sure you include control plants. Turn the outline in to your instructor, so he or she will know what materials to prepare for next week. Also remember that next week one or more members of your group will need to return to collect the leaf area data. Plan your schedule now.

Materials Available

These items are routinely available. If you need something that is not on this list, ask your instructor whether it can be made available.

- Mung bean seedlings in small pots
- Sodium chloride solutions at various osmolarities
- Solutions of inert solutes at various osmolarities
- High concentration solutions of mineral nutrients (nitrates, phosphate, potassium)
- Solutions of micronutrients (calcium, sulfate, iron, or others; ask your instructor which ones will be available)
- Buffer solutions to change soil pH
- Low, high humidity chambers
- High temperature incubators
- Electric fans
- Light boxes
- Shade cloths of different densities
- Plastic wrap and resealable bags
- Top-loading balances

UNIT 12

Transpirational Control

Teacher Pages

INSTRUCTORS' NOTES

Background and Key Concepts

Many undergraduates are quite knowledgeable about animal physiology, and humans in particular. Yet it has been the author's experience that students know surprisingly little about plants. Most can name the major inputs and outputs of photosynthesis, and they know that water and soil nutrients are required for growth (but not necessarily which nutrients, or how they are used). Depending on prior background, some also know that water moves by evapotranspiration, but overall the majority have little or no prior knowledge of specific mechanisms by which root cells create a selective barrier between soil and xylem fluids, how roots transport nutrients and water to xylem tubes, or which abiotic factors in the soil and air might alter rates of flow in the xylem.

Given these limits in background knowledge, a conscious decision was made when designing this unit to eliminate as many potentially confusing elements as possible. To complete this unit as written, students do not need to know the anatomy of roots beyond the fact that the xylem tubes begin in the roots and extend in an unbroken path to the leaves. Similarly students do not need to know details of the symplast and apoplast pathways or the mechanisms for ion selection by root parenchymal cells. This unit treats the root as a monolithic structure that undergoes a fairly simple, two-compartment process: moving water and selected nutrients from the outside compartment (soil, which includes water and ions) across the root surface (the barrier layer) to the inner compartment (the xylem tubes). The question that students must answer is, Do abiotic conditions like soil salinity affect the rate at which this transfer occurs?

The exercise described here is a stand-alone unit, but it can be combined with Unit 6, Energetics and Photosynthesis and Unit 14, Resource Allocation in Plants, to provide students with a series of explorations of plant physiology and ecology. Unit 8, Animal Hormones also can be modified by substituting bean seedlings for Drosophila and treating them with various plant hormones. Similarly, the methods for measuring biological diversity described in Unit 16, Measuring Biological Diversity, can be applied to any vegetative habitat; all the instructor would need to do is provide students with locally appropriate plant field guides.

General Teaching Strategy and Common Problems

This unit was designed for a first semester freshman course in organismal physiology that serves both biology majors and nonmajors fulfilling general science requirements. The model system is extremely versatile and works equally well for a purely non-majors course. Typically, novice students become more engaged when the exercise is presented in light of a practical problem relating to horticulture or agriculture; hence the choice of soil salinization in California's Central Valley. If desired, an instructor can use a more theoretical approach to the model with students in advanced courses who have gained a deeper understanding of root structure and function.

This unit assumes there are multiple lab sections that will share duties. For example, students in a lab meeting on Monday afternoon will set up the initial exercise, then students in a Tuesday lab will measure the final outcome. The data are pooled, then shared by both sections. Another alternative is to have students return after 24 hours and measure their own plants.

Initial setup of the main exercise can be completed in one hour, and measurements can be completed in two hours. Theoretically there should be an observable effect in two to four hours, but in practice students have not seen clear-cut results. Results are more consistent if mung bean seedlings sit 18–24 hours in continuous light between the initial setup and final measurements.

Recommended Prelab Skills

No prior skills are needed, making this a good unit with which to begin a semester. Students can use a spreadsheet program for calculations and graphing, but it is not essential.

Assessment

As written, the goal in this unit is for students to make predictions about root functions *before* they know specific details of how the root actually handles solutes. Thus, for a prelab quiz, students should not be expected to know the exact mechanisms by which solutes are transported via the symplast and apoplast paths across the Casparian strip. Instead students should be asked to make hypotheses, then predict how various abiotic factors might affect transpiration rate.

The postlab quiz can draw heavily from the Unit Summary Questions on pages 384. It is reasonable to expect students to be able to summarize and explain their results from the first week. A typical data graph is shown in Figure 12.5.

Figure 12.5

Sample data for a postlab quiz

The apparent rise in activity is a false positive, caused by shriveling of the severely desiccated leaves. Initially students may misinterpret this to mean transpiration restarts; if so, ask them to think about what physically happens to desiccated leaves and then to look again at the physical evidence.

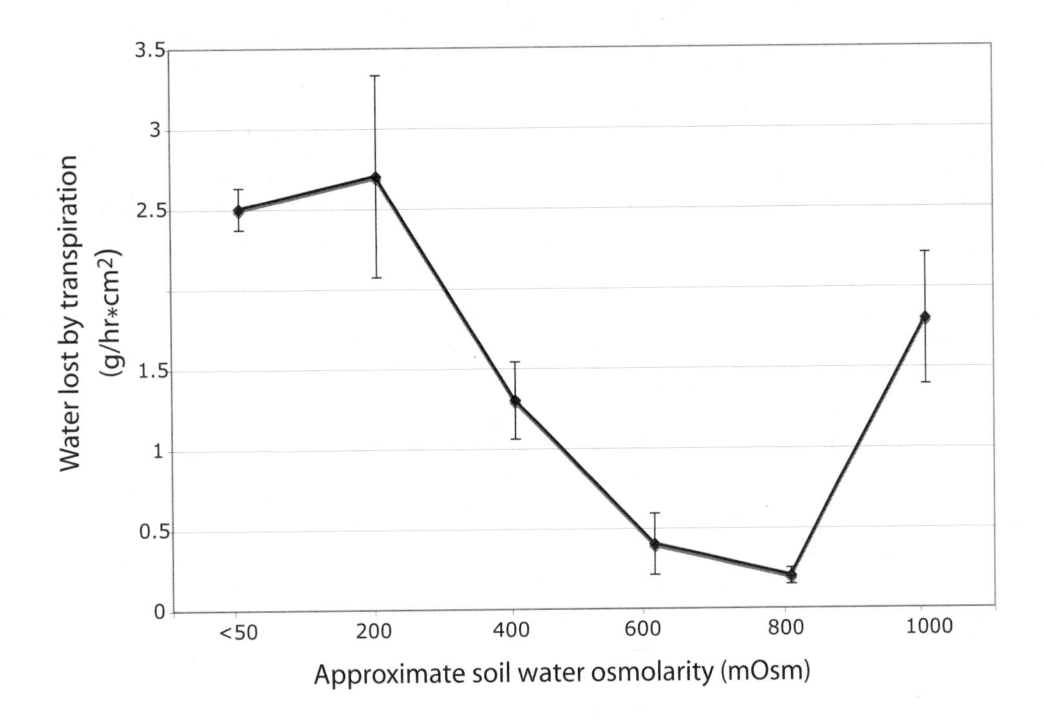

Most students can explain why transpiration rate decreases initially, but many are surprised to see that the rate apparently increases once more at the highest soil salinity. What they fail to recognize is that, at the upper end of the range, leaves wilt as water is pulled from the plants back into the salt-laden soil. The weight of the leaves drops, and apparent surface area declines. These changes are not obvious when students look just at the numerical data. This is an excellent opportunity to discuss with students why graphical or numeric data alone may not be sufficient for understanding a complex process; making direct observations is equally important.

Safety and Housekeeping

As written, this unit uses no hazardous chemicals or materials. Wet vermiculite is slippery, so it is important to sweep or mop the lab floor frequently when this lab is in progress.

Other Tips

- Regularly remind students to handle their plants by the pots only. Warn them not to lift or handle their bean plants by the stems; bending, pinching, or bruising the stem in any way dramatically reduces transpiration. After plants have been bagged, students often lift or carry them by the edges of the bags. The leaves slide down inside the bag, blocking transpiration. Again, handle the pots only.
- If students are overly confused by the apparent rebound in transpiration rate that occurs at high salinity, consider switching to a narrower range of soil salinities. Usually, a range of 100–500 mOsm is sufficient to demonstrate a significant drop in transpiration rate.
- Most student experiments have shown that brighter light, low humidity, and elevated temperature (up to 28°C) increase transpiration. Conversely, transpiration slows if plants are watered with 0.5 M sucrose, placed in high humidity, or shaded. Experiments in which students simulate drought by comparing plants that are or are not watered tend to produce poor results. Similary, there usually is little or no difference in transpiration rate between plants that are fertilized versus those that are not fertilized.
- Students who want to test the effects of wind often overstimulate their plants. When they use wind in an experiment, it is not necessary to turn fans up on high. A very gentle draft or breeze is sufficient. In still air, leaves have a thin layer of air near their surface called the **boundary layer** where the actual transpiration process occurs. Anytime this boundary layer is disrupted, transpiration accelerates. It takes very little breeze to disrupt the boundary layer; if leaves move just a little, the boundary layer is probably broken.

Supplemental References

Kirnak, H., T. H. Short, and R. C. Hansen. 2002. Studies on the relationships among moisture tension, microclimate, and transpiration rate of container grown *Acer rubrum*. *Journal of Applied Horticulture* 4: 65–69.

Munns, R., J.B. Pasioura, J. Guo, O. Chazen, and G. R. Cramer. 2000. Water relations and leaf expansion: Importance of time scale. *Journal of Experimental Botany* 51: 1495–1504.

PREPARATORY NOTES

Quantities listed are for a lab section of 20 students working in groups of three or four.

Week 1: Shared Materials

- 36 mung bean seedlings (3–4 weeks old), grown in vermiculite in plastic seed trays
- Distilled water, large carboy
- Sodium chloride solutions: 500 mL/lab section

    ◊ 200 mOsm NaCl (100 mM)
    ◊ 400 mOsm NaCl (200 mM)
    ◊ 600 mOsm NaCl (300 mM)
    ◊ 800 mOsm NaCl (400 mM)
    ◊ 1000 mOsm NaCl (500 mM)

- Clear acetate transparencies, with photocopied 0.5 cm × 0.5 cm grid
- Mop and wringer bucket for cleanup

Week 1: Materials at Each Work Station

- Top-loading balance
- 4 large weigh boats
- 2–3 dishes for soaking potted plants and for catching runoff from pots
- 200 mL beaker
- Sandwich-sized resealable bags
- Clear tape
- Permanent marker

Week 2: Shared Materials

- 60–70 mung bean seedlings (3–4 weeks old), grown in vermiculite in plastic seed trays
- Distilled water, large carboy
- Clear acetate transparencies, with photocopied 0.5 cm × 0.5 cm grid
- Mop and wringer bucket for cleanup

Week 2: Materials at Each Work Station

- Top-loading balance
- 4 large weigh boats
- 2–3 dishes for soaking potted plants and catching runoff from pots
- 200 mL beaker

- Sandwich-sized resealable bags
- Clear tape
- Permanent marker

Optional Materials for Experiments (make available only if requested)

- Sodium chloride solutions at various osmolarities
- Solutions of sucrose or other inert solute at various osmolarities
- High concentration solutions of mineral nutrients (nitrates, phosphate, potassium)
- Solutions of micronutrients (calcium, sulfate, iron, and others; see Table 12.5)
- Buffer solutions to change soil pH
- Low, high humidity chambers
- High, low temperature incubators
- Electric fans
- Light boxes
- Plug-in timers (for fans or light boxes)
- Shade cloths of different densities
- Colored acetate film

Table 12.5

Composition of single nutrient solutions

Ion or Condition of Interest	Acceptable Range in Cropland Soils	To Match Acceptable Range, Water Vermiculite With This Solution
Ammonium	2–10 ppm (Excess burns roots)	10 mg/L ammonium chloride (NH_4Cl) in water
Boron	0.5–2 ppm	2 mg/L boric acid (H_3BO_3) in water
Calcium	1000–2000 ppm	2 g/L calcium carbonate ($CaCO_3$) in water
Copper	>0.6 ppm	1.5 mg/L cupric chloride monohydrate ($CuCl_2 \cdot H_2O$) in water
Magnesium	60–180 ppm	45 mg/L anhydrous magnesium chloride ($MgCl_2$) in water

(cont. on p. 393)

Table 12.5 (cont. from p. 392)

Manganese	>1.5 ppm	5 mg/L manganese chloride tetrahydrate ($MnCl_2 \cdot 4H_2O$) in water
Molybdenum	>0.1 ppm	0.2 mg/L ammonium molybdate tetrahydrate ($(NH_4)_6Mo_7O_{24} \cdot 4H_2O$) in water
Nitrate	10–30 ppm (Excess burns roots)	25 mg/L sodium nitrate ($NaNO_3$) in water.
pH	6.0 to 7.0 (Fertile acid soils may reach 5.0, but are rarely any lower)	Prepare 20 mM sodium citrate; adjust to desired pH (from 5.0 and 7.0) with 0.1 M HCl or NaOH. Dilute to final concentration of 10 mM citrate, & recheck pH.
Phosphate	20–100 ppm	50 mg/L phosphorus pentoxide (P_2O_5) in water
Potassium	150–800 ppm	400 mg/L potash (potassium superoxide, KO_2), OR 900 mg/L KCl in water
Sodium and related salts	640–1280 ppm	1.3 g/L sodium chloride (NaCl) in water.
Sulfate	2–10 ppm	6 mg/L sodium sulfate (Na_2SO_4) in water
Zinc	>1 ppm	2 mg/L zinc chloride ($ZnCl_2$) in water

Growing Bean Plants

Mung beans are used for the exercise because the seeds are widely available locally and sprout quickly. Also, seedlings are large enough for students to handle and measure easily. However, most any bean variety works equally well. Beans should be soaked overnight in tepid water before planting. Unsoaked beans still germinate, but less reliably and more slowly.

Three weeks prior to the start of the unit, fill a standard 12 × 24 in. nursery flat with 48 1 × 2-in. (i.e., six-well) plastic seedling inserts. Larger plants are fine for this lab, so in cooler winter months, it may be better to start flats four weeks in advance. Loosely fill inserts with agricultural grade vermiculite, then top water thoroughly to settle the medium in each insert. Fill the flat with 1–2 in. of tap water, and allow the inserts containing vermiculite to hydrate thoroughly (1–2 hours).

Remove inserts from the flat and allow to drain. Gently tamp down the vermiculite in each insert. Place three or four soaked mung beans into each insert, for a total of 150–200 seeds per flat. Cover the beans with twice their own diameter of dry vermiculite, then top water gently to settle the new medium. Do not tamp the vermiculite down on top of the seeds. Place flats in a sunny window or greenhouse bay and keep flats evenly moist until most of the seeds have germinated. Germinating beans break the surface in 3 to 10 days, depending on temperature.

Once the majority of seeds break the surface, water each flat thoroughly with liquid complete fertilizer mixed according to the manufacturer's directions. One week before the unit begins, water plants a second time with complete fertilizer. In between fertilizer treatments, water the flats when the top one-half inch of vermiculite is dry. To encourage more even growth, rotate flat positions every three to five days.

On average, one out of three plants in each pot will be stunted or deformed. This is normal. Students are instructed to trim out these damaged plants before starting the exercise. If leaves in a large area of a flat begin to curl or brown, the beans likely are being scalded by too much sunlight. Move the flats to a less brightly lit position, or hang a piece of shade cloth overhead to protect them from the midday sun.

Other Solutions

Full instructions for preparing single nutrient solutions and buffers for changing vermiculite pH are in Unit 14, Resource Allocation in Plants.

100–1000 mOsm NaCl

For each solution needed, dissolve the quantity of sodium chloride listed in Table 12.6 in 950 mL distilled water, then bring the volume to 1000 mL.

Table 12.6

Preparing sodium chloride salt solutions

Osmolarity Required	NaCl (g/L)		Osmolarity Required	NaCl (g/L)
100 mOsm	2.92		600 mOsm	17.52
200 mOsm	5.84		700 mOsm	20.44
250 mOsm	7.30		750 mOsm	21.90
300 mOsm	8.76		800 mOsm	23.36
400 mOsm	11.68		900 mOsm	26.28
500 mOsm	14.60		1000 mOsm	29.20

UNIT 13

Animal Behavior

Student Pages*

BACKGROUND

Ethology, which is the study of animal behavior, combines the observational skills of a natural historian with modern insights from ecologists, geneticists, and especially evolutionary biologists. Modern ethologists see an animal's current behavioral repertoire as a product of natural selection acting on interactions between that species and its environment. Ethologists try to perceive the world "as the species sees it," not as human beings see and interpret it. For example, dogs lack broad-spectrum color vision. As a result their view of the world looks similar (but not identical) to what we see in black and white photos. However, a dog's visual acuity is greater than that of humans for small movements. Likewise, dogs smell and hear much more than humans.

All these differences make evolutionary sense, given that dogs are hunting carnivores that must track and locate prey of various sizes. Color vision can be fooled easily by camouflage coloration and countershading, but odor, sound, and movement will betray the location of almost any prey animal. Differences in perception of the world also help explain why your pet dog wants to stop and smell what seem to be random patches of grass for several minutes. Dogs are able to gather considerably more sensory information than humans can, and they will spend considerably more time collecting that information than we might think is necessary. For humans the equivalent would be a color photograph: While humans can extract considerable information from it, to a dog a color photograph offers little or no significant stimuli.

*Teacher Pages begin on page 405.

To understand and describe the complete behavioral repertoire of a species, a researcher must spend considerable time patiently observing it, usually in both the field and the laboratory. Observational fieldwork is both demanding and enjoyable; many ethologists have spent most of their adult lives in the field with the species they are studying. Long periods of observation help the ethologist recognize repeated patterns of behavior and learn the limits of that species' behavioral repertoire.

Ethology also is an experimental discipline. Once a species' behavioral repertoire has been defined, it is important to find the cues and stimuli that trigger those behaviors. Some cues are obvious; for example, for dogs the sight of a running prey item (like a squirrel) triggers them to chase it. Other cues are not so obvious. We can hypothesize about the possible stimuli all we want, but the only way to determine the actual stimulus is experimentation. Knowing the behavioral and evolutionary literature adds depth and appropriateness to these experiments and allows us to make comparisons among species with similar or different ecological backgrounds and histories.

Natural History of Bettas

Betta splendens, also known as Siamese fighting fish, are common pets (Suborder Anabantoidei, Family Belontiidae). Their wild ancestors are indigenous to Thailand and Southeast Asia, where they live in ponds, ditches, drains, rice paddies, and other sluggish or muddy water. At rest wild males and females alike are inconspicuous, dull green or grayish-brown fish. However, bettas are highly territorial. When males see another male of their own species, they flare their gills, flick their fins, and change color to an intense dark blue or red.

Since about 1850, wild bettas have been bred in captivity. Bettas arrived in the United States through the aquarium trade in 1910, and, by 1930, selective breeding had produced males that retained their "fighting colors" even if another male was not present.

Unlike many fish, bettas are not wholly dependent on their gills. They have an accessory labyrinth organ, which can extract oxygen from air that they gulp at the water's surface. This adaptation apparently allows wild fish to tolerate oxygen-poor water and provides the males with additional oxygen during protracted displays or grappling contests (which in the wild may last for hours).

To prepare for spawning, males create bubble nests from a mixture of mucus and air. After releasing eggs and sperm, both males and females pick up their fertilized eggs and spit them into the nest, where they remain suspended. After mating, the male drives off the female and guards the bubble nest. For two to four days, the male fans the nest, returns any eggs and newly hatched fry that sink out of it, and adds bubbles to the nest as old ones dissipate. Male parental care is essential for the eggs to develop; if males are removed, the eggs become hypoxic and die.

Basic Betta Behaviors

An **intraspecific behavior** is any behavior that an organism performs for, or directs toward, members of its own species. Male bettas have a wide range of intraspecific behaviors. Near a female, the male's behaviors revolve around courtship and mating. Conversely, in the presence of a second male, aggressive and territorial behaviors predominate. These behaviors range in intensity. Some are very subtle and are early responses to challengers. Other behaviors occur only as the aggressive display intensifies.

Aggressive behaviors (referred to by ethologists as **agonistic behaviors**) of bettas can be divided into two main types of movements: broadside movements and facing movements. Broadside movements are always performed with the fish's side turned toward its opponent.

Here are some examples of broadside behaviors, listed in order of increasing intensity.

- Pelvic fin flickering, in which the fish moves its pelvic fin vertically opposite the opponent
- Tail beating, where the fish makes sudden movements with its tail fin toward an opponent
- Tail flashing, when the fish closes and reopens its caudal fin

Here are some of the more common facing movements. Again, they are listed in order of rising intensity.

- Erecting pectoral fins and spreading gill covers (opercula)
- Extending the branchiostegal membranes (located underneath the opercula), to create a red "mane"
- Raising the dorsal fin, lowering the head, arching the back, darkening the skin, and darting toward an opponent
- Nipping and pulling off an opponent's fins

The preceding lists of behaviors are not all-inclusive. There are many others remaining for you to discover as you work through this unit. Also, agonistic behaviors do not always occur in a clear order. They may blend from one to the other, and often intermediate level behaviors may be skipped. In other situations, multiple agonistic behaviors may even occur simultaneously. Despite these caveats, it is possible to make some general conclusions based on observations of behaviors. Suppose a male betta were presented with two stimuli. Stimulus A causes the betta to start pelvic fin flicking, while Stimulus B causes the male betta to alternately raise its dorsal fin then turn sideways and flash its tail. Based on the previous descriptions, it could reasonably be concluded that Stimulus B causes the greater agonistic response.

Ethologists Must Resist Being Anthropomorphic

Anthropomorphism is the act of applying human traits or characteristics to non-human organisms or to inanimate objects. When observing animals, people are tempted to interpret their behaviors in human terms and to apply human emotions, motives, and standards. Observers may even use terms like *love, hate, spite, kindness*, and *nastiness*. Ethologists, on the other hand, want to view the world as their animal subject perceives it. To do so they must not use terms that interpret behaviors, only terms that describe them.

For an example of this, look back at the descriptions of each behavior performed by bettas. Each of them is described in empirical terms that have no emotions attached to them, such as "upright posture," "fin waving," "darting movements," or "biting." As you describe the behaviors you see today, try to use the same kind of emotionally neutral language that is used by professional ethologists.

Goals for This Unit

In this unit you will uncover more of the behavior repertoire of betta fish by observing their responses to several stimuli. You will compile a set of observations that allow you to answer these questions:

What are the main behaviors that bettas perform or exhibit in the presence of other males versus females of their own species (intraspecific interactions)?

What are the main behaviors that bettas exhibit in the presence of fish from other species (interspecific interactions)?

What is the minimal stimulus that is required to trigger the various intraspecific displays in male bettas? Does the same stimulus always trigger the same response? Is one stimulus enough to cause escalating agonistic behavior or are multiple stimuli required?

Based on your observations, you will try to describe the bettas' world as they sense it, and within that context explain the evolutionary basis for their behaviors. After you have completed the exercises, you will have the opportunity to develop and carry out an experiment of your own that explores some other aspect of behavior in greater detail.

General Safety Precautions

Bettas are very strong swimmers and may jump out of an observation tank. Cover all tanks or holding containers that you are not using. Domesticated male bettas are very aggressive to one another. Unlike their wild kin, they may kill each other. *Do not* let male bettas actually fight; they will rip off each other's fins, gill arches, and scales. Always have two nets on hand to separate males before they can actually attack one another.

On rare occasions, people working around aquariums develop lumps or nodules on their hands, especially if they put their hands in the water frequently. This condition is caused by an opportunistic bacterial infection, but other than itching, it is not harmful. To reduce the risk of such an infection, wash your hands thoroughly before you leave the lab.

UNIT EXERCISES

Exercise 1: Intraspecific Interactions Among Bettas

For the exercises in this unit you will be concentrating on the behavior of the male bettas. Each work group will use a small aquarium that is divided into three compartments. Two compartments contain males, and the third, a female. All three fish have been visually isolated from one another for several hours by an opaque plastic partition.

Procedure

Behavior of Isolated Males

1. Quietly observe both males for a minimum of five minutes. Don't tap the glass, bang on the table, or intentionally disturb the fish. Take your time; you are observing the baseline behaviors to which you will compare behaviors in other conditions.
2. Describe the position of the fish in the aquarium, its color, the position of its fins, tail and gill covers, and the frequency and direction of any swimming or fin movement. Write down your descriptions. Make sketches where appropriate.

Behavior of Physically Separated Males

3. Remove the opaque partition between the two males. Leave the glass partition in place. Observe both fish for about five minutes.
4. Observe their initial reactions.

   How quickly do the fish react to one another?

5. Describe any changes in the movements of the gills, pectoral fins, or tail.

   Are there changes in posture, color, or apparent size?

6. Replace the opaque partition between the fish and give them five minutes to settle down before you proceed.

Behavior in Response to Self

7. Place a mirror outside the tank, in front of one male betta.
8. Describe the male's reaction initially, after 5 minutes, and after 10 minutes.

   *Is there any sign of **habituation**, that is, waning interest in the reflection?*

   What is your evidence for waning interest?

   Why might habituation be adaptive?

9. Remove the mirror and let the male fish rest for five minutes before you proceed.

Behavior of Two Males Together

10. Remove both the glass and opaque partitions separating the two males.
11. Describe how their display patterns and behaviors differ from resting behaviors and from the behaviors you described above. Do not leave the fish together for more than five minutes.

    Are the interactions more or less intense now that the two males are not separated by the glass? How can you tell?

    Are more stimuli available now than when there was a clear glass partition between the males? Why or why not?

12. If the two males physically attack each other before the five minutes have elapsed, use two dip nets to gently separate them. Then replace the partitions. Let the fish rest at least five minutes while you complete your notes.

Behavior of One Male in the Presence of a Female

13. Remove the partitions separating the female from the closer of the males. Observe the male's behaviors for five minutes.
14. Describe the behaviors of the male that shares space with the female.

    Is there any difference in the behavior of the male in the presence of a female compared with the behavior of a male in isolation?

15. Describe any changes in the movements of the gills, pectoral fins, or tail.

    Are there changes in posture, color, or apparent size?

16. Place the opaque partition between the male and female fish and give them five minutes to rest before you proceed.

Behavior of Two Males Together in the Presence of a Female

17. Remove the partitions isolating the female and the two males.
18. Observe the males for a maximum of five minutes. Describe the interactions as before.

Is there any difference in the intensity of behavior of the two males in the presence of a female?

19. If the fish make physical contact, isolate all three fish again. While they are resting, look back at your notes describing the displays and other activities observed and the relative intensity of the behavior.

    Can you come up with a hypothesis to explain why bettas fight? In other words, can you explain the adaptive advantage these fish gain by having so much of their behavioral repertoire devoted to aggressive behavior?

Exercise 2: Interspecific Interactions

Procedure

1. Gently transfer your less active male betta to another container. Replace it with a blue gourami.
2. Remove the partitions between the gourami and male betta.
3. Observe the initial reactions.

   How quickly do the fish react to one another?

4. Describe any changes in the movements of the gills, pectoral fins, or tails.

   Are there changes in posture, color, or apparent size?

5. After five minutes, replace the partitions between the male betta and gourami and give them five minutes to rest. While they are resting, record your observations. Note especially any behaviors that differed from what you observed between two male bettas.
6. Repeat the test, except this time remove all the partitions between the male and female betta and the blue gourami. Make notes of your observations of posture, color, movement, fin position, and orientation of the male betta.

   Does the male betta behave differently toward a gourami in the presence of a female? If so, what might be the adaptive advantage in behaving differently to the blue gourami now that a female is present?

7. Replace the partitions so that all three fish are separated once again. Give them all five minutes to rest before continuing to Exercise 3. While they are resting, record your observations. Note especially how the behaviors have changed, relative to what you observed between two male bettas.

Exercise 3: Stimulus Filtering

Background

Sometimes animals do not require all the information available from a stimulus in order to trigger the corresponding response. Sometimes just one or a few parts of the whole stimulus are sufficient. This phenomenon is often referred to as **stimulus filtering.** Stimulus filtering can be a very powerful instinctive trigger for behaviors, and may even override learned behaviors. Consider this situation: A well-trained dog with no prior history of aggressive behavior suddenly kills another pet or attacks a child. From the perspective of ethology, these attacks are not surprising. Remember that dogs were derived from predatory ancestors. Studies of wild canids (wolves, coyotes, and foxes) have reported that rapid movement of a smaller, prey-sized animal triggers specific hunting, attack, and killing behaviors. It does not matter if the animal is black, brown, or even green, just that it is prey-sized and moving correctly. When these same motion and size stimuli are presented suddenly, the dog responds instinctively by biting and killing the prey item.

In Exercises 1 and 2 you observed several agonistic behaviors of male bettas. The question is, What stimulus is required to trigger those behaviors? Do different stimuli trigger different agonistic behaviors? Is there stimulus filtering? Is the whole fish required or just one anatomical part of the other fish? Does size matter? Are the movements or behaviors of the other fish important?

These questions can be addressed empirically using simple models. Certain characteristics are added or removed, then the model is shown to male bettas. The intensity of their responses determines the relative ability of the stimulus to trigger a particular behavior.

Your goal in this exercise is to determine the minimum stimulus required to elicit an agonistic response from a male betta.

Procedure

1. Remove the female from the observation tank. Leave the opaque and glass partitions in place.
2. Obtain a piece of thin foam sheeting and cut out several shapes (and sizes) with a pair of scissors. You want to determine the minimum stimulus to which the male betta will respond, so start with simple shapes (squares, ovals, circles, or triangles).
3. Tape one shape to a wire and lower it into the empty part of the tank slowly. Hold the model still, and remove the opaque partition so that the male betta can see the model.
4. Let the betta see the model broadside for one to two minutes, then turn the model "end-on" for another minute. Make notes of any changes in behavior that you see.
5. If your fish does not respond to the first shape, try a new shape that is just slightly more fishlike in appearance. Present each shape both broadside and "end-on," for at least one minute each. Note any changes in behavior.
6. After you have tested several different shapes, dry off the shape that stimulated the greatest response and the shape that stimulated the least response. If no shape elicited responses, select two different shapes at random.
7. Using black and colored permanent markers, start adding features of a real

fish, one at a time, to each shape. Add the same features to both shapes. Use your ingenuity and imagination in selecting features to add. Just be sure they are features found on the actual fish. Some other hints:

- Add features one at a time. For example, you might add eyes in black; then body colors, and, finally, fin outlines. Test each feature before you add new ones.
- Make sure that the ink on your models has dried *thoroughly* before you place the model into the observation tank. Loose ink flakes and solvent residues can sicken or kill fish.
- Think about the size of features, not just presence or absence of a feature.
- Test each new feature on both shapes.
- As you test, check whether movement or different angles of view affect the betta's response.
- Write down your observations *as you go along*. Do not wait until you complete your entire set of tests.

Unit Summary Questions

1. What were the main behaviors that you actually observed bettas performing in the presence of other members of their own species? What is their rank order; in other words, which behaviors did you regularly see early in agonistic displays and which behaviors did you only see once the displays intensified?
2. Overall, how did the presence of a female change the behaviors of males?
3. What were the main behaviors that bettas exhibited toward fish from other species? Given what you saw during intraspecific interactions, what would you hypothesize is the purpose of these interspecific behaviors? Is there an adaptive advantage in behaving this way? If so, what might the advantage be?
4. Rather than undergo an elaborate display, a male betta could immediately attack an intruder. What are the adaptive advantages of preceding an attack with a complex display?
5. Bettas expend a part of their metabolic energy and physical resources on behaviors associated with aggression. What might be the adaptive advantages and disadvantages of this allocation strategy?
6. What is the minimum stimulus that is required to trigger an intraspecific aggressive display by a male betta? What is your evidence for that conclusion? What might be the selective advantages and disadvantages to stimulus filtering?
7. As was stated in the Background section to this unit, ethologists try to perceive the world "as the species sees it," not as human beings see and interpret it. Based on your observations, how would you describe a betta's "view" of the world around it? (Remember, an ethologist does not use emotional terms. Focus on a mechanistic viewpoint.)

Your Own Experiment

Now you have the opportunity to explore some aspect of betta behavior in greater depth. Unlike previous units, there are no obvious independent and dependent variables for which you can take numerical measurements. So your behavioral experiment will require some additional thought and planning.

First, think about the behaviors that you observed and the data you collected. What questions about betta behavior do you have that your current data do not answer well? Is there a question you have that is not addressed at all by your current data? These questions would be good starting points for your experiment. Once you have decided on a general question that interests you, break that question down into one or more specific, testable hypotheses. Select one hypothesis to test further.

Next, think about the types and format of data you might collect as you test this hypothesis. In the unit exercises you collected descriptive data only, but you certainly are not limited to just descriptions. Is there some way to quantify your data? Can you put agonistic behaviors on a scale and report the intensity of responses to stimuli? Also think about how many animals you plan to test and how long it will take you to complete the testing.

Finally, think about how you will report and interpret your results. Will you report it as a narrative, or is there a way to graph the data? Should you have photos or illustrations of specific behaviors? How do you intend to place your results into an evolutionary and behavioral context? Remember that any behavior you see has a specific trigger stimulus and a specific function and is beneficial to the betta for some reason. You must try to explain these elements.

Materials Available

If you'd like to use something different, ask your instructor whether the appropriate materials are available.

- Male and female bettas
- Tropical fish, mixed species
- Observation tanks with partitions
- White and colored foam, wire, and other materials for making models
- Mirrors
- Colored acetate
- Brown, white craft paper
- Aluminum foil

UNIT 13

Animal Behavior

Teacher Pages

INSTRUCTORS' NOTES

Background and Key Concepts

Many lab courses have an exercise that is similar to the one described here, but often it is organized so that students look for examples of a predefined set of behaviors. As a result, students mistakenly think behavioral research is not open to hypothesis testing. A traditional cookbook exercise has been modified here so that it uses a more inquiry-based approach.

In preparing students for this unit, instructors should introduce three central principles:

- Most complex behaviors can be separated into several smaller, stereotyped behavioral elements.
- Many behaviors proceed in a specific order or pattern and have specific triggers.
- Even highly stereotyped behavior may be influenced by environmental conditions.

If students do not understand these principles initially, compare them to the central tenets of physiology and homeostasis. Students likely already know that the goal of a physiological system is to maintain homeostasis and allow the organism to respond to its environment. A complex process like gas exchange requires several organs to perform several tasks together in a coordinated way to achieve a desired response to a particular stimulus. Complex behaviors are simply an extension of the physiological paradigm; several smaller discrete behavioral events occur together and in sequence, in response to a specific environmental stimulus. The outcome of the complex behavior is also the same as a physiological process: The organism responds to its environment.

When students realize behaviors are a quantifiable biological response to stimuli, they are more likely to view ethology as both an observational and an experimental discipline.

General Teaching Strategy and Common Problems

This two-week unit was developed for a freshman premajors ecology course, but it works extremely well with nonmajors as well. In the first week, groups of three to four students place male and female bettas into several controlled situations, then observe and record the animals' behavioral responses. As they work, students will see several complex behaviors, which can be broken down into smaller stereotyped behavioral elements. In a single two- to three-hour lab period, students can see most (if not all) of this species' agonistic behavioral elements. In the second week, students design and execute experiments to identify the external stimuli that trigger one or more specific behavioral elements.

As written, this unit requires about three hours to complete. If time runs short, or to accommodate a two-hour lab period, drop Exercise 2, which examines interspecific interactions with a blue gourami. Exercise 3: Stimulus Filtering can also be dropped to shorten the lab. If Exercise 3 is not part of the routine lab procedure, consider making it available as an option for student experiments in the second week.

This unit is very simple to set up and execute, but the quality of learning hinges on the instructor taking the time to prepare students properly. Stress to students that they must make careful observations of betta behaviors; watching casually for a few seconds is not enough. Students must watch for subtle changes in behavior and for the individual behavioral elements (like broadsiding, fin raising, spreading of the opercula), not just the big agonistic displays that consist of several behavioral elements at once. Also stress to students that they *must* write down specific observations *as they work*. Summarizing their observations at the end of class is not sufficient, and is not how ethological research is conducted normally.

Assessment

If pre- and postlab quizzes are used as part of formal assessment, the prelab quiz should focus on the basic life history of bettas and on general terms such as *agonistic behavior or interspecific interaction*. After students have completed the first week of the lab, quiz questions should focus on specific details of their observations, such as when a particular behavior occurs or what behavioral elements it contains.

While students can report the results of their experiments in formal scientific reports, this lab is particularly suited to using oral reports and roundtable discussions as assessments, since all students will have observed the same baseline behaviors. Many students have cameras that capture simple movies; consider having students build a mini movie library of betta behaviors.

Safety and Housekeeping

Male (and in some cases female) bettas show visual displays first, then physically attack each other. Males bite and tear off the fins of opponents, until one or both males are unable to display. Fin barbering is not fatal, and healthy males will regrow their lost fins and tails within a month or so. However, fin barbering prevents animals from being used more than once. Also, intentionally letting the fish injure each other is a violation of AAALAC (Association for the Assessment and Accreditation of Laboratory Animal Care) and most institutional animal care guidelines. Students should watch males carefully, and separate them with aquarium nets before they actually make physical contact.

Other Tips

- In Exercise 3, students try to find the minimal stimulus needed to trigger a behavioral response by a male betta. Students often want to start with a full replica of a fish, then remove components, which leads to habituation. Remind them to start with a simple geometric shape, then add individual traits until they get a response.
- Students must give bettas sufficient time to rest (at least five minutes) between encounters with other fish, or with stimulus models; if the fish are not given sufficient time to rest, they may habituate and ignore all stimuli.
- When a betta stops responding to stimulus, even after resting, it probably has habituated to its reflections in the tank glass. Move the fish to a tank or container with less reflective sides, or line the tank with a plastic diffusion filter from an aquarium supplier.
- Most fish can be used twice a day and will still display most of their behavioral repertoire. Ideally, though, there should be sufficient fish on hand so that individual animals are used only once each day. With a once-daily rotation, fish may be used for up to five days before generalized habituation develops. If they are given a weekend to rest, bettas can be used for a second experimental week with no ill effects.

Supplemental References

Clotfelter, E. D., L. J. Curren, and C. E. Murphy. 2006. Mate choice and spawning success in the fighting fish *Betta splendens*: The importance of body size, display behavior and nest size. *Ethology* 112: 1170–1178.

Dzieweczynski, T. L., A. M. Bessler, D. S. Shelton, and W. J. Rowland. 2006. Effect of a dummy audience on male-male interactions in siamese fighting fish, *Betta splendens*. *Ethology* 112: 127–133.

Peake, T. M., R. J. Matos, and P. K. McGregor. 2006. Effects of manipulated aggressive interactions on bystanding male fighting fish, *Betta splendens*. *Animal Behaviour* 72: 1013–1020.

Snekser, J. L., S. P. McRobert, and E. D. Clotfelter. 2006. Social partner preferences of male and female fighting fish (*Betta splendens*). *Behavioural Processes* 72: 38–41.

PREPARATORY NOTES

Quantities listed are for a lab section of 20 students working in groups of three or four.

Week 1: Shared Materials

- Handheld mirrors
- Craft foam, sheets, both white and mixed colors
- Scissors (to cut foam)
- Aluminum foil
- Permanent colored markers
- Galvanized 16-gauge, single-strand utility wire (for holding foam models)
- Additional male bettas, housed in individual 500 mL plastic containers or jars
- Additional female bettas (may be housed together in an aquarium on a side bench)

Week 1: Materials at Each Work Station

- 1 observation tank (2.5 gal.), divided into three compartments with both opaque and clear partitions
- 2 male bettas, in two adjoining chambers of tank
- 1 female betta, in third chamber of tank
- 1 blue gourami or other tropical fish of similar size, stored in a 500 mL container or jar
- 2 small aquarium fishnets

Week 2: Shared Materials

- 2.5 gal. observation tanks, half-filled with aged tap water
- Spare empty aquariums
- Male bettas, housed in individual 500 mL plastic containers or jars
- Female bettas, held in an aquarium on a side bench or in individual containers

- Blue gouramis or other tropical fish of similar size, stored in a 500 mL container or jar
- Large carboy of aged tap water for filling aquariums
- Hand-held mirrors
- Craft foam
- Scissors
- Permanent colored markers
- Galvanized 16-gauge, single-strand utility wire
- 2 small aquarium fishnets

Note that during the second week students must arrange their own tanks. Other materials may be provided, but must be requested in advance of the lab.

Maintaining Organisms

Large numbers of bettas can be purchased from a wholesale aquarium supply company. With proper care their life span averages two to three years. Male bettas *must* be kept in individual containers and should be separated visually from one another by placing cardboard between the containers. Males can tolerate small containers, but for best health should be housed in a minimum volume of 2 L per animal, with some type of bottom covering, such as gravel or aquarium sand. Individual 2–4 L aquarium jars with lids are ideal. Filtration is not essential, but two-thirds of the water should be exchanged weekly. Cover all containers to prevent bettas from leaping out, but leave head space for air to circulate. Female bettas and blue gouramis can be housed together in 10 gal. aquariums.

Equipment

Preparing Observation Tanks

Most aquarium suppliers sell dividers that fit standard tanks. These directions let instructors retrofit tanks when appropriate dividers are unavailable (see Figure 13.1, p. 140). A standard 2.5 gal. glass tank is large enough to house three fish for an entire week, which minimizes the number of times they must be netted and moved. A finished tank has three approximately equal size compartments, and fish should not be able to move between compartments unless the inserts are removed. It may be necessary to adjust dimensions slightly for your particular brand of tank, so prepare one unit first to ensure they are correct. For each tank purchase or prepare:

- 1 glass aquarium tank (2.5 gal.), 12 in. wide × 8 in. high × 6 in. deep
- 2 clear, one-eighth-inch-thick glass inserts, ~5.5 × 7.5 in.
- 2 white or other opaque one-eighth-inch-thick acrylic inserts, ~5.5 × 7.5 in.
- 8 strips of one-quarter-inch-thick plexiglass, 7 in. long × approximately one-half inch wide
- Small tube of silicone aquarium sealant (do NOT substitute silicone caulk)

Figure 13.1

Assembling a partitioned observation aquarium

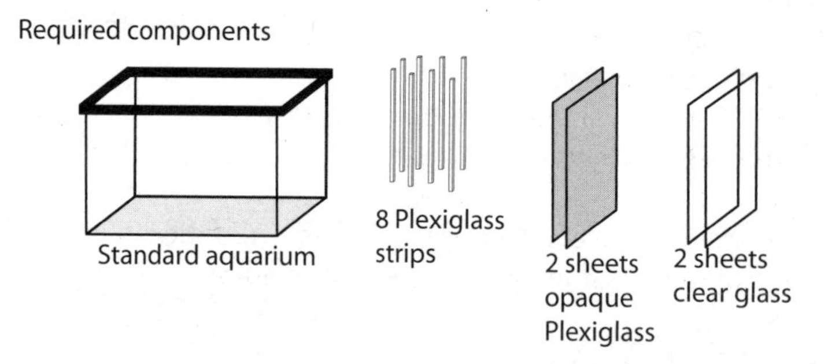

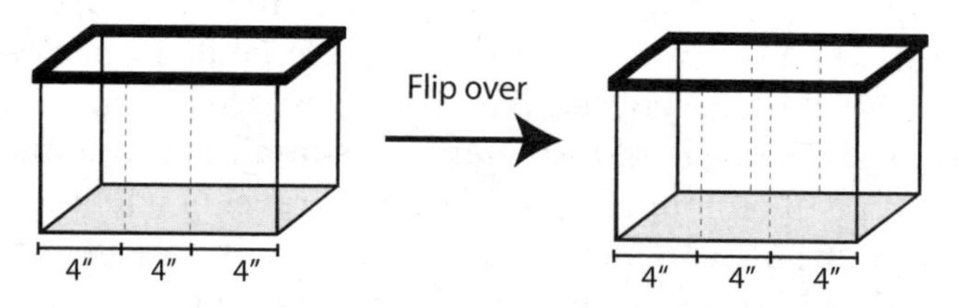

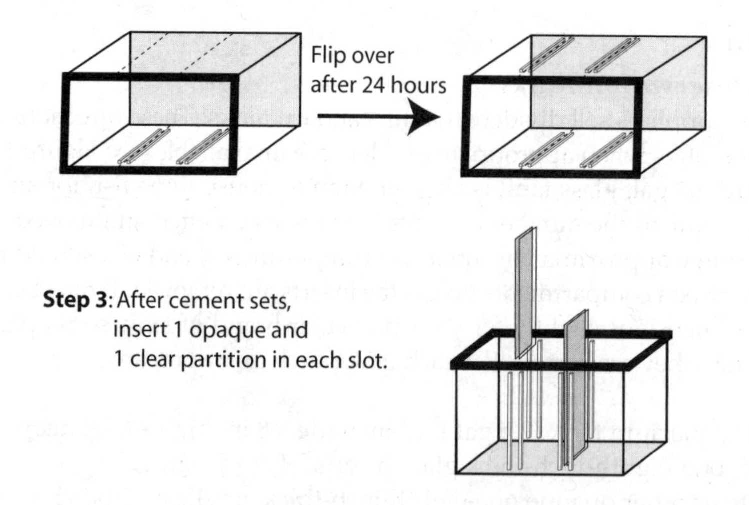

1. Lay the tank on a counter, long side facing up. Using a Sharpie, draw two vertical lines on the glass on the outside of the tank, approximately 4 and 8 in. from one edge, that divide the tank into thirds.
2. Flip the tank over. Make the same marks on the other long side.
3. Spread a thin layer of aquarium sealant on the long edges of two strips of Plexiglass. Place the strips on the inside surface of the tank, one-half inch apart and centered over one of the marks on the outside glass. When attaching these strips, be sure they are flush with the bottom of the tank. If they are a bit short and do not extend to the lip of the tank, do not be concerned; simply do not fill the tank completely with water. It is more important that the bettas not be able to swim between the plates and the wall at the bottom of the tank.
4. Place a strip of tape over the Plexiglass strips so they will not slip out of place.
5. Cement the second set of strips over the second line on the same side of the tank.
6. Allow the sealant to cure for at least 24 hours, then flip the tank over and repeat the process on the other long side of the tank. Tape the strips in place again.
7. Allow the sealant to cure for the time listed on the tube, then remove the tape.
8. Stack together one piece of glass and one piece of opaque acrylic. Slide the stack into the tank between the Plexiglass guides. Repeat with the second pieces of glass and acrylic. The glass and acrylic sheets should slide in and out of the tank easily.
9. Check the spacing between the sides and bottom of the aquarium and the glass and acrylic inserts. There should not be enough room for fish to squeeze through and move between compartments. Close off any spaces with additional aquarium sealant.

UNIT 14

Resource Allocation in Plants

Student Pages*

BACKGROUND

In an ecological community, plants and animals must interact with and adapt to each other and to the abiotic environment around them. Among animals, intra- and interspecific behavioral interactions often include some sort of movements. Plants also interact with other species in the community but are less obvious because they do so where they are rooted. Yet plant interactions shape the entire community and in large part determine the number and types of animals that are present.

Many intra- and interspecific plant interactions center on obtaining essential resources: sunlight, nitrogen or other nutrients, and water. When these resources are limited, it affects what species are present in a community. For example, an abandoned field has abundant sunlight but fairly dry soil. Pine seedlings are drought-tolerant and grow well in full sun. So they will colonize an open field fairly quickly. After pine trees become established, though, oak and other hardwood tree seedlings appear that eventually grow up and create a shady canopy. Most hardwood seedlings are shade-tolerant but require moister soil than pines. They cannot easily colonize an open field, but they thrive once pines or other hardwoods have appeared. In contrast, pines cannot tolerate the shade produced by maturing hardwoods and are not replaced as they die out.

*Teacher Pages begin on page 421.

So why hasn't natural selection led to a tree species that grows equally well everywhere? It has to do with **resource limitation**. Organisms can extract a finite amount of resources from their environments for growth and reproduction. Any species with the ability to outgrow all others in every conceivable set of conditions likely would need more resources than are available. This physical limitation is a potent selection force; in response, most species have evolved to allocate their resources in a particular pattern that determines where they are most likely to grow.

Each species' pattern of **resource allocation** is largely genetically determined, but it is not completely fixed. Individuals can modify their typical allocation pattern somewhat to allow them to adapt to different abiotic conditions and to the presence of other species. The extent of **phenotypic plasticity** differs between species. Some have little and so do not tolerate any change from optimum growing conditions. Other species have significant plasticity, so they can adapt to a variety of conditions. *How* resource allocation changes may differ as well. For example, some plants respond to drought by shifting resources to rapid downward growth of existing roots, while others reallocate their resources to forming a waxy protective cuticle on leaves.

Where Are Resources Allocated?

A plant's limited resources can be spent on growth either above or below ground. Above ground, plant shoots (leaves and stems) are the source of carbon compounds such as sugars derived from photosynthesis. These sugars get converted into molecules needed for energy storage (starches and lipids), for information storage (nucleotides), as structural components (phospholipids, proteins, and cellulose), and for metabolism (glycolytic intermediates). Belowground structures take up water and essential nutrients like nitrogen, phosphorus, potassium, and other minerals. Thus, allocation above or belowground is not just a matter of increasing body size; it affects the plant's ability to continue gathering resources.

In terrestrial environments, nitrogen is often the limiting nutrient that determines growth rate for plants. In response, plants differ in how they allocate nitrogen among their roots, stems and leaves. **Leaves** are the principal photosynthetic organs of the plant. Many rapidly growing weeds like knotweed, burdock, and dandelions favor placing more nitrogen in leaves, thereby increasing the photosynthetic ability of the plant and consequently the growth rate. However, these plants can only effectively grow in high light environments such as old fields that provide plenty of nitrogen. They do not perform as well if light or nutrients are scarce.

Stems are the principal supporting structures of the shoot. They are involved in light capture and competition by lifting the leaves above surrounding plants. In many plants there appears to be a trade-off between investment in stem and in leaves. For example, trees invest heavily in woody stems that support the leaves for a long time. The cost of this expenditure is slower growth. In contrast, many vines rely on neighboring plants for support and have flexible, metabolically inexpensive material in their stems, unlike self-supporting plants such as trees. Vines have additional resources they can partition into leaves and fast growth. Good examples of species using this strategy are kudzu, wisteria, and honeysuckle.

Roots not only serve as the primary absorptive organ, but they also anchor the plant to the soil. In many environments like the desert or the sandhills, soil water can be scarce and plants must allocate most of their resources to root growth and maintenance to survive. For example in the high desert of Utah, some shrubs like the creosote bush have deep roots that are able to access water 20 *meters* below the surface. Yet the stems and leaves of the shrub may only be half a meter tall.

In this lab, you will be comparing allocation strategies among plant species and within a species under different abiotic conditions. You will quantify how seedlings of several common crop plants allocate their carbon and nitrogen resources by measuring root and shoot biomasses, then expressing them as a **Root:Shoot (R:S) ratio**. The R:S ratio is simply a measure comparing relative root size (weight, length, etc.) to shoot size. The larger the value, the larger the roots are relative to the shoots, and vice versa.

Think about the relative contributions of roots, stems, and leaves to the survival of a plant. Which plant would be better suited for living in the diffuse light of a forest understory: one with a high root:shoot ratio or one with a low ratio? Why?

What would be the best allocation strategy for a plant living on a windy summit near the tree line?

What would be the best allocation strategy for a plant living in a seasonally dry climate like that found in southern California, where a rainy winter is typically followed by a dry summer?

Goals for This Unit

By the end you should be able to answer these questions about several crop plant species.

1. What is the normal allocation pattern for each species?
2. Does the normal allocation pattern change for a species when nitrogen becomes a limiting nutrient?
3. If the allocation pattern changes for a species, in what way does it change and by how much?

Based on your findings, you will design an experiment of your own that explores a question related to resource allocation.

General Safety Precautions

None of the plants you will use are harmful. If you request chemicals for treatments in your own experiment, ask your instructor about their associated hazards and proper handling precautions. You will be using vermiculite as a low-nutrient soil substitute. Vermiculite spilled on the floor can be very slippery. Clean up all spills promptly.

UNIT EXERCISE

Phenotypic Plasticity and Allocation

Background

In the lab there are trays of seedling plants grown in vermiculite, a soil-free mixture that contains few minerals. When the seeds were planted, the vermiculite was supplemented either with standard strength liquid fertilizer (1x) or with fertilizer diluted to one-tenth normal concentration (0.1x). You and your classmates will measure the root and shoot lengths and weights for several different plant species, which have been grown in either low or normal nutrient conditions. Once everyone is finished, you will compile your results into a large shared dataset.

The number of plants you must harvest is not the same for all species. Large plants like beans are heavy enough that you can measure the height and weight of a single plant accurately. For small or light plants like rye, you must harvest and measure 5–10 plants simultaneously to get an accurate measurement.

When you take multiple measurements of the same trait or feature, you should report the **mean**, which is almost always more accurate an estimate than a single measurement. Taking multiple measurements also allows you to calculate the **standard deviation**, which can be used to statistically test whether the mean for one treatment group is significantly different from the mean for another group. So that you can perform statistical analyses of your data, you must harvest and measure three single plants, or three sets of 5–10 plants.

Procedure

1. Your instructor will assign each work group one or two species to analyze. Harvest 1 (large) to 10 (small) plants by carefully pulling plants with adherent medium out of the tray.
2. *Gently* wash the plants in a container of water to remove the vermiculite and help separate the roots. Pat the plants dry with a paper towel.
3. Weigh the 1–10 whole plants together and record the weight in Table 14.1.
4. Separate the roots from the shoots by cutting each plant apart at the original level of the vermiculite.
5. Weigh the shoots and record the data in Table 14.1, page 417.

6. Calculate and record root mass, which is merely the total mass minus the shoot mass.
7. Measure the maximum shoot length and root length to the nearest mm, then record the mean value for each group of 1–10 plants in Table 14.1.
8. Repeat these steps for your second species. Record your data in Table 14.2 for the second species.

Results

If you had to collect several plants for each replicate, record the weight of all of the plants together in the *total wt.* column (T). Determine the average weight of a single plant by dividing total weight by the number of plants in the group. Enter the result in the *individual wt.* (I) column. Calculate the root:shoot weight ratio for each of the three replicates, using the *individual* root and shoot weights, and enter those values in Tables 14.1 and 14.2.

Next, calculate the root:shoot length ratios for each of the three replicates, using the maximum lengths. Enter your results in the tables. Finally, calculate the mean and standard deviation for the R:S weight ratios and the length ratios. Report these values to the rest of the class.

Table 14.1

Data for species 1

Species: ______________________ **Nutrient Concentration:** ______________________

Replicate and # of Plants Used	Roots (g)		Shoots (g)		R:S Wt. Ratio (Use I)	Max. Root Length (cm)	Max. Shoot Length (cm)	R:S Length ratio
	T	I	T	I				
1 ___								
2 ___								
3 ___								
			Mean				Mean	
			s.d.				s.d.	

Table 14.2

Data for species 2

Species: ______________________ **Nutrient Concentration**: ______________________

Replicate and # of Plants Used	Roots (g) T	Roots (g) I	Shoots (g) T	Shoots (g) I	R:S Wt. Ratio (Use I)	Max. Root Length (cm)	Max. Shoot Length (cm)	R:S Length Ratio
1 ____								
2 ____								
3 ____								
			Mean				Mean	
			s.d.				s.d.	

Class Summary
Be sure to record both mean and standard deviation for each species and treatment in Table 14.3.

Table 14.3

Class summary data

	R:S Weight Ratio Mean+s.d.		R:S Length Ratio Mean+s.d.	
Species	**Low Nitrogen**	**High Nitrogen**	**Low Nitrogen**	**High Nitrogen**

Using either computer or graph paper, create bar graphs that show root:shoot ratios for each species. For each bar in the graph, include an error bar to show the variation. Put ratios based on weight on one graph and ratios based on length on a second graph.

Unit Summary Questions

To answer these questions, refer back to your own and the class summary data, and give specific evidence to support your statements and conclusions. State what you saw, not what you would have expected.

1. Did you see a general pattern that describes how allocation varies between nutrient conditions? What was it?
2. Is there a relationship between plant growth rate (total plant size) and patterns of allocation when species are compared? In other words, do faster growing plants have a particular allocation pattern?
3. Plants can adjust their root:shoot ratio somewhat (i.e., demonstrate some **phenotypic plasticity**). Suppose there are insufficient nutrients in the environment. How might a plant change its root:shoot ratios to acquire nutrients when they are limiting?
4. Based on your observations, are differences in the root:shoot ratios between species primarily caused by changes in root size or by changes in shoot size? What is your reasoning?
5. The range of variation observed between the two nutrient treatments is a measure of the amount of phenotypic plasticity. What types of environmental conditions would favor plants with high phenotypic plasticity? Low plasticity?
6. What factors other than nutrient or light conditions might affect the plants' growth rate?
7. What factors other than overall size of the organisms might affect belowground competition? What factors other than size might affect aboveground competition?
8. When nitrogen was limited, did you find any species for which the R:S weight ratio changed, but the R:S length ratio did not? Did you find any species for which the length ratio changed, but the weight ratio did not? How would you interpret a difference in response of the length ratio and weight ratio?

Your Own Experiment

Caution: Do not finalize the details of your experiment until you have completed your analysis and answered the questions for the preceding exercise.

The unit exercise demonstrated how different plants change resource allocation in response to nitrogen levels. However, many other abiotic and biotic factors can affect resource allocation. The following lists summarize the materials available and the independent and dependent variables that can be manipulated easily. They are not inclusive; if you want to examine different independent or dependent variables, ask your instructor. Most likely, the necessary arrangements can be made. Just remember, your experiment in some way should explore how plants allocate resources.

Materials Available

- Seed of various crop species
- Vermiculite or similar nutrient-free planting mix
- Greenhouse flats and inserts
- Support stakes
- Shade cloth
- Colored acetate
- Growth boxes with light timers
- Balanced liquid fertilizer solution
- Nutrient-deficient fertilizer solutions
- Nutrient supplement solutions

Possible Experimental Variables to Test

Independent Variables

- Light intensity and wavelength. Shade cloth tents can be used to reduce total light, and colored acetate can be used to change the spectrum of light reaching the plants.
- Temperature. Plants can be grown in warm incubators versus a cold refrigerator.
- Soil composition and particle size (high sand, high clay, loam, etc.)
- Macronutrient levels (can adjust nitrogen, phosphorus, or potassium)
- Micronutrient levels (calcium, sulfate, copper, zinc, molybdenum, etc.)
- Water availability (ample, excessive, restricted)
- With versus without nitrogen-fixing symbionts (alfalfa +/- *Rhizobium* sp.)
- Seed size (white clover or alfalfa [small] versus. peas or beans [large])
- Planting density (ample room versus crowded)
- Herbivory (mimicked by occasional *light* clipping)

Dependent Variables

- Total, root, leaf, or stem mass, or root:shoot ratio
- Growth rate or plant height
- Shape or architecture of plant
- Germination rate or seedling mortality

UNIT 14

Resource Allocation in Plants

Teacher Pages

INSTRUCTORS' NOTES

Background and Key Concepts

All organisms extract mass and energy resources from their environments for growth, survival, and reproduction. However, not every organism extracts the same resources, or uses them in the same way. The pattern of resource allocation for organisms is determined in part by genetically determined differences in how they allocate similar resources. However, a species' general pattern of resource allocation is not fixed. Environmental factors influence allocation too.

In this unit students compare resource allocation among different crop plant species and within a single species under different abiotic conditions. There are several advantages to using plants as models. First, it is fairly simple to estimate relative allocation. In young herbaceous plants that are not yet flowering, nearly all resources go either to root growth or to shoot (stem and leaf) growth. A second advantage is that plants grow more rapidly than animals, so differences in allocation between species or treatment conditions can be measured sooner. Finally, plants are affected by a wide variety of environmental conditions, making it possible for students to design many different experiments.

In preparing students for this unit, stress these concepts:

- Organisms cannot obtain sufficient resources to dominate in all possible conditions. They must allocate the limited resources they do obtain to specific structures.
- Each organism within a species has a similar, genetically predefined pattern of resource allocation. All members of one species will emphasize similar physical features or processes over others.
- The genetically defined pattern of allocation is not completely fixed. Environmental factors can modify the basic pattern of resource allocation in specific ways. This is called phenotypic plasticity.
- The allocation strategy that helps plant species survive and reproduce under one set of environmental conditions may make it *less* likely to survive in a different set of conditions.

General Teaching Strategy and Common Problems

This unit was developed for a freshman majors course in ecology, but the methods have worked well in a nonmajors course. They have also been expanded and used in a more advanced class. The main exercise can be completed in one two-hour class period. Typically students require about one hour to set up their experiments, which they can do on the same day as the main exercise or during a later lab period. Students will need another two-hour block of time 14–17 days after setting up their experiments when they can collect their results.

In this unit relative resource allocation is expressed using a simple root-to-shoot (R:S) ratio, which is calculated from wet weights as well as from overall lengths; generally, the ratio calculated using weights is more informative. When the R:S ratio is ~ 1, resource allocation to roots and shoots is approximately equal. This pattern of allocation is typical of generalist plants like ryegrass that are moderately successful under a variety of conditions. When roots predominate, the R:S ratio for a species is >1. These species excel at pulling nutrients and water from dry or poor quality soils. Alfalfa is one species that uses this allocation strategy. Legumes that have nodules of symbiotic nitrogen-fixing bacteria also tend to have high R:S ratios. Plants also may have a R:S ratio >1 if their roots act as storage organs; good examples are radishes, potatoes, carrots, and other root crops. These species often dominate where growing conditions vary seasonally, because the plant can go dormant when conditions are unfavorable. When shoots predominate, plants have a R:S ratio <1. These species have a relatively high rate of solar energy capture. They often are annual plants, like oats, that grow quickly to reproductive maturity.

Students also compare R:S ratios between plants of the same species that have been grown in the presence of high versus low nitrogen. By the end of the main exercise, it should be very clear that some species are very plastic, while others are not. From this, students should be able to predict which crop species will adapt easily to a range of soil conditions, and which will not. For example, radish plants usually do not show a significant change in their R:S weight ratio between low- and high-nitrogen conditions. From this, students could rightly predict that radishes are less adaptable than the other species. In fact, gardening references confirm this prediction, stating that radishes require well-worked soil and supplemental nitrogen to thrive and form usable roots. In comparison, there usually is a very large change in R:S weight ratio for alfalfa grown in high- versus low-nitrogen conditions. As would be predicted, alfalfa adapts extremely well to a variety of growing conditions. Usually the cereal grains and beans exhibit less plasticity than alfalfa but more than radishes, and are moderately adaptable.

Recommended Prelab Skills

Students can perform the repetitive calculations and construct graphs more quickly if they know how to use the basic arithmetic and statistical formulas in a spreadsheet program. It is helpful if they have discussed energy capture in a lecture course, but this lab is sufficiently intuitive that students can complete it successfully without ever having heard about allocation previously.

Assessment

This unit is not particularly suited to using prelab quizzes. There is minimal factual content on which to base the prelab quiz. For a postlab quiz, it is reasonable to expect students to know what a R:S ratio means and why R:S ratios can change. It is difficult for students to remember the specific outcomes for every species tested, so questions about the behaviors of particular species in low- versus high-nitrogen conditions are not particularly informative. Instead, provide students with sample data such as a graph of R:S ratios from several species; ask them to interpret the graph or make predictions about which species would predominate under a given set of conditions.

Students can write a formal report that summarizes the results of the experiment they performed. However, a poster or oral presentation may be a better option. This particular model system lets students ask many different questions. It is unlikely that two groups in the same lab section will conduct similar experiments or make similar presentations. Hearing what their peers have discovered provides students with a rich learning context for interpreting their own results.

Safety and Housekeeping

There are no significant hazards associated with this unit. Leftover plant material from the main exercise can be composted, but plants or vermiculite treated with salt, metals, phosphate, or highly acidic or alkaline buffers should be disposed of in general trash rather than composted.

Other Tips

- The large amount of numerical data generated by this lab unit makes it ideal for teaching students basic biostatistics. If students struggle to understand the concept of phenotypic plasticity ask them to imagine two identical human twins that both weigh 170 lbs. One is a trained runner, and the other is a couch potato. Because they are genetically alike, their faces and general builds will be similar. However, the athletic twin will be more muscular, carry less body fat, and have denser bones from regular exercise, while the sedentary twin will have more body fat, less dense bones, smaller muscles, and more skin. The life stresses their bodies experience (constant versus rare exercise) have created significant phenotypic differences.
- The preceding tip points to another common misunderstanding that students have about how phenotypic plasticity works. In animals, body shape changes by rearrangement of existing structures—for example, bones get denser or lighter and muscles shrink or enlarge. In plants, phenotypic plasticity is expressed mainly by changes in current growth patterns, not by modifications of the plants' existing body structures. When a plant suddenly is exposed to nitrogen, its existing leaves rarely grow larger. Instead, the plant grows taller or grows more leaves. Plants do have the ability to reduce body mass by leaf, limb, and root abscission. However, in seedlings, changing the rates of root and shoot growth are the primary means of expressing plasticity, not tissue restructuring.
- This unit is difficult to complete when there are fewer than 12 students in a lab section. Smaller sections' datasets are supplemented with results from previous years or from other lab sections. Another option is to divide species to be tested into smaller sections, then compile a shared dataset. For example, one lab section might measure just monocots grown in low- and high-nutrient conditions, while another section measures just dicots.
- If time runs short, have students calculate and graph just the root:shoot weight ratios in class, then complete the root:shoot height ratios for homework.
- Most of the seeds used for this unit will germinate faster and more uniformly if they are presoaked in lukewarm tap water. Beans should be soaked overnight; other seeds should be soaked for 20–30 minutes.

Supplemental References

Hegazy, A. K., G. M. Fahmy, M. I. Ali, and N. H. Gomaa. 2005. Growth and phenology of eight common weed species. *Journal of Arid Environments* 61: 171–183.

Johnston, F. M., and C. M. Pickering. 2004. Effects of altitude on resource allocation in the weed *Achillea millefolium* (yarrow, Asteraceae) in the Australian Alps. *Australian Journal of Botany* 52: 639–646.

Tolvanen, A., J. M. Alatalo, and G. H. R. Henry. 2004. Resource allocation patterns in a forb and a sedge in two arctic environments—short-term response to herbivory. *Nordic Journal of Botany* 22: 741–747.

Yang, Z., and D. J. Midmore. 2005. Modeling plant resource allocation and growth partitioning in response to environmental heterogeneity. *Ecological Modeling* 181: 59–77.

PREPARATORY NOTES

Quantities listed are for a lab section of 20 students working in groups of three or four (except for the preplanted flats; each 12 × 24 in. nursery flat has sufficient material for five to six lab sections).

Week 1: Shared Materials

- Standard nursery flats filled planted with the following (one type or treatment per flat):
  - ◊ Radishes, fed twice with a high level of balanced nutrients
  - ◊ Radishes, fed twice with a low level of balanced nutrients
  - ◊ Alfalfa, high nutrients
  - ◊ Alfalfa, low nutrients
  - ◊ Mung beans, high nutrients
  - ◊ Mung beans, low nutrients
  - ◊ Annual rye, high nutrients
  - ◊ Annual rye, low nutrients
  - ◊ Oats, high nutrients
  - ◊ Oats, low nutrients
  - ◊ Winter triticale, high nutrients
  - ◊ Winter triticale, low nutrients

Week 1: Materials at Each Work Station

- 30 cm ruler
- Scissors
- Plastic boxes or beakers for rinsing plants
- Paper towels
- Large weigh boats
- Top-loading balance able to read 0.01 g
- Waste bin for organic matter

Optional Materials
(only provide if students are conducting experiments)

- Empty 6 × 6 in. plastic inserts for nursery flats, with drain holes, 4–6 per work group
- 1 empty nursery flat, without drain holes, per work group
- Bulk vermiculite
- 1x balanced liquid fertilizer
- 0.1x balanced liquid fertilizer
- Various crop seeds
- Large bowls or dishes that can hold 6 in. square inserts (for presoaking vermiculite)
- Small bowls or weigh boats (for soaking seeds)
- Shade cloth, three grades (e.g., 10%, 25%, 50% shade)
- Plant stand with lights (or arrange for space in a greenhouse)
- Labeling tape
- 24 in. bamboo or other small plant support stakes
- Colored acetate film, on rolls (available from craft stores)
- Single nutrient supplement solutions

Week 2: Materials at Each Work Station

- 30 cm ruler
- Scissors
- Plastic boxes or beakers for rinsing plants
- Paper towels
- Large weigh boats
- Top-loading balance (0.01 g sensitivity)
- Waste bin for organic matter

Sources of Materials

High-quality seeds are available in bulk from several vendors, including Seeds of Change (*www.seedsofchange.com*), Seed Savers Exchange (*www.seedsavers.org*), Nichols Garden Seed (*www.nicholsgardennursery.com*), and Johnny's Seed (*www.johnnysseed.com*). Frequently seeds can be purchased locally. Check farming suppliers and garden centers first, as they will carry the largest variety and freshest seeds. Home improvement stores sell annual rye, fescues, and other lawn grass seed. Specialty grocers and organic food stores regularly sell alfalfa, mung bean, radish, and other seeds for sprouting. Also, the seeds listed above are not the only ones that will work. Do not be afraid to experiment with other species, particularly cover crops such as buckwheat or clover.

Avoid purchasing seed from educational supply houses; often it has not been stored properly, so germination is uneven.

Growing Plants for Week 1

Start flats of seedlings two weeks in advance for summer or fall courses and three weeks ahead for spring semester courses. Twelve nursery flats of seedlings (six species × two nutrient levels) are ample for six sections of 20 students each to complete this unit; scale up as necessary for your course.

If students are going to perform their own experiments, expect to use twice the amount of seed and vermiculite listed below. Unused seeds can be refrigerated in closed plastic bags for up to one year.

1. To speed germination, presoak mung beans overnight in two changes of warm tap water. Soak all other seeds 30 minutes. For every two flats to be sown, soak:
   Beans—8 oz. (1 cup) of seed
   Oats, wheat—4 oz. (1/2 cup) of seed
   Alfalfa, radish, rye—2 oz. (1/4 cup) of seed
2. While seeds soak, fill 12 standard nursery flats (12 × 24 in. × 2 in. deep) with agricultural-grade vermiculite. Typically, one large (~18 lb. or 4 cu. ft. size) bag is sufficient.
3. Label each flat with the species to be planted and then label each flat "High Nutrients" or "Low Nutrients."
4. Prewet vermiculite by setting the filled flats in a tray of water. Alternatively, wet down the vermiculite from the top using a hose fitted with a fine misting nozzle.
5. Seeds should be sown about 3 "seed lengths" deep and 3 "seed lengths" apart. For example, to plant mung beans, scrape 3/4 in. (the length of three beans end to end) of vermiculite off the tops of the two flats marked "Beans." Divide the soaked mung bean seeds in half, and spread them out evenly on the exposed surfaces, about 3/4 in. apart. Press the beans down onto the surface, then cover them with the removed vermiculite.
6. Sow the other seeds in their flats, adjusting depth and spacing for seed size.
7. Water all flats again with a misting hose to settle the vermiculite around the seeds.
8. Place flats in a sunny room, greenhouse, or growth chamber set for a 12 hr.:12 hr. light/dark cycle. Keep all flats evenly moist until the seeds germinate.
9. Once ~75% of the seeds have germinated in a flat, water each flat thoroughly with liquid fertilizer. For the flats labeled "High Nutrients" use a 1x solution of liquid 10:10:10 plant food. For flats labeled "Low Nutrients" use plant food diluted 1:10.
10. When the first true leaves appear, reduce watering frequency. Wait until the top 1/2 in. of the vermiculite becomes dry. Watering more frequently encourages damping off and overgrown plants.
11. One week before the lab, water each flat again with 1x or 0.1x liquid fertilizer.

Solutions, Reagents, Equipment

Liquid Fertilizer Solutions

Purchase a balanced 10-10-10 liquid fertilizer; local generic equivalents of a national brand usually work fine. Follow package directions (typically 1 tablespoon per gallon of water) to prepare a 1x solution of fertilizer. For the 0.1x solution, dilute the 1x solution 1:10 with tap water. For convenience, 20 L of 1x and 0.1x fertilizer solutions can be prepared then stored in separate carboys. Check the listed formulation carefully before purchasing fertilizer; a balanced 20-20-20 mix can be diluted an additional 1:2 and used, but other liquid plant foods may have a 10-20-10 or other unbalanced ratio that can adversely affect the allocation pattern.

Materials for Students' Experiments

Students have dry seed, dry vermiculite, 6-inch square inserts for nursery flats, and an empty nursery flat to transport them. Each work group of three or four students will need two to four inserts for their experiment. Unless their experimental design specifically prohibits it, students should add 50 mL of 1x balanced fertilizer to the vermiculite when they sow their seeds. If students intend to manipulate specific macro- or micronutrients, they still must water their flats *at least once* with 50 mL of 0.1x balanced fertilizer. If they do not add at least some fertilizer, the seedlings are likely to die before the end of their experiment.

To ensure sufficient light, students' experiments can be kept in a departmental greenhouse. Alternatively, 24 in. fluorescent shop fixtures mounted 6–8 in. above the seed flats will provide ample light.

It is impossible to provide a comprehensive list of items that might be used for experiments, but the most common choices for your student population should become obvious after two or three semesters. In the author's experience, the most common student experiments are variations on two-species competition, where students measure rates of germination, growth, and root:shoot ratios of two plant species grown alone versus together in the same insert. For variety, seed can be kept on hand for several monocot and dicot species. If students want to use a species that is not already on hand, they must provide the seed.

Students often want to look for effects of light intensity or wavelength on allocation. Individual 6 in. inserts can be fitted with tents of shade cloth or colored acetate film supported on bamboo stakes. The colored acetate craft film is available at most craft stores or from educational suppliers. Small pieces of shade cloth may recycled from a greenhouse; larger rolls are available from greenhouse suppliers.

Other students have examined the effects of acid rain, salt deposition, or other environmental contaminants on allocation. Students should be asked to provide a reasonable contaminant concentration range (both EPA and USDA websites are good resources for this information). The instructor or preparator then can prepare a solution that matches the highest concentration in the range, which students can apply directly to the vermiculite in the same way as regular fertilizer. For lower concentrations, students can make serial dilutions of the stock contaminant solution.

Single Nutrient Supplement Solutions

Students can test the effects of specific single nutrients on allocation or plasticity using the solutions listed in Table 14.4. First students water their vermiculite once with 0.1x fertilizer solution when the seeds are first planted. After germination, they water plants two to three times with the appropriate solution listed in Table 14.4. If students want to look at the effects of excess nutrients, increase the concentration five- to tenfold.

Table 14.4

Composition of single nutrient solutions

Ion or Condition of Interest	Acceptable Range in Cropland Soils	To Match Acceptable Range, Water Vermiculite With This Solution
Ammonium	2–10 ppm (Excess burns roots)	10 mg/L ammonium chloride (NH_4Cl) in water
Boron	0.5–2 ppm	2 mg/L boric acid (H_3BO_3) in water
Calcium	1000–2000 ppm	2 g/L calcium carbonate ($CaCO_3$) in water
Copper	>0.6 ppm	1.5 mg/L cupric chloride monohydrate ($CuCl_2 \cdot H_2O$) in water
Magnesium	60–180 ppm	45 mg/L anhydrous magnesium chloride ($MgCl_2$) in water
Manganese	>1.5 ppm	5 mg/L manganese chloride tetrahydrate ($MnCl_2 \cdot 4H_2O$) in water
Molybdenum	>0.1 ppm	0.2 mg/L ammonium molybdate tetrahydrate ($(NH_4)_6Mo_7O_{24} \cdot 4H_2O$) in water
Nitrate	10–30 ppm (Excess burns roots)	25 mg/L sodium nitrate ($NaNO_3$) in water
pH	6.0 to 7.0 (Fertile acid soils may reach 5.0, but are rarely any lower)	Prepare 20 mM sodium citrate; adjust to desired pH (from 5.0 and 7.0) with 0.1 M HCl or NaOH. Dilute to final concentration of 10 mM citrate and recheck pH.
Phosphate	20–100 ppm	50 mg/L phosphorus pentoxide (P_2O_5) in water
Potassium	150–800 ppm	400 mg/L potash (potassium superoxide, KO_2) OR 900 mg/L KCl in water
Sodium and related salts	640–1280 ppm	1.3 g/L sodium chloride (NaCl) in water
Sulfate	2–10 ppm	6 mg/L sodium sulfate (Na_2SO_4) in water
Zinc	>1 ppm	2 mg/L zinc chloride ($ZnCl_2$) in water

UNIT 15

Population Ecology

Student Pages*

BACKGROUND

Two questions that ecologists frequently ask are (1) What factors limit where a species can occur? and (2) What factors limit the abundance of that species? These seem like simple questions, but answering them can be extremely difficult. One problem is that ecologists cannot account for every individual of a particular species within a particular area. So instead they work to understand the factors that control the size and growth of **populations** of that species. A population is defined as all individuals of one species that live together in similar conditions and usually in a single physically contiguous area. Populations are not static groups, but change dramatically over time. A combination of **biotic factors** (those associated with other living things) and **abiotic factors** (climate, temperature, and other factors that are *not* due to living things) affect a population's **intrinsic rate of growth**, that is, the rate at which it grows or shrinks. These factors also determine the **carrying capacity** of the habitat, that is, how many members of a particular species can live in a given space. If abiotic or biotic factors change, the rate of growth and carrying capacity of the affected populations change too.

The questions asked by population biologists are not just academic curiosity; entire economies can rise and fall on them. For example, the population of North Atlantic codfish was once believed to be inexhaustible. Yet in the mid to late 20th century, cod populations plummeted because of overfishing. As a result, many communities along the northern Atlantic coast that had relied on cod for over a

*Teacher Pages begin on page 445.

century were financially ruined. Current data show several more pelagic fish populations following the same downward trend as cod. In response, marine fisheries ecologists now monitor the populations of several commercially important species very closely. They use mathematical models to predict how these populations will change, then set catch limits and recommend fishing practices that will (hopefully) allow the fishing fleets to stay in business without decimating additional species.

How Do Abiotic and Biotic Factors Affect Populations?

The most obvious abiotic factor to affect population growth is physical space. If there is less space available, the carrying capacity for a particular species will be smaller than when more space is available. However, abiotic conditions can affect both carrying capacity and the rate of population growth in other ways. For example, day length is an abiotic condition. Longer days may allow a particular organism to find more food and, ultimately, produce more offspring. So for this species, day length is an abiotic factor that affects its population growth rate. Water availability is another obvious and powerful abiotic factor, especially in arid regions. Abiotic factors also can have effects on one population that in turn become biotic factors that affect other populations. Imagine an isolated valley in the southern Rockies, where groundnuts are a common forage plant. Total groundnut biomass depends on the amount of solar energy the plants capture, nitrogen content of the soil, average summer temperature, rainfall, and other variables. The size of the wood rat population depends on the number of groundnuts, which are the rats' main food. The size of the bobcat population in the valley in turn depends on the availability of wood rats, which are their preferred prey. Ultimately, the carrying capacities and rates of growth for all three populations depend on the same set of abiotic factors.

Biotic factors affecting population growth and carrying capacity can be subdivided into **intraspecific interactions** and **interspecific interactions**. Intraspecific interactions occur between members of the same species. One example of the many intraspecific interactions that affect populations is mate distribution. Suppose a solitary bobcat in the valley is searching for a mate. If mates are hard to find (sparsely distributed), the population growth rate will be lower than if mates are easy to find. This same limitation exists for plants that cannot self-pollinate. A member of the same species must be close enough for pollen exchange to occur; the closer a mate is, the more readily the plants reproduce, and the faster their population grows.

There also are many different forms of interspecific interactions, the best known being **competition**. Competition occurs anytime two species need the same limited resource and usually lowers the population growth rate and carrying capacity of both species. However, the effects of each species on the other are not always equal. Theoretically, if one species has a competitive edge over the other (no matter how slight), in time the species with the advantage will eliminate the other. This is called **competitive exclusion**. In the real world, though, there are few clear-cut examples of pure competitive exclusion. The potential for elimination by competitive exclusion puts tremendous selective pressure on the less competitive species. Selection will favor accumulation of any traits that make the less "fit" species a better competitor or that let that species use other resources that its competitor cannot. Going back to the southern Rockies valley example, the wood rats may compete with voles and marmots for groundnuts. In response to this competition, each species may have specialized to seek different parts of the plant; for example, voles might focus on tender roots, while marmots eat shoots, and wood rats eat the hard underground tubers.

Population growth rates also are affected when one species eats another. If the population or species being modeled is a plant, then its consumers are called **herbivores**. If the species of interest is an animal, then its consumers are either called **parasites** or **predators**. There are positive interspecific interactions as well. Two species may live as **mutualists**, where the presence of each species directly benefits the other. For example, recent studies suggest that most (if not all) forest trees have symbiotic fungal colonies on their roots. The trees provide sugars to the fungi, and the fungi transport and concentrate phosphate and trace minerals for the roots. When these fungi are present, the trees grow and reproduce more quickly than when the fungi are absent. In turn, additional roots from larger trees provide habitat for expansion of the fungal population. In this case, the interspecific interaction increases the total population size and rate of reproduction of both species involved.

Goals for This Unit

In this unit you will explore how biotic and abiotic factors affect population growth. You will explore the effects of competition on growth of different species of molds. You will be given actively growing cultures with which you will make an inoculation mix of mold spores (which are reproductive gametes) and mycelium (vegetative cells). You will inoculate sterile plates containing nutrient agar with a single mold species or with a combination of two molds. Over the next seven days, you will measure the density and types of molds growing in each dish.

Based on your initial results and outside reading, you will formulate one or more hypotheses about the abiotic, intraspecific, or interspecific factors that favor growth of the population of one mold species over another. Then you will design and set up your own experiment to test one of your hypotheses.

General Safety Precautions

The species of molds used for this exercise are not pathogenic, but they do produce light, tiny spores that can spread quickly and contaminate other students' experiments. To limit the spread of spores, keep plate lids closed as much as possible, wash your hands thoroughly before leaving the lab, and promptly clean up all spills or waste from your bench.

Close or repeated contact with molds can induce an allergic reaction in sensitive individuals. If you know that you are severely allergic to molds, or have serious reactions to penicillin or related antibiotics, inform your instructor. You can still record and help interpret data for your group.

UNIT EXERCISE

Competition Among Molds

Background

The typical life cycle of a mold is shown in Figure 15.1.

Figure 15.1

The life cycle of *Aspergillus* sp., a common mold

Most species of *Aspergillus* molds can form colonies either by vegetative growth of fragments of mycelium or by germination of spores. In this unit, the presence of spores will be used as a marker of mature mycelium.

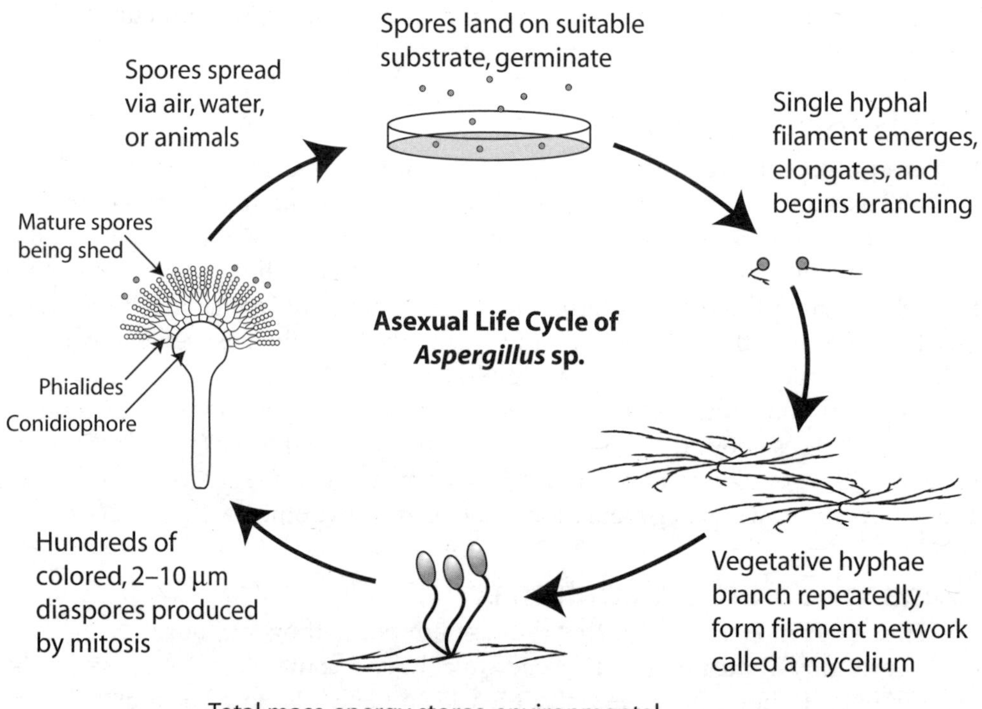

You will use several species of molds in the lab today:

- *Aspergillus flavus*: A common mold species found worldwide. It produces yellowish brown spores on a stalk with a star-shaped end bulb. Opportunistic, it will grow just about anywhere it has the chance.
- *Aspergillus niger*: Similar to *A. flavus*, except with black spores.
- *Penicillium camemberti*: Dull blue-green mold responsible for the veins in Camembert bleu cheese. Again, it can grow on a variety of substrates.
- Wild molds: Collected by leaving PDA plates uncovered for one to two hours in several locations around the building. Some of them may have been subcultured for one or two generations.

All molds are growing on plates of **potato dextrose agar** (PDA for short). This general nutrient media can support the growth of a variety of molds and fungi, but PDA is not the optimum growth medium for any particular species.

Procedure

Setting Cultures

1. Work in groups of two to four. Select two species of mold that you could easily tell apart if they were growing together. Take one plate of each mold culture you plan to use back to your bench.
2. Obtain three empty, 1.5 mL microcentrifuge tubes and nine plain, sterile PDA plates. Label the first tube and three plates with the name of Species #1. Label the second tube and three plates with the name of Species #2. Label the third tube and three plates "1:1 MIX." Also put your group name and today's date on the nine PDA plates. Set the plates aside for now.
3. On your bench is a small tube of dissociation medium (2% glucose in water). Use a sterile graduated plastic transfer pipet to place 1 mL of dissociation medium into each of the two tubes labeled with species names. Do not put dissociation media into the tube labeled "1:1 MIX" yet. Throw away the transfer pipet.
4. Uncover the autoclaved beaker and remove one glass Pasteur pipet. IMMEDIATELY replace the cover on the pipets. *Do not touch the large, open end of the pipet; hold it by the narrow end only!*
5. Open your stock plate of Species #1, and find an area that is completely covered with colored, spore-bearing growth. Use the large end of the glass pipet to punch out a round piece of the culture from this area. Close the plate lid.
6. Using the pipet only, transfer the plug of agar with mold to the corresponding labeled microcentrifuge tube containing dissociation medium. If the plug becomes lodged, gently blow it out with a plastic transfer pipet placed over the narrow end of the glass pipet. Close the cap on the tube.

7. Use a second sterile glass pipet to punch out a sample from the plate of Species #2. Again, place the plug in the appropriately labeled microtube, and close the cap.
8. Return the two stock culture plates to the front of the lab.
9. Shake your two microtubes vigorously for 10–20 seconds to wash the spores and mycelium from the PDA plug, and suspend them in the dissociation media. Put the two tubes containing single-species mold slurry back in the plastic tube rack at your bench.
10. Use a clean, sterile plastic graduated transfer pipet to transfer 0.5 mL of the slurry for Species #1 to the tube labeled "1:1 MIX."
11. Use a second transfer pipet to transfer 0.5 mL of the slurry for Species #2 to the tube labeled "1:1 MIX."
12. Close the "MIX" microtube, and invert it several times to mix. When you are finished, you will have two tubes containing 0.5 mL of mold slurry for a single species and a third tube with 1.0 mL of mixed slurry from two species.
13. Using a sterile plastic pipet, place ONE drop of mold slurry for Species #1 into the center of each of three corresponding, prelabeled PDA plates. Use a second clean pipet to inoculate the next three plates with one drop of mold slurry for Species #2. Finally, place one drop of the 1:1 mixed mold slurry onto each of the three prelabeled plates labeled "1:1 MIX." Replace the covers on all nine plates.
14. Remove a bent glass spreader rod from the dish of 95% alcohol, and wave it gently in the air until the alcohol dries off. Immediately open the first plate, and use the glass rod to spread the drop of mold slurry as evenly as possible on the three PDA plates inoculated with Species #1. Re-cover the plates.
15. Put the spreader back in 95% alcohol for at least 30 seconds to sterilize it.
16. Repeat Steps #14 and #15 for the three plates inoculated with Species #2. Sterilize the spreader in alcohol for at least 30 seconds. Then spread the drop of slurry on the three plates inoculated with the 1:1 mix.
17. When all nine plates have been spread, place the glass spreader back into alcohol.
18. Secure the lid to each plate with two pieces of clear tape. Then place the nine plates face up in a drawer or on the side bench where your instructor indicates.

Quantifying Mycelial Growth

You will estimate how much of the culture plate is covered by each species using a method called point-counting (see Figure 15.2). This method is often used in plant ecology to estimate species occurrences in natural communities. You must practice this procedure in lab with a sample culture plate before you begin collecting data on your experimental plates. After you have practiced it a few times, counting nine plates will take about 15 minutes each day.

Figure 15.2

Using point-counting to estimate mold density in petri plates

Tape the three PDA plates containing mycelium and mold face down onto an acetate grid with a preprinted grid of 0.5 cm squares. When the plates on the acetate sheet are turned upright, the spore-bearing mold should be visible through the acetate sheet and lid of the plate.

A. Mounting plates

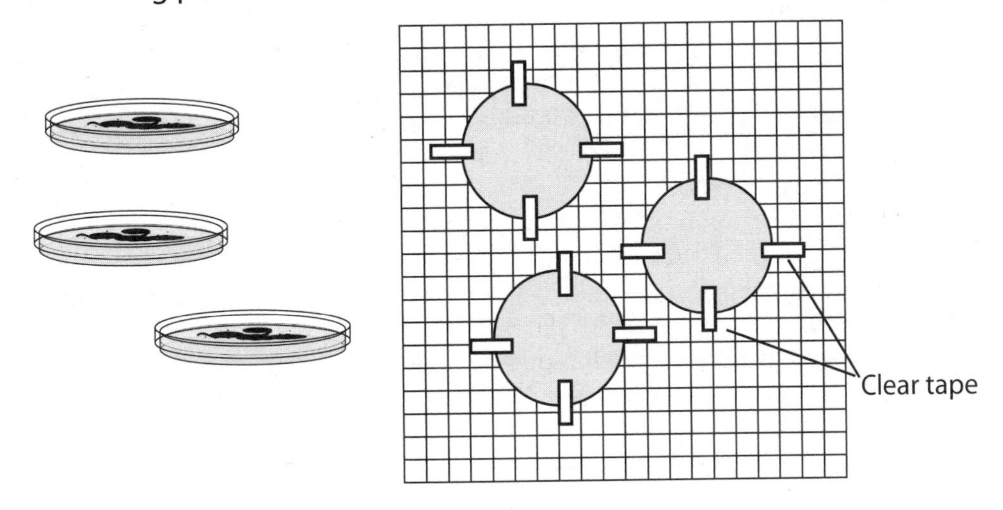

B. Marking and counting plates

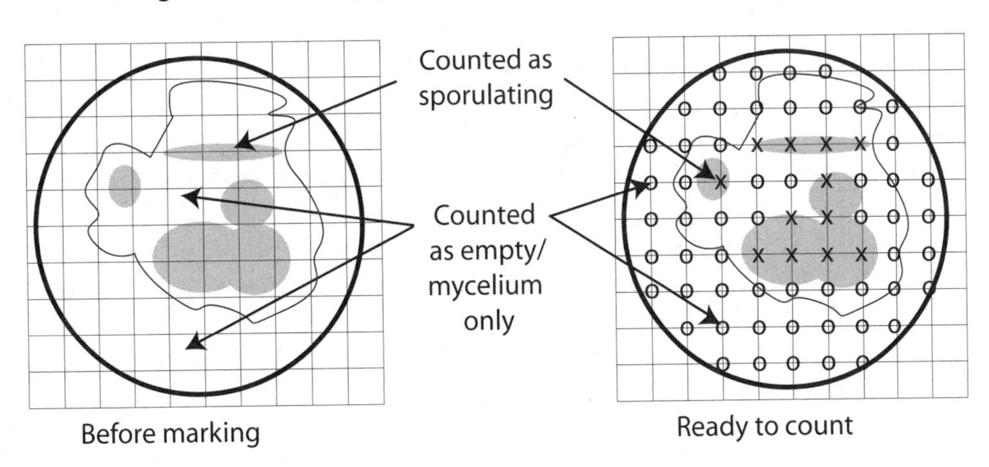

19. On the bench are clear acetate sheets that have grids printed on them. You will need one of these sheets and a dry erase marker.
20. You need be able to see whether there is anything growing on the culture surface at each of these points. If there is too much condensation to see clearly, open the lid of the plate, and wipe away the moisture with a Kimwipe. Then use two small pieces of transparent tape to attach the grid sheet to the lid of your three single-species plates.
21. Look down through the clear grid sheet at the mold on the surface of the plate. Focus on the surface *directly beneath* the point where two lines intersect.
22. Starting at the top of the plate, use the marker to make a small dot at every point at which intersecting lines overlay a spot where mature (i.e., colored, spore-containing) mold culture is present. Continue down the plate until you have marked every intersection that overlies mature, spore-bearing mold on the plate.
23. Now use a small open circle to mark every intersection that either is over an area of the plate that has no mold growing under it or is over an area that has whitish mycelium without any visibly colored spore bodies. Be careful not to erase any of the marks you made in the previous step.
24. When you are finished marking the grid, count the number of dots that are over spore-containing areas. Then count the number of circles that are over areas of the plate that are empty or have mycelium but no spores. Write both counts in Table 15.1, page 440.
25. Wipe off the grid, and repeat the previous steps with your second plate inoculated with Species #1. Write the counts in Table 15.1. Do the same for the third plate.
26. When you have finished counting and recording data for the three plates inoculated with Species #1, wipe the sheet clean and repeat for the three plates inoculated with Species #2.
27. For the competition (1:1 MIX) plate, you will need to count three things: number of intersections overlying spores for Species #1, the number of intersections over spores from Species #2 and the number of intersections that overlay areas that are either bare or have mycelium that cannot be assigned to either species. *(Hint:* A simple way to mark the mixed-species plate is to use a solid dot for Species #1, an X for Species #2, and an open circle for mycelium or empty space.) Write your results in Table 15.1.
28. When you have finished counting, wipe the acetate clean with a paper towel so it is ready to reuse.

29. To calculate the **percent coverage** for each mold species on each day, take the number of mature mold spots you counted, and divide it by the *total* number of spots counted (both mold and empty/mycelium only). Then multiply by 100%. The formula is:

$$\text{Percent coverage} = \frac{\text{\# spots w/spores}}{(\text{\# spots w/spores} + \text{\# empty spots})} \times 100\%$$

Record your results in Table 15.2, page 441.

30. Use graph paper or a spreadsheet program to plot growth curves for each species of mold you tested. Plot four separate lines:

- Average growth in the three plates by Species #1 alone
- Average growth in the three plates by Species #2 alone
- Average growth in the three plates by Species #1 in presence of Species #2
- Average growth in the three plates by Species #2 in presence of Species #1

Results

Table 15.1

Daily point-counting data

		Species #1 Name: ____________		Species #2 Name: ____________		Competition (1:1 Mix)		
Day	Plate #	# Spots Over Spores	# Spots Over Empty/ Mycelium Only	# Spots Over Spores	# Spots Over Empty/ Mycelium Only	Spots Over Species #1 Spores	Spots Over Species #2 Spores	Spots Over Empty/ Mycelium Only
1	1							
	2							
	3							
2	1							
	2							
	3							
3	1							
	2							
	3							
4	1							
	2							
	3							
5	1							
	2							
	3							
6	1							
	2							
	3							
7	1							
	2							
	3							

Table 15.2

Percentage of plate containing mature (spore-bearing) mold each day

Day	Plate #	Species #1 Name: ______	Species #2 Name: ______	Competition (1:1 mix)	
				Species #1	Species #2
1	1				
	2				
	3				
	Mean				
2	1				
	2				
	3				
	Mean				
3	1				
	2				
	3				
	Mean				
4	1				
	2				
	3				
	Mean				
5	1				
	2				
	3				
	Mean				
6	1				
	2				
	3				
	Mean				
7	1				
	2				
	3				
	Mean				

Unit Summary Questions

1. In your individual cultures where the molds are growing separately, what abiotic or biotic factors are present that could affect their rates of population growth? What abiotic and biotic factors determine the carrying capacity? What is your rationale for choosing these?
2. Look at the graphs for each species alone. The slope of the line connecting any two days indicates how fast that species of mold is growing over that time period; a small slope means growth is slow, while a large slope means the mold is rapidly taking over space in the plate.

   Looking at your own graphs, is the slope the same from day to day, or does it fluctuate? What might cause the pattern you observe? Suppose the slope was small early in the experiment, then became larger as time passes. How would you explain this observation? Now suppose the slope was steep in the middle of the curve, then flattened out toward the end; what would that mean? What might cause the growth curve to flatten out?
3. Look at the lines showing mold growth when both species were together. Pick a day during the experiment when spores from *both* molds were present. On this day, did each species cover the same percentage of the plate as it did on the same day when grown by itself? If so, why do you think the presence of a second species did not affect the rate of growth of each mold? If each species covers less of the plate, why do you think that happened?
4. Compare the rate of growth for Species #1 alone and in the presence of Species #2. Do the same for Species #2, in the presence of Species #1. Were the growth rates the same, faster, or slower for each species in the presence of a second species? Go back and review the theories behind interspecific competition. Now, what do you think your results mean?
5. Imagine that equal numbers of spores from Species #1 and Species #2 were inoculated simultaneously onto a PDA plate as described in the exercise. After seven days, a plug of media is removed from three places on the plate. The plugs are taken randomly, so may contain one or both species of molds. Each plug is washed with dissociation buffer, and the slurry is inoculated onto fresh PDA plates. Based on your own data for Species #1 and #2, what do you predict will happen on these "2nd generation" plates over the next seven days? Will both species be present, or only one, and why? Do you expect to see the same thing in all three of the new plates? Why?
6. Fungus beetles are specialized to eat molds and other fungal mycelia. Each day, a single fungus beetle can eat all of the mycelium and spores from up to 10% of the area of a standard 100 mm diameter petri plate. Suppose you performed an experiment in which you inoculated three replicate plates with slurry from your Species #1. You allow mycelium to grow for

one, two, or four days after inoculating the plates with spores. Then you add one fungus beetle to each plate. Each day for seven days after adding the beetle, you determine percent mold coverage on each plate. Describe what you would expect to see in each plate over the seven days. Be sure to explain your rationale. (*Hint:* Look at your own data graphs, then add in the additional variable of the beetle.)

Your Own Experiment

Once again, you have an opportunity to design and execute an experiment that extends your observations and results for the exercise or that addresses an entirely different question about population ecology. You may look at abiotic factors, interspecific factors, or intraspecific factors. When planning your experiment, remember to keep it simple. Do not design an impossibly difficult study. Just test one variable at a time. Also remember that your experiment must be hypothesis-driven. Develop your hypothesis first, then design the experiment to test it. Do not forget to include proper controls and replicates.

Materials Available

In addition to the materials you used in the main exercise, most of the items on the following list will be available. Remember, if you want to use something different you need to ask your instructor whether it will be available before finalizing your experimental design.

- Colored cellophane, shade cloth, artificial lights on timers (for modifying ambient light)
- Various temperature-controlled chambers (incubators and refrigerators)
- Alternative substrates: breads, cheese, old fruits and vegetables
- Supplemental nutrient mixtures (to provide additional nitrogen, phosphorus, or other ions required for growth)
- Buffers of differing pH (used to change pH of PDA plates)
- Antimycotics (create negative selection pressure)
- Low- and high-humidity chambers

Additional Suggestions

1. Do not limit yourself to the lab strains of mold. Colonies of opportunistic local fungi often appear in cultures. These can be sampled and compared to lab strains, or to each other.
2. In the exercise you inoculated plates by spreading out a mold slurry on them. However, you do not have to spread the mold slurry; you can cut plugs from mold-containing plates and place them in wells on clean plates, or add drops of slurry without spreading them. Think about how

the distribution of the mold spores could affect population growth or competitive ability.

3. You are not limited to measuring just one dependent variable. The type of data you can collect is not limited to spore coverage or to just using point-counting. For your experiment you might measure the distance of migration or the growth of the vegetative mycelium instead of the mature spores. Think about other types of data you might collect or different measurements you could make.

UNIT 15

Population Ecology

Teacher Pages

INSTRUCTORS' NOTES

Background and Key Concepts

Most instructors introduce students to the logistic growth model that was first developed by Vito Volterra in 1926 and then tested experimentally by G. F. Gause in the early 1930s. By examining the assumptions and components of this model, students can gain a deeper understanding of how and why population sizes change over time. Despite its predictive power, though, the logistic growth model has not been used widely in inquiry-based learning. In many lab courses students derive the model as a paper exercise or apply it to an historic dataset. This unit gives students an opportunity to see the principles of the model in action, to make predictions, then experimentally test them with living organisms.

To prepare students for this unit, instructors should emphasize these concepts:

- Growth of a population of a single species is not random. It usually follows a specific pattern that can be described using the logistic growth model.
- The overall shape of the logistic growth curve is the result of interactions among individual members of a single species (intraspecific interactions).
- Changes in abiotic conditions or interactions with a second species can change the shape or absolute numerical values for the logistic growth curve.

As a model system for this unit, common molds (Deuteromycetes) offer several advantages, the most obvious being rapid reproduction in a small space. In optimum conditions, germinated spores can complete their entire mycelial cycle and sporulate again in as little as four days. Many different species are available that can be distinguished from one another easily when they reach reproductive maturity. Finally, mold growth is very sensitive to environmental conditions; small changes can have dramatic effects. This last trait is particularly useful when students begin designing their own experiments.

General Teaching Strategy and Common Problems

This unit is used routinely for both nonmajor and majors courses. The nonmajors' version is presented here; majors also complete a prelab exercise in which they derive the logistic growth model for themselves. This unit is relatively simple to set up and execute, it can be scaled up to accommodate large numbers of students, and there are many different options for robust student experiments.

In the first week, students inoculate nine 100 mm petri dishes of potato dextrose agar (PDA) with mold spores. Two sets of three plates contain spores from two separate species, while the third set of plates contains a mixture of spores from both species. Over the next seven days, students estimate percent coverage of the plates with mature spores from each species using point counting.

This is an ideal inquiry unit for institutions with short lab periods. If students stay on task, plate setup can be completed in a one-hour lab session. However, students will need to return for about 20 minutes daily for the next week so that they can take point-counting measurements. Alternatively, one member of each work group can take the plates home, but they must be willing to make all the daily measurements for their group. At the start of the second lab meeting, students graph their group's point-counting data and discuss their initial results. Again, the required lab work can be completed in one hour or less.

Under optimum conditions, the graph showing the percentage of area with reproductively mature (i.e., sporulating) mold will fit a traditional logistic growth curve. Depending on inoculation density, though, some mold plates grow more slowly than others. It is not unusual for the graphs to lack a clear plateau. Some students choose to measure their plates for several more days until the curve plateaus, but this is not essential. It is more important that they see the relative rates of log-phase growth for the two species they tested. When students compare the relative rate of growth for each species alone against the rate for that species in the presence of another, there should be clear differences in the growth rate for at least one mold type. *A. niger* usually grows at the same rate regardless of whether another species is present or not. In contrast, the presence of *A. niger* delays log-phase growth of *A. flavus*, *P. camemberti*, and most other wild species of molds; it is not unusual for *A. niger* to completely eliminate other species. Usually,

A. flavus and *P. camemberti* mutually inhibit one another, while wild-type molds delay growth of the lab strains of *A. flavus* or *P. camemberti* to varying degrees.

During the second week, students also can set up an experiment of their own. Students should try to develop an experiment that builds on a question raised by their initial results rather than develop an entirely new one. After the instructor approves the experimental proposals, students set up, label, and inoculate the plates from which they will collect data during the coming week. At the third lab meeting, students collate and analyze their data with their group and make a brief informal presentation of their results to the rest of the class. Usually these informal oral presentations improve the written reports dramatically. However, instructors may prefer to have student groups collect data and complete the analysis entirely on their own time, freeing the third week for another topic.

Often students will not make a distinction between interspecific interaction and competition. Remind them frequently that there are *many* different forms of interspecific interactions, including herbivory, predation, mutualism, commensalism, parasitism, allelopathy, and succession. Pure head-to-head competition for one limited resource by two species is quite rare, because it results in the extinction of one of the competitors. For that reason, encourage students to use the more general term *interspecific interaction,* until they have data to indicate a more specific type of interaction is occurring.

It also is very difficult for students at this level to compare their curves statistically. Fortunately mold growth usually is quite different alone versus in competition. Typically, one species dominates the other or eliminates it completely, making the growth curve for one of the two species in the competition plates nearly or completely flat. Thus the difference in growth rates is obvious without requiring statistical analysis. However, there are situations in which a statistical comparison would be useful. In cases where the students' graphed data from replicate plates do not show such an obvious trend, students can determine whether there are any points on their curves where the means of the two different treatments are significantly different from one another on the same day. Each group will have sufficient data to calculate a mean and standard deviation for percent coverage at each time point for each species or treatment, which they can compare using a simple t-test. The author is painfully aware that this approach violates statistical rules, but most freshman-level students do not have the background to attempt more rigorous analyses. Fortunately, this particular approach is more likely to indicate that groups are not different when in reality they are, rather than the converse. Another suggested option is to have students conduct a regression on the log phase to generate a slope, then compare the slopes. Students who are confident in their abilities to interpret statistical outcomes can use more advanced techniques such as ANOVA.

Recommended Prelab Skills

Some experience with basic sterile or microbiological technique is helpful, but not required. Prior experience using a spreadsheet program to construct data graphs is useful as well.

Assessment

If the instructor uses pre- and postlab quizzes routinely as part of formal assessment, there are several general concepts and terms in bold that make good content questions for a prelab quiz. Generally, specific questions about methods should be reserved for a postlab quiz. When designing a postlab quiz, the instructor should remember that not all students select the same two species for their competition assays in the first week. The quiz questions should be relevant to all possible combinations of mold species. These are examples of questions from a typical postlab quiz:

- Based on the data from your own plates, which species appeared to predominate over the other? What is your rationale?
- What biotic or abiotic factors might have caused your dominant species to cover a larger part of the petri dish?
- If you repeated this past week's exercise, what would you change about the conditions within the plates to make the less successful species dominant? Why?

Other postlab quiz questions can focus on interpretation, application, and skills demonstration. Students reasonably could be expected to use point counting to estimate percent coverage of some photocopied images of plates, then graph the data. Alternatively, students can be given a set of fictitious percent coverage data and be asked to graph and interpret it.

Safety and Housekeeping

None of the lab strains of mold are biohazardous. Regardless, instructors should collect all PDA plates that contain mold growth and autoclave them prior to disposal.

Some strains of *A. flavus* produce aflatoxins, which are potent carcinogens. The lab strain sold by science education suppliers has inactivating mutations in the aflatoxin synthesis pathway. The strain does not produce aflatoxins, even when grown on peanut powder (the substrate most likely to induce toxin production [Wei and Jong 1986]). Also, aflatoxin production by wild-type strains is very low on potato-dextrose agar. Instructors can use *A. flavus* safely in teaching labs as long as they use commercial stocks. If plates appear to have been contaminated by wild-type *A. flavus,* they should be sealed and autoclaved to destroy them.

All molds can induce allergic reactions in sensitive individuals. More serious reactions can occur in students who are highly sensitive to penicillin. Students who know they are severely allergic to penicillin should avoid direct contact with the mold plates.

Students will be handling *Penicillium camemberti* and other molds used to flavor foods. Remind them that these are laboratory strains, not food-grade material. Students should *never* sample the molds to see what they taste like.

If the laboratory is equipped with gas burners, the instructor may opt to have students flame their glass spreaders instead of letting them air dry. This will not change the exercise significantly. Instruct students to be careful with open dishes of alcohol near any flames.

Other Tips

- A simple 20-minute oral exercise/informal assessment can significantly improve learning in this unit. During the wrap-up at the end of the first week, draw one or two fictitious logistic growth curves on the board, similar to those in Figure 15.3. Ask the students to describe any trends, then predict what is likely to happen to each species over time. Next ask them to predict which species would predominate in a mixed population. Students who have this early opportunity to practice evaluating logistic growth data usually make more meaningful interpretations of their own data at the start of the second week and design better experiments.
- Each semester one or two student work groups will set up experimental plates in which, for some unexpected reason, the mycelium fails to sporulate or mold does not grow at all. Usually this happens in the first week; most likely, students are not letting the ethanol-soaked spreader dry. Regardless of the reason, these students will not have data. Table 15.3, page 450, contains a dataset that two students granted permission to reproduce here for other students to use. These are the original data used to create Figure 15.3.

Figure 15.3

Sample population growth data

Undergraduates collected these data as part of their own experiment. PDA plates were inoculated with a long-standing laboratory strain of *A. niger*, a locally common wild mold that secretes prodigious red pigment (species unknown) or a combination of both. They hypothesized that the secreted pigment was allelopathic and predicted that the pigment would inhibit growth of the *A. niger*. What they observed in their experiment was that *A. niger* was unaffected by the presence of red mold. In contrast, growth of red mold was dramatically reduced by *A. niger*, and eventually *A. niger* overgrew it and filled most of the plate.

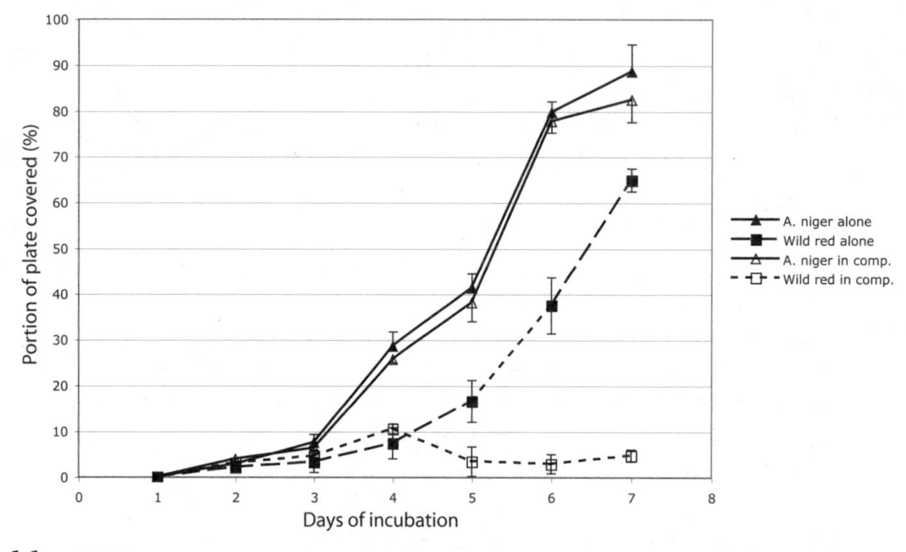

Table 15.3

Sample data from a student experiment

All values are mean ± standard deviation for percentage covered by spore-bearing mold, three replicate plates.

Day	Species #1: *A. niger*	Species #2: Wild Red Mold	Competition (1:1 mix)	
			Species #1	Species #2
1	0	0	0	0
2	2.8 ± 0.2	2.1 ± 0.3	3.9 ± 0.4	3.1 ± 0.4
3	7.5 ± 1.8	3.3 ± 2.3	6.3 ± 0.4	4.6 ± 0.6
4	28.6 ± 3.1	7.4 ± 3.4	25.7 ± 0.7	10.5 ± 1.1
5	41.3 ± 3.3	16.6 ± 4.6	38.1 ± 4.1	3.4 ± 3.2
6	79.7 ± 2.4	37.5 ± 6.1	77.7 ± 2.5	2.9 ± 2.1
7	88.6 ± 5.9	64.9 ± 2.5	82.4 ± 4.8	4.7 ± 1.3

- If students struggle with how to think through Unit Summary Questions 3 and 4, pose the questions to them graphically. Draw three two-row by three-column tables as shown in Table 15.4. Each table represents the outcome of a different experiment or pair of mold species. The "X" shows the relative growth rate for each species relative to its growth alone in a plate.

Table 15.4

Sample data for Unit Summary Questions 3 and 4

A

	Mixture		
	Less	**Equal**	**More**
Sp.1		X	
Sp.2			X

B

	Mixture		
	Less	**Equal**	**More**
Sp.1		X	
Sp.2	X		

C

	Mixture		
	Less	**Equal**	**More**
Sp.1	X		
Sp.2	X		

- Students should be able to see that, in Panel A, Species #1 is not affected but growth of Species #2 has increased, meaning it is gaining some benefit from growth with Species #1. In Panel B, the interpretation is that the presence of Species #1 is detrimental to growth of Species #2. Panel C suggests that both species mutually interfere with the others' growth; whether the cause is competition, allelopathy, or some other interaction is unknown. This matrix method can help students envision all possible outcomes.
- Opportunistic molds often appear in students' PDA plates during the first week. At the author's institution several "local" species pop up so regularly that they have earned informal names like red thumb, green dusty, black agar, or pink ooze. Encourage students to save the local species and test them in their experiments.
- *A. niger* is unusual in that it secretes large amounts of citric acid, which reduces the pH of its microenvironment to <3.5 and blocks growth of competing fungi. This is a textbook example of allelopathy and is why *A. niger* dominates over most other species. When a PDA plate is buffered so that pH remains around 5–6, growth of *A. niger* is inhibited and other molds can grow successfully, while at pH 7.5 or higher, bacteria begin to appear and displace molds.

Supplemental References

Alvarez-Vasquez, F., C. Gonzalez-Alcon, and N. V. Torres. 2000. Metabolism of citric acid production by *Aspergillus niger:* Model definition, steady-state analysis and constrained optimization of citric acid production rate. *Biotechnology and Bioengineering* 70: 82–108. This is one of many references describing optimiza-tion of industrial-scale production of citric acid by this species.

Wei, D. L., and S. C. Jong. 1986. Production of aflatoxins by strains of the *Aspergillus flavus* group maintained in ATCC. *Mycopathologia* 93: 19–24. This study identifies *A. flavus* strains that do not produce the toxin.

PREPARATORY NOTES

Quantities listed are for a lab section of 20 students working in groups of three or four.

Weeks 1 and 2: Shared Materials

- 1–2 confluent 100 mm plates of *Aspergillus flavus*
- 1–2 confluent 100 mm plates of *Aspergillus niger*
- 1–2 confluent 100 mm plates of *Penicillium camemberti*
- 3–4 plates (100 mm) of subcultured "wild" molds
- 5–6 subconfluent training plates
- 60 potato dextrose agar (PDA) plates (keep refrigerated prior to use)

Weeks 1 and 2: Materials at Each Work Station

- 5 mL dissociation medium in a closed tube, sterile
- 8–10 graduated plastic transfer pipets (1–2 mL), sterile (singly wrapped is best)
- 4 microtubes (1.5 mL), in a beaker covered with foil and autoclaved
- Plastic microtube rack
- 4–6 sterilized glass Pasteur pipets, standing in a beaker with their wide ends *down*, covered in aluminum foil, and autoclaved
- Bent glass spreader
- Dish of 95% alcohol (ethanol or isopropanol)
- 1–2 transparency acetates, photocopied with a 0.5 cm grid
- Dry erase marker, preferably fine point

Optional Materials for Experiments (provide only if requested)

- Incubators or similar temperature-controlled chambers
- Refrigerator
- Low- and high-humidity chambers
- Sliced bread, with no preservatives (alternative substrate for molds)
- Sliced provolone cheese (alternative substrate for molds)
- Colored acetate film (to change light wavelength)
- Shade cloth, aluminum foil, artificial lights on timers (to modify the ambient light and to control day length)

- Supplemental nutrient mixtures (to provide additional nitrogen, phosphorus, or other required ions)
- Buffers of differing pH (used to change pH of PDA plates)
- Disinfectant, antimycotic sprays

Sources of Materials

Stock cultures of *A. niger, A. flavus,* and *P. camemberti* can be purchased from education supply vendors. To capture local "wild" species, place six to eight PDA plates at various locations, and remove the lids for one hour. Reclose the plates, and incubate them at room temperature in the dark for five to seven days, or until colored, sporulated molds are visible. Plates of sporulated molds can be sealed with Parafilm and stored refrigerated for six months to a year.

Approximately two weeks before the first week of this lab, expand the commercial stock strains and wild-type molds. For each lab section of 20 students, inoculate two PDA plates with each species of mold that students will use. It is not necessary to amplify every species of wild mold that was captured; only amplify wild species that can be distinguished from *A. niger, A. flavus,* and *P. camemberti.* Detailed directions for inoculating plates are in the student exercise (p. 435).

Solutions, Reagents, Equipment

Potato Dextrose Agar (PDA) plates

Students use a large number of PDA plates in this unit. To ease the time crunch, start making plates two to three weeks in advance. Mix 39 g of PDA powder (e.g., Fisher, #DF0013-17-6) with 1 L of distilled water in a 2 L Erlenmeyer flask. Stir for 10 minutes to dissolve salts (the agar will not dissolve). Autoclave for 20 minutes. Swirl the flask to mix melted agar, allow the media to cool to 55°C, then pour into 100 mm plates. Replace lids, then let plates solidify and dry out on the counter overnight. Each liter of media makes 35–40 100 mm plates. Plastic sleeves of finished plates can be stored refrigerated for six weeks. Use oldest plates first, and discard any that develop mold spots.

Dissociation Medium

Mix 10 g of glucose with 500 mL distilled water. Dispense 5 mL aliquots into screw-cap glass tubes, and put caps on loosely. Autoclave for 20 minutes. When completely cooled, tighten caps and label.

High pH, Low pH, Single-Nutrient Plates

Unbuffered PDA plates average ~pH 5.5. Instead of preparing plates of differing pH or with different supplements in advance, let students prepare them as needed.

For plates of different pH, prepare acetate or phosphate buffer at the desired pH (see Unit 5, Applied Enzymology for detailed instructions) and filter sterilize.

Have students overlay PDA plates with 5 mL each of the appropriate pH buffer, and allow the plates to soak for 30 minutes. Invert the plates and allow them to drain thoroughly, then inoculate as usual.

Instructions for preparing solutions of individual ions or nutrients are in Table 12.6 on page 392. Overlay individual plates, allow to soak, and drain as described for changing pH of plates.

Alternative Substrate Plates

To make cheese plates, place a thin slice of provolone or similar block cheese onto a clean piece of plastic wrap. For bread plates, place one slice of deli or other fresh bread between two sheets of plastic wrap, flatten it with a rolling pin or book, then remove the top piece of plastic wrap. Open a sterile PDA plate. Then, using it like a cookie cutter, punch out a round from the cheese or bread. Turn the plate back over, and use an ethanol-sterilized, dry spreader to push the cheese or bread down so it contacts the PDA (which will supply the necessary moisture). Inoculate the cheese or bread surface with the desired mold slurries or with plugs from confluent plates.

Avoid using singly wrapped American cheese food. It contains sufficient salt and preservative to inhibit all mold growth. Similarly, preservatives in commercial white bread make molds grow poorly; use fresh deli bread instead.

Subconfluent Training Plates

These plates may be any species, as long as they are *not* completely confluent; one set of four or five plates usually is sufficient for an entire week. Rub a countertop with two fingers, then touch the surface of a PDA plate four or five times. Incubate the plates at room temperature in the dark for three to five days until sporulated molds cover one-third to one-half of the plates. To limit further growth, store subconfluent plates in the refrigerator between lab sections.

Disinfectant and Antimycotic Sprays

Students are very interested to find out how well various commercial compounds affect fungi. Aerosol cans or spray bottles of a general disinfectant such as Lysol, two or three over-the-counter antimycotic foot sprays for treating athlete's foot, and two or three garden fungicides can be kept on hand routinely. Remind students to spray away from anyone's eyes and to use a light touch; the spray should barely cover the plate surface.

UNIT 16

Measuring Biological Diversity

Student Pages*

BACKGROUND

Biological diversity is a hallmark of life on Earth. The fossil record indicates that a variety of species evolved very soon after life first appeared on this planet. In its long history, there have been several global catastrophes that wiped out most of the species living at that time. Each time, a small fraction of species survived then underwent natural selection and adaptive radiation, resulting in a wide variety of new species once more.

One reason this planet supports so many different species is that they are not distributed evenly. Each has a suite of adaptations that allows them to inhabit a specific ecosystem or microenvironment within an ecosystem. For example, brown trout and smallmouth bass are both predatory fish found in freshwater; some rivers even contain both species. However, brown trout and smallmouth bass rarely are present in the same sections of a river. Differences in temperature, oxygen content, and flow rate separate the river into distinct ecosystems. Trout will inhabit the colder, more highly oxygenated areas, which usually are nearer to the headwaters of the river, while smallmouth bass will be found in warmer, slower-moving areas downstream. While both species live in the same river, they are not in direct competition with one another for space, food, and other resources.

*Teacher Pages begin on page 471.

Every ecosystem has a set of species that can typically be found there. If we know what the typical species of an ecosystem are in one location, we can predict what species should be found together in that same ecosystem at another location. Going back to the previous example, smallmouth bass usually live in rivers that have stony or gravel bottoms, aquatic grass beds (where the bass can forage and hide to ambush prey), and a large population of crayfish (a major food item for bass). If we netted a random sample of fish from a river and found that the majority were smallmouth bass, we could reasonably predict that the river will contain crayfish, aquatic grasses, and other species associated with gravel-bottomed rivers.

There is a catch to such predictions: Just because certain species *could and should* be in an ecosystem, it does not guarantee they *will* be there. The number of potential species actually present depends on the **quality** and **stability** of the ecosystem. These two factors are closely intertwined but not exactly the same thing. *Quality* describes how closely a particular ecosystem matches a theoretical ideal. For example, a river that has every one of the features favored by smallmouth bass would be high quality. Suppose the river bottom is mostly sand, and gravel is only present in patches. This second river can still support smallmouth bass, but it is not as high quality a habitat.

Stability refers to how long the defining features of a particular ecosystem stay in place. Imagine two rivers, the Abel and the Baker, each with identical bottom structures that have the potential to support smallmouth bass populations. However, the Abel River maintains a relatively constant temperature throughout the year. In contrast, the temperature of the Baker River fluctuates because it has a shallower channel and heats more quickly in the sun. In addition, the Baker receives more warm summer runoff than the Abel River. In this scenario, the Abel River is the more stable ecosystem because the temperature does not fluctuate as much.

These fundamental principles point to some important questions:

- ***What determines the quality and stability of an ecosystem?*** There are many different factors that affect quality. Some are physical features of the landscape or are imposed by the local climate. Other factors that affect quality and stability are imposed by human activities. For example, nitrogen-rich runoff from agricultural fields or animal pastures or hydrocarbon-laden stormwater from a city can dramatically affect the quality of an aquatic ecosystem. The frequency at which these events happen affects stability.
- ***How do quality and stability affect diversity?*** As a general rule, species diversity in an ecosystem increases as the quality and stability increases and falls as the ecosystem becomes unstable or of lower quality. This relationship holds true regardless of the reasons why the ecosystem changes. A high-quality, stable ecosystem will contain a larger percentage of the species that are adapted to living there (high biodiversity). If an

ecosystem is unstable or of low quality, then a smaller percentage of the species that could potentially live there will actually be living there (low diversity).

- ***Suppose abiotic and biotic factors change so much that the previous species inhabiting an ecosystem cannot adapt to it anymore; what happens then?*** Typically, the original ecosystem collapses, and the species that cannot adapt disappear from the area. New species enter that are better adapted to the conditions, and a new, different ecosystem develops. How long the new ecosystem remains in place depends again on how stable conditions are and on the quality of the habitat for the newly arrived species.

  Ecosystems change, become unstable, or collapse for many reasons. Often it is a natural event resulting from changes in climate or other abiotic factors. Ecosystems also may collapse due to biotic factors such as the introduction of new exotic species or as a result of environmental impacts by human activity.

Why Measure Biological Diversity?

How do we know whether an ecosystem is stable or high quality? We measure the overall biodiversity within the ecosystem. What can we learn by doing this? If we compare the relative number, variety, and types of organisms in one location at two different times, we can determine if and how that ecosystem is changing over time. We can also calculate and compare biological diversity indices for two locations that should be similar and see if one is more or less diverse than the other. Using this information we can start to assess the impact of human activities, climatic changes, and other events on the ecosystem in question.

The mathematical methods for estimating biological diversity can be applied to almost any type of ecosystem. So why not compare two locations in a terrestrial ecosystem instead? We focus here on aquatic ecosystems because they act as amplifiers for small, often unnoticed changes in their surrounding terrestrial ecosystems. Small creeks, streams, and ponds collect stormwater and other runoff from a much larger area of surrounding land. In addition, most natural bodies of water are inhabited by many small, thin-bodied invertebrates. These organisms absorb the materials in runoff and are directly affected by it. Changes in aquatic ecosystems serve as early indicators of events happening in the surrounding terrestrial ecosystem.

Goals for This Unit

This week you will be estimating biological diversity in two different aquatic ecosystems. First you will go out and collect water and bottom detritus from two locations. Your instructor will tell you more about the particular sites you will be visiting. You also will make field notes about the locations. If lab time is short, your instructor may have already collected the sample material for you.

Normally, field ecologists collect several different samples for each site they want to survey over weeks, months, or even years. For this exercise, each work group will collect and examine one sample from each of two different locations, then pool their data for analysis.

In lab you will locate invertebrate organisms that are present, use a dichotomous key to identify them to phyla, and tabulate the number of species in each phylum. Then the entire class will tabulate the total numbers of phyla and species present. Using these pooled data you will calculate two estimates of biodiversity for each location and compare the two locations.

General Safety Precautions

There may be sharp sticks, stones, or thorns in the collected material. If you are stuck or cut, wash the wound thoroughly and report the injury to your instructor.

There is an extremely slight chance that aquatic insects in the samples might bite you. The offenders usually are toad bugs, water boatmen, or other Hemipterans. They are difficult to see, but when disturbed defend themselves with a needle-like beak. Their bite is painful (it is similar to sticking your finger with a straight pin or staple) but not dangerous. If you find a large insect, do not pick it up carelessly. Gently move it into another container with forceps or a plastic spoon for observation.

UNIT EXERCISES

Exercise 1: Field Collection

Procedure

Your instructor will take you to the two sites and show you exactly where they want you to collect your samples.

1. At Site #1, use a dip net to fill one-half of a 1 pt. jar with bottom detritus. Try to collect a sample that represents the typical bottom material. For example, if the bottom is mostly covered with decaying leaves, do not collect a detritus sample that is mostly sand and silt; collect decaying leaves.
2. Fill the remaining half of the jar with water from the site and close the lid.
3. Now look around Site #1, and make notes about what you see.

    - What is the source for the water here?
    - What type of terrestrial ecosystem does it pass through before reaching your sampling location?
    - Are there trees, shrubs, or herbaceous plants next to or in the water?
    - How fast is the water flowing? Is it highly aerated?

- Is there evidence of larger animals (tracks, partly eaten plants or prey, or scat) in the area?

4. Go to Site #2, and repeat the sampling procedure. Take notes about the site as well. Before leaving Site #2, write down a preliminary prediction of which location (Site #1 or Site #2) you think is going to have greater biodiversity. What is your rationale?

Exercise 2: Identifying Invertebrates

Procedure

You will work in pairs for this exercise.

1. Transfer 10–15 mL of water and a small amount of the detritus from your sample of Site #1 to a glass dish or white plastic plate. Start with a pile about the size of a ripe cherry. When you finish searching through this first batch of material, empty it into the waste container provided and start searching through a second batch.
2. One partner should use forceps, a dissecting needle, and his or her fingers to separate and move the detritus around. Look for movement both in the water and between pieces of debris. Use a hand lens or, better, a dissecting microscope. Proceed carefully and methodically; most invertebrates will be well camouflaged.
3. When you find an organism, use forceps or a plastic spoon to transfer it to a plastic weigh boat or small dish. Use the dichotomous key to identify the group to which it belongs. If you cannot see the organism clearly, gently transfer it to a depression slide. Add a coverslip and use a microscope to see additional detail.
4. The other partner should place a sample of silt or other fine material from the Site #1 sample directly onto a microscope slide. Using your brightfield microscope on its lowest magnification, search through the material for any very small invertebrates.

    - The amount of material you can place on a slide and search effectively is much smaller than in Step 2. Place only a few drops of silty debris on the slide.
    - Do not just take a quick look at your slide. Methodically scan back and forth through the material. Look for movement to indicate an organism; you are not likely to see the whole organism all at once.

5. Record each phylum that you identify—using either the dissecting or brightfield microscope—in Tables 16.1 and 16.2, pages 460–461. Also make a note of each obviously different species you find within each phylum. If you find more than one individual of a particular species, make a note of how many individuals of that species you find.

6. After about 30 minutes, switch over and start searching the material from Site #2. Try to use the same searching pattern and methods as you did for material from Site #1.

Results

Individual Results

Table 16.1

Site 1 individual student data

Phylum/Class of Invertebrate	Your Descriptive Name for This Particular Species	Number of Individuals You Found of This Species

Table 16.2

Site 2 individual student data

Phylum/Class of Invertebrate	Your Descriptive Name for This Particular Species	Number of Individuals You Found of This Species

Class Summary Data

Table 16.3

Site 1 class summary data

Phylum/Class of Invertebrate	Names of Distinct Species Found at Site by the Class	Total Number of Individuals Found of Each Species

Table 16.4

Site 2 class summary data

Phylum/Class of Invertebrate	Names of Distinct Species Found at Site by the Class	Total Number of Individuals Found of Each Species

Exercise 3: Calculating Diversity Estimates

Background

There are many ways to estimate relative biodiversity. Today you will compare the two sites based on species richness, phylum/class richness, and dominance. These methods assume you sampled both habitats in the same way, that equal amounts of time were spent searching each one, and that species were characterized in the same way in both habitats.

Richness

This is the most intuitive measure of biodiversity; it is just the number of different groups in a defined area. Richness can also be estimated at any taxonomic level by counting the number of genera, families, or even phyla present. Regardless of the level used, the question is the same: Which has greater biodiversity?

Dominance

Sometimes an area will have many species, but there are only a few individuals representing most of the species present. In such cases, species richness would be misleading. For example, suppose all the plants in a cornfield and in a meadow were counted. Following are the results. Which has greater biodiversity?

- Cornfield: 10,000 corn plants and four plants each of five wildflower species.
- Meadow: 6000 plants, but there are 1000 plants of each of six flower species.

Intuitively, you would say that the cornfield is less diverse, because most of the plants are one species and because there are very few individual plants for the remaining species. However, if just the number of species are counted, both plots have six species. So according to the diversity index described above, the two sites have equal species richness.

To account for the number of representatives of each species, **dominance** (d) is calculated using the Berg-Parker Index. The equation is:

$$d = \frac{\text{total number of organisms } (N_{total})}{\text{No. of organisms in most abundant species } (N_{abund})}$$

Going back to the cornfield versus meadow example:

- Cornfield: d = [10,000 + (4 x 5) total plants] / 10,000 = 1.002
- Meadow: d = [6000 total plants] / 1000 = 6.00

The larger the value for d, the greater the diversity, so based on the Berg-Parker Index, the meadow has greater biodiversity.

Procedure

1. Using the class summary data in Table 16.3, add up the total number of different phyla/classes that your lab section identified in samples from Site #1 and enter those values in Table 16.5. Do the same for the number of different species your lab section found. Then count up the number of species and phyla/classes that your lab section found in samples from Site #2 (these data are in Table 16.4). Enter these results in Table 16.5 as well. You now have your estimates of species richness for both sites.
2. To calculate dominance, add up the total number of organisms, regardless of species, that your entire lab section observed in samples from each site (data are in Tables 16.3 and 16.4). Record these values in Table 16.6.
3. Determine the species for which your lab section found the most individuals in each site (again, data are in Tables 16.3 and 16.4). Record the number of individuals your lab section found of that species in Table 16.6.
4. Use the numbers in Table 16.6 to calculate the dominance for both sites.

Results

Table 16.5

Relative site richness

Site	Number of Species Present	Number of Phyla/Classes Present
#1		
#2		

Table 16.6

Relative site dominance

Site	Total # Organisms (N_{total})	# of Most Common Species (N_{abund})	Dominance $(N_{total})/(N_{abund})$
#1			
#2			

Unit Summary Questions

1. Which of the two sites has the greatest species richness? Phyla/Class richness? Go back to your field notes describing the two sites. Think about how the surrounding environment would affect the water quality at each location. Based on this, are your results for species richness or phylum richness what you would have predicted, or are they surprising? Why?
2. Which site has the larger value for dominance? What does that mean? Based on dominance, which site has more biodiversity? Did you get the same result for species richness? For phyla richness? If not, how might you explain the differences?
3. Which phylum/class is the most common at each site, in terms of total number of individuals? Do you think this will be true year-round, or will other organisms come to predominate? What is your reasoning?
4. If the most common phylum/class is the same at both sites, why do you think that is so? If they are different between the two sites, why do you think are they different?
5. Do you think that you have observed all of the phyla that are present in this particular ecosystem? Why or why not? If not, how could you ensure that you have seen all the phyla that are present?
6. Relying on just one sampling technique may prevent you from seeing all of the organisms present in a particular system. This is referred to as **sampling bias**. Think about how you collected your samples; what might be some sources of sampling bias? Which species might you have missed?
7. In general, what external factors or events could cause one freshwater body to have greater biodiversity than another? Out of all the factors you just listed, which ones are most likely affecting the overall biodiversity in the two particular sites you sampled?

Dichotomous Key of the Major Freshwater Invertebrate Phyla

Table 16.7

Key to common aquatic invertebrates

	I. Overall size of the organism	
A.	Organism is microscopic, usually <1mm in size; may be possible to see moving with unaided eye, but must use a microscope to see overall body layout.	Go to Category II
B.	Organism is macroscopic, usually >1mm in length or overall size; usually can see overall body layout with unaided eye. May require a microscope to see details of body structure.	Go to Category III
C.	Body micro- to macroscopic. Under microscope, individual cells are obvious, and contain bright green membrane-bound organelles called chloroplasts (a microscope is needed to see them). Organism is not swimming or moving. May occur as single blocklike cells, strands, clumps, or ball-like colonies.	Freshwater algae (Do not count)
	II. The common microscopic pond organisms	
A.	Body shape variable; may be round, oval, slipper-shaped, or amorphous. One or two large dark nuclei may be visible in center, indicating a single-celled organism. Vacuoles often present in cytoplasm. Organism moves by crawling, by whipping 1–2 long flagella, or by waving multiple rows of cilia. Often see cytoplasmic streaming inside cell.	Phylum Protozoa (protists)
B.	Organism has vase- or wine bottle-shaped, elongated body. Swims rapidly but erratically. Larger end of the body (oral end) has ring of beating cilia that appear to be rotating. Narrow end usually tapered, and often has "toes."	Phylum Rotatoria (rotifers)

(cont. on p. 468)

Table 16.7 *(cont. from p. 467)*

III. Subgroups of common macroscopic pond organisms		
A.	Shape variable; there is no clear body plan. Body mass arranged around a water-filled space. Does not move when touched, and there are no well-defined or specialized body parts visible. Pale yellow to pale brown or greenish gray.	Phylum Porifera (sponges)
B.	Body plan resembles a jellyfish or sea anemone, with similar body parts arranged like the spokes of a wheel (radial symmetry). Often has flexible tentacles. Usually withdraws suddenly when touched.	Phylum Cnidaria (hydras, jellyfish)
C.	Wormlike. Body is soft, long, and wormlike. May or may not appear to be segmented. When present, segments number more than three.	Go to Category IV
D.	Body is encased in a stonelike, rigid shell. Aside from shell, body is very soft.	Phylum Mollusca (clams, snails) Go to Category V
E.	Organism has internal skeleton or has a jointed exoskeleton. Often it has a well-developed head region with obvious eyes.	Go to Category VI

IV. Wormlike pond organisms		
A.	Body strongly flattened and ribbonlike. Usually moves with a wavelike or gliding motion. Often sufficiently transparent to see the internal organs.	Phylum Platyhelminthes (flatworms)
B.	Body is thickened and round. Body is not divided into obvious segments. Usually less than 1 cm long. Swims with a whiplike or thrashing motion.	Phylum Nematoda (roundworms)
C.	Body is thickened and round. Body is obviously divided into segments, resembling earthworms. Usually greater than 1 cm long. If present, appendages are bristles and are found on more than three body segments.	Phylum Annelida (earthworms and kin)
D.	Body is thickened and round. Body is obviously divided into segments, resembling earthworms. Usually less than 1 cm long. Often brightly colored. May rest near water surface, using a tubelike siphon extending through the water film to obtain air.	Phylum Arthropoda Class Hexapoda (larvae of aquatic insects)

(cont. on p. 469)

Table 16.7 *(cont. from p. 468)*

V. Classes of Mollusca		
A.	Shell covering body is composed of two approximately equal halves.	Class Pelecypoda (clams and kin)
B.	Shell on body is composed of a single spiraling unit.	Class Gastropoda (snails and slugs)
VI. Organisms with eyes and a distinct head		
A.	Internal bony skeleton. Body fusiform, usually macroscopic. Two large eyes that are not compound. Number of legs: zero, two, or four. Legs (if present) have webbed toes or feet. Animal appears to be a fish or tadpole.	Phylum Vertebrata (vertebrates)
B.	Body encased in plates or sheets of exoskeleton. Legs (if present) are clearly jointed. Number of legs >4. Compound eyes often present.	Phylum Arthropoda Go to Category VII
VII. Classes of Arthropods		
A.	Body size (without legs) ranges from 1 mm to 5 cm. Eight legs. Two body segments (head may not be distinct). Exoskeleton soft. No gills. Antennae not visible.	Class Arachnida (spiders, mites, and kin)
B.	Body up to 15 mm long. More than eight legs present. Body segments not distinct; exoskeleton always hard. Legs can be obscured by a carapace (an over-hanging extension of the exoskeleton). Gills are frequently present, but may not be visible from above. Legs and antennae may be forked and often emerge from under carapace.	Class Crustacea (crabs, shrimp, and kin)
C.	Body is minute to 5 cm long. Six legs, three body segments (may not be obvious). All legs emerge from middle segment. Wings may be present. Exoskeleton may be soft in larval forms. Three pairs of jointed legs on thorax. Gills may be present on abdomen, but will be obviously different from jointed legs. Antennae usually visible, vary in their length.	Class Hexapoda (the insects)

UNIT 16

Measuring Biological Diversity

Teacher Pages

INSTRUCTORS' NOTES

Background and Key Concepts

Often the emotions and rhetoric surrounding preservation of biodiversity mask the fact that most students lack even a basic understanding of the issue. Many students agree it is important to maintain biological diversity, but they support their position primarily with emotional arguments. Others have a more anthropocentric view: The goal of preservation is to protect potential biological resources that might be valuable for humans ("There are plants in the jungle that may have drugs we can use.") Few students can explain the meaning or inherent value of biodiversity from a strictly biological point of view.

The debate on how best to preserve biodiversity has fostered a major misconception as well. Students mistakenly think biodiversity applies only to preservation of international hot spots containing rare charismatic megafauna or unique habitats. They do not realize areas of high and low diversity exist all around them, nor do they see that conservation and preservation are local as well global concerns.

This unit was designed with three goals in mind: (1) to help students build a functional understanding of biodiversity from the empirical point of view, (2) to help students understand that differences in biodiversity can be seen all around them, and (3) to show students how biodiversity is measured and evaluated. Discussion of hot-button issues has been avoided intentionally, so that students could focus on the underlying biological principles. Before and during this unit, instructors should emphasize these concepts:

- Every habitat or ecosystem has a wide range of species that could potentially be found in it. Which species are adapted to a location depends on several abiotic and biotic factors.
- Some habitats are able to support a larger number and variety of species than others. Habitats that support a larger variety are said to have high biodiversity, while habitats that support only a few species have low diversity.
- For any habitat, there is a difference between its maximum *potential* diversity and its *actual* diversity. Species can be absent because of abiotic factors, biotic factors, or both.
- By regularly measuring the biodiversity of an area, we can estimate its overall stability, and make predictions about its future behavior.

Readers with backgrounds in field ecology or conservation biology may argue that this unit oversimplifies or ignores contextual factors that affect biodiversity measurements and interpretation. Given that the target audience is nonmajors (many of whom are phobic about science in general), the author opted for greater understanding of a smaller number of core concepts. However, other instructors may want to revise this unit and introduce one or more of the following ideas:

- **Measurements of stability and quality are sensitive to timescale.** Biological diversity changes seasonally. In the temperate zones, estimates of aquatic diversity may be very low during winter, when many organisms are inactive or are present only as eggs. Similarly, diversity estimates may be low immediately after heavy runoff scours a stream, or they may change dramatically after the midsummer metamorphosis of large numbers of aquatic larval insects. The point is, diversity estimates vary considerably over short timescales. More accurate estimates of quality and stability require repeated measurements made over longer timescales.
- **Measurements of stability and quality vary over geographic scales.** Suppose the state office of the Natural Resources Commission seeks to determine if catch pond ecosystems in a particular county are stable. Thirty diversity estimates are obtained over three years for three catch ponds, all of which are in an area subject to routine flooding by irrigation runoff. The resulting analysis is likely to indicate that catch pond habitats are unstable.

Yet if 1000 diversity estimates were obtained over three years for 100 catch ponds within the county (including the three described above), the data may indicate that catch pond habitats in the county are quite stable over time.

- **Greater diversity is not always better**. All estimates of diversity are relative; there is no magic number that indicates an ecosystem or habitat is stable or is high quality. In some cases, increases in diversity estimates indicate a decline in quality or stability. For example, prior to the arrival of chestnut blight, the great majority of hardwood canopy in Appalachia consisted of American chestnut. When blight eliminated the dominant species, three to five other tree species moved into the opened niches. Although the loss of chestnuts was an ecological disaster, mathematically hardwood biodiversity increased. A similar scenario occurs when a stable endemic habitat based on a few species is overrun by nonnative invasive species. Estimates of overall diversity are higher, despite the environmental decline.
- **Estimates of biodiversity based on a single method are prone to error**. No one method for estimating diversity is ideal. Each method has advantages and limitations. The most robust estimates of quality and stability are obtained by combining data collected with multiple methods.

General Teaching Strategy and Common Problems

This unit was designed for a freshman nonmajors course, but it can be adapted for more advanced courses in ecology. Unlike other units, there is no option for students to design experiments of their own. The goal in this unit is for students to evaluate existing habitats, not test new ones. That said, students can use the basic methods to compare any number of different habitats. Instructors are encouraged to adapt this unit and allow students to explore areas besides just aquatic habitats. Students often are surprised by the amount of life they find in habitats they walk or drive past every day.

The entire exercise can be completed in one three-hour class period. During the first hour students walk to two preselected sites on campus, which should be located within 200 yards of one another. The first site should be a stormwater creek that collects considerable runoff and silt from the campus. The second should be a small, spring-fed stream, a small pond, or some other minimally disturbed waterway. At each site students make field notes describing the types of surrounding vegetation, appearance of the banks, and differences in the color or flow rate of the water. Students collect their own water and bottom detritus samples to carry back to the lab for comparison. Before leaving the second collection site, students are asked to write down a preliminary *field* prediction of which location is likely to have greater biodiversity and their rationales for that prediction.

In the laboratory, pairs of students spend the second hour searching their collected materials for invertebrates. Students are reminded that most organisms will be found in the detritus and silt, not in the water. Each time a student finds an organism, they identify it to phylum or class, show it to their partner, and give it a simple descriptive name. It is *vital* that students work methodically and spend the same amount of time on material from each site.

Instructors should try to keep the mood light for this lab. Microscopic surveys can be tedious and tiring. Students also get disheartened quickly if they find few organisms, so when one student locates something unusual, announce it to the class and make sure everyone gets to see it. Consider providing students with small specimen containers with lids, so they can hold onto specimens for later demonstration.

In the last hour of lab, the instructor will tally the number of species and individual animals that students found in each phylum in a table on the board. When tallying the number of different species in a phylum, two or more groups will likely discover that they have found the same species, but duplicates are not a major problem. The common names students assign usually are sufficiently descriptive to eliminate most duplicates. Additionally, differences in diversity indices between sites tend to be large enough to offset any species counted twice.

After the class summary data have been compiled for each site, students calculate diversity indices as described in Exercise 3. These require only basic math skills, so if time is short, students can complete them and the Unit Summary Questions as homework assignments.

This lab can be split into two separate, two-hour lab periods. In the first session students collect samples, search for invertebrates, and collate their data for Site 1. The following week, students collect and collate data for Site 2. Final calculations of diversity estimates are completed as a homework assignment.

This unit also can be used as the starting point for a multiweek inquiry lab sequence focusing on aquatic ecosystems. Subsequent units would focus on quantifying the effects of abiotic and biotic factors identified here on energy capture via photosynthesis (Unit 6, using algae or aquatic plants as the model system), homeostasis (Unit 10), and metabolism (Unit 11).

Recommended Prelab Skills

Students must be comfortable using a brightfield microscope to find microorganisms. Prior experience with a dissecting microscope is helpful as well.

Assessment

If the instructor uses pre- and postlab quizzes routinely as part of formal assessment, the prelab quiz questions should focus primarily on general definitions of high and low biodiversity and on factors affecting overall diversity in an ecosystem. It also is appropriate to ask what specific factors could raise or lower diversity in an aquatic ecosystem. Wait until the postlab quiz to ask which specific indices are being used to estimate biodiversity and about differences between richness and dominance.

On the postlab quiz, students should be able to summarize the results they obtained the previous week (Unit Summary Questions 1–3 prepare them for this).

They should be able to summarize their observations of the two field sites visited and to explain how what they observed might relate to measured differences in biodiversity between the sites.

If students are required to use the dichotomous key rather than match specimens to photos or drawings, the instructor should include practical questions on the postlab quiz. Preserve five examples of invertebrate phyla in 70% ethanol, and label them "A" through "E." Provide students with a *text-only* key and ask them to key out any three of the five examples; to receive full credit, they should be able to explain each step they took through the key to reach the correct identification.

Safety and Housekeeping

Pond detritus samples may contain sharp sticks, stones, thorns, or litter such as broken bottles. We recommend having forceps, antiseptic, and small bandages on hand to administer first aid. Bites from aquatic insects are very rare, with little chance of an adverse reaction. Make sure students dispose of glass slides and coverslips in a dedicated glass waste receptacle.

When they finish working, students can empty waste samples into a 5 gal. plastic bucket near a sink. When the bucket is two-thirds full, cover it and allow the contents to settle. Pour off the water, then dispose of the solid matter in general trash, or return it to the collection site.

Other Tips

- This unit can degenerate into a cookbook confirmation lab very quickly if the instructor tells students what to expect at the two sites. *Do not tell students which location is more diverse*; let their results drive that discovery. It is important to not correct students if their data indicate the apparently more degraded habitat is actually more diverse. Focus on interpreting the data and evidence at hand properly, not on obtaining a "correct" answer. The author learned this lesson firsthand when this unit was first developed. Students sampled several streams on campus that, given their physical features and locations, were expected to be significantly different. Yet students' data showed quite clearly that almost all of the sites had essentially the same range of species and equivalent diversity indices. Subsequently it was determined that both sites were fairly degraded and unstable, despite their appearances.
- If convenient collection sites are not available, or the lab must be shortened to fit available time, water and bottom detritus can be collected two to seven days in advance of the lab meeting and kept in aquariums. Detailed instructions for collection are in the Preparatory Notes.
- The general format of this unit can be adapted to compare other habitats and employ other sampling methods. The key requirement is that samples be collected from the two sites and examined in the same way. For example,

students could easily compare biodiversity in soil detritus from two sites by placing equal wet weights of collected detritus into two separate, identical Berlese funnels.

- For a multisemester project, the instructor can set up artificial ponds in 300 gal. stock watering tanks at various locations (in sun versus shade, near other water versus isolated) with different starting conditions (e.g., with addition of bottom detritus versus without addition of bottom detritus or with low versus high organic nitrogen load), then let students follow the development of the artificial pond ecosystems over time.
- Inexperienced students will not know how to search methodically for organisms using a microscope. An effective search strategy is to use dissecting forceps and a spoon to transfer about 20 mL packed volume of larger debris (sticks, leaves) to a 6 in. white Styrofoam disposable plate. Spread the debris thinly across the white plate, then place the plate on the stage of a dissecting microscope. Set the microscope to its lowest magnification. Starting at the top of the plate, scan the field of view for any movement. If there is none, move one field to the left or right, and scan that field. Each time, move over by only one field of view. At the edge of the plate, move down one row, and move back field by field in the opposite direction as the previous row. Scanning one plate of debris should take four to six minutes, not including time spent identifying the organisms found. After scanning one plate, empty the debris into the waste bucket, and repeat the process with fresh material.
- To search for even smaller organisms, mix an equal volume of clear pond water with some silt and small debris. Run a bead of liquid and suspended debris down the middle of a microscope slide, then using a plastic pipet, spread the liquid out to cover most of the slide; do not add a coverslip. Place the slide onto the stage of a compound microscope, and focus on the material using the lowest power objective. Scan the slide methodically moving left to right, then a second time moving from top to bottom. After making two passes, dump the slide, rinse it with water, and prepare another. As before, one slide should take four to six minutes to scan.
- Advanced students can be required to prepare a collection of "voucher specimens" from the two sites, to reduce the likelihood of species being counted twice. As students search for invertebrates, they transfer the first example of each species they find into individual glass shell vials containing either dechlorinated water or 70% ethanol. The instructor photographs the organism in each vial, transfers the images to a laptop, and displays the images as class members collate the class data. Serviceable photographs can be taken simply by holding the camera up to the eyepiece of the microscope. Some instructors may wish to keep ethanol-fixed voucher specimens and digital photographs as references for later semesters.

Creating a Dichotomous Key for Local Organisms

At the end of the student exercises there is a brief key to major freshwater invertebrate phyla. This key is based on simple field characters and does not represent phylogenetic relationships. Individual instructors must decide how much emphasis they want to place on using the key for making identifications. The primary learning objective for this unit is biodiversity, not field biology. Therefore, instructors at the author's institution make the key available but let students make identifications by matching specimens to drawings or photos directly, without working through the key.

The key provided is text only, because the most common representative species for each group differs among geographic regions. Readers will need to supplement it with diagrams or photos of locally common representatives from each group or find keys that are specific to their areas.

PREPARATORY NOTES

Quantities listed are for a lab section of 20 students working in groups of three to four.

Shared Materials

(*Omit if students will be collecting their own sample materials)

- 1 aquarium containing bottom detritus from a high-diversity creek or pond site*
- 1 aquarium containing bottom detritus from a low-diversity creek or pond site*
- 1–2 waste buckets (5 gal.), near a sink

Materials at Each Work Station

(**Omit if instructor is supplying precollected sample materials)

- Clipboard with paper (for field notes) **
- 2 glass jars (pint) with screw lids**
- Large aquarium dip net, with long handle**
- 2–3 disposable white Styrofoam plates (6 in.)
- Dissecting microscope
- Compound microscope
- Glass microscope slides
- Coverslips
- Dissecting needles
- Medium dissecting forceps

- Fine dissecting forceps
- 3–4 plastic transfer pipets
- Paper towels
- White plastic spoon (for moving organisms; can substitute weigh spatula)
- Small glass dish or plastic weigh boat (for observing isolated organisms)
- Optional: 15–20 glass shell vials with dechlorinated water (if students will be collecting voucher specimens)

Sources of Materials

Significantly different sampling sites are needed if this unit is to succeed. Ideally there will be obvious physical and biotic differences between them. For example, the low-diversity site that the author uses is a stormwater stream. The site is open to full sun, and its banks are mostly red clay. About a third of the banks are eroded bare spots; the remainder is covered with woody brambles and saplings. The stream bottom is predominantly silt and sand and is routinely scoured by heavy runoff from the campus. Typically students find large numbers of annelids and mollusks—and a smaller number of flatworms and various arthropods—but overall biodiversity is low. The stable, high-diversity site is a small, spring-fed feeder creek that is a tributary of the larger stormwater stream. The feeder creek has a stone and gravel bottom, with overlying layers of leaf and twig debris from canopy hardwoods. The banks are mainly dark forest duff covered by mosses and shade-tolerant forbs. The dominant organisms in this stream year-round are arthropods, with seasonally variable numbers of caddisflies, midge larvae, and dragonfly/damselfly naiads. Diversity of the two sites is about the same at the phylum level, but much higher at the species level for the spring creek.

If suitable collection sites are not available within walking distance of the teaching lab, this unit can be done using "miniponds" set up in 20 gal. aquaria with precollected materials. Six miniponds (three each for two sites) can hold more than enough material for 200 students working in pairs to complete the unit.

1. One week before lab, collect one 5 gal. bucket of bottom debris and one 5 gal. bucket of water from a low-diversity stormwater pond or stream. Collect the same amount of material from a high-diversity natural pond or wooded stream.
2. In the lab, divide the bucket of bottom debris from the low-diversity site equally between three clean, 20 gal. aquaria. Add one-third of the collected water from the site to each tank. Each tank will be less than one-third full. Mark the water level in each tank, and label these three tanks "Site 1."
3. Divide the bottom debris from the high-diversity site equally between the remaining three tanks, add the collected water, mark the initial water levels, and label these tanks "Site 2."
4. Place two air stones in each tank for aeration. Keep tanks on a ~16 hr. light:8 hr. dark cycle.

5. Allow the material and sediment to settle for at least two days before use. Replace any water lost to evaporation with dechlorinated tap or springwater. Keep all tanks approximately one-third full.

Notes:

- In the spring dragonfly naiads may start emerging after three to five days in the lab. They can be captured with a butterfly or other soft sweep net and released outside.
- To reduce the effects of variation between tanks, have all students in a single lab section take their samples from the same two tanks. Rotate which tanks are being used.

APPENDIX A

The Instructional Methods Inventory

Directions

This survey instrument is designed to measure the relative amount of class time an instructor devotes to teaching methods that promote inquiry and active learning (Bohrer et al. 2008). The results can be used to (1) track changes in teaching practices over time and (2) compare teaching practices between instructors.

There are 31 different instructional methods listed. Beside each one is a scale from 1 to 5. Read each description, then circle the number that best corresponds to the frequency with which you use that method. Mark only one number on the scale.

There are no right or wrong answers, so please answer honestly. If you do not recognize a particular teaching method, or the description is confusing, indicate that by circling the number of that item in the *left* column. Leave the scale for that item blank and go to the next question.

Reference

Bohrer, K., A. Ferrier, A. D. Johnson, and K. Miller. 2008. Teaching the teacher: Introducing and training TAs in inquiry-based learning methods. *Proceedings of the 29th Conference of the Association for Biology Laboratory Education* 29: 67–126.

THE INSTRUCTIONAL METHODS INVENTORY (IMI)

How frequently do your students engage in each of the following behaviors when you are teaching in a lab setting? Please rate them using the following scale:

(1) Never. I do not use this method at all when I teach.
(2) Rarely. I use this method in less than 25% of labs I teach .
(3) Sometimes. I use this method in about 50% of labs I teach.
(4) Frequently. I use it in about 75% of labs I teach.
(5) Extensively. I use it in 90–100% of labs I teach.

Item #	Description	Rating
1	Completing worksheets	1 2 3 4 5
2	Listening to the instructor lecture	1 2 3 4 5
3	Taking multiple-choice/true or false/fill-in-the-blank tests	1 2 3 4 5
4	Reading assignments in a textbook or manual	1 2 3 4 5
5	Engaging in experiments with predetermined outcomes	1 2 3 4 5
6	Engaging in experiments with predetermined, written procedures	1 2 3 4 5
7	Memorizing concepts	1 2 3 4 5
8	Writing lab reports on experiments with set procedures and results	1 2 3 4 5
9	Receiving factual information from the teacher	1 2 3 4 5
10	Passively watch a demonstration of a principle or process	1 2 3 4 5
11	Identifying variables and designing appropriate controls for experiments	1 2 3 4 5
12	Answering questions about prior knowledge	1 2 3 4 5
13	Asking clarification questions during or after class	1 2 3 4 5
14	Participating in an in-class simulation or group exercise	1 2 3 4 5
15	Participating in a class discussion	1 2 3 4 5
16	Developing new examples of a specific concept or process in action	1 2 3 4 5
17	Making predictions based on prior knowledge	1 2 3 4 5
18	Giving individual presentations or participating in group presentation in class	1 2 3 4 5
19	Writing formal lab reports on novel results	1 2 3 4 5
20	Reviewing or critiquing another student's work	1 2 3 4 5
21	Searching outside primary literature sources to learn what is already known	1 2 3 4 5

22	Designing and implementing new procedures or models	1	2	3	4	5
23	Exploring alternative methods for solving problems	1	2	3	4	5
24	Identifying questions/concepts that guide scientific investigations	1	2	3	4	5
25	Comparing data or otherwise collaborating with other groups	1	2	3	4	5
26	Communicating findings to the rest of the class	1	2	3	4	5
27	Using graphs, basic statistics (mean, st. dev., t-test, etc.) to summarize and analyze results	1	2	3	4	5
28	Explaining unexpected results and considering potential sources of error	1	2	3	4	5
29	Asking new questions based on data analysis from a previous experiment	1	2	3	4	5
30	Reflecting on their own work or learning	1	2	3	4	5
31	Explaining data from experiments without a predicted outcome or using other evidence to make and defend conclusions	1	2	3	4	5

Scoring the Inventory

Which specific methods are used is not important for scoring this inventory. What *is* important is the relative amount of time that an instructor spends using low-, medium-, and high-level teaching methods. Items 1–10 are low-level teaching methods that do not foster an inquiry mind-set well. Items 11–21 are teaching methods that can foster inquiry when they are used correctly, while Items 22–31 are methods that strongly foster inquiry.

Determining Relative Time Distribution

Calculate the average score for each section using Table A1, page 484. First, add up the total score on all items in each section and place that total in Column A. Next count the number of questions answered in each section and place the total in Column B. The maximum number of responses for each section is given in parentheses. To determine the average scores for low-, medium-, and high-level teaching activities (C), divide A by B in each row. Values for C can range from 1.0 to 5.0 (inclusive).

Table A1

IMI calculations

	Sum of All Scores for Section (A)	Total Number of Items Scored (B)	Average Score for Section (A/B) (C)
Items 1–10 (Low level)		(10)	
Items 11–21 (Medium level)		(11)	
Items 22–31 (High level)		(10)	

Most Common Teaching Methods

Below, list all teaching methods that were given scores of 4 or 5 in each section. These represent the instructor's most frequently used teaching methods.

Low:

Medium:

High:

Suggestions for Interpreting Inventory Results

Relative Time Distribution

Ideally, the inquiry-oriented instructor spends less class time using low-level teaching methods and more time using higher-level methods. The midpoint score on this inventory is 3.0, which corresponds to using a particular group of classroom activities 25–50% of the time. Table A2 shows how the author would interpret various combinations of average scores from Table A1, assuming it is the first time that a novice instructor has completed the inventory.

Table A2

Interpreting the results of the IMI calculations

Scores	Interpretation
$C_{Low\ Level} > 3.2$ $C_{Mid\ Level} + C_{High\ Level} < 6.0$	Instructor is not inquiry-oriented; spends majority of time teaching with methods that prevent or inhibit development of inquiry mind-set.
$C_{Low\ Level} = 3.0$ $C_{Mid\ Level} + C_{High\ Level} = 6.0$	Instructor is spending equal class time using low-level teaching methods and methods that foster inquiry. Some effort to foster inquiry.
$C_{Low\ Level} > 3.2$ $C_{Mid\ Level} + C_{High\ Level} > 6.5$	Instructor is trying to use inquiry instructional methods on top of traditional didactic methods. Students are likely to feel stressed or overworked.
$C_{Low\ Level} < 2.7$ $C_{Mid\ Level} + C_{High\ Level} > 6.5$	Instructor is spending majority of class time using teaching methods that promote the inquiry mind-set. Class format fosters inquiry.

If the instructor has completed the inventory previously, the author would look for several additional trends. First, if the instructor being assessed is trying to increase his or her use of inquiry, the average score for low-level teaching methods should decline over time. Also, the sum of the two scores for mid- and high-level methods should rise over the course of two to four years, and remain above 6.5 once the novice instructor gains confidence and skill in using inquiry-oriented instructional methods.

Most Common Teaching Methods

The point of this evaluation question is not to define which specific methods are used, but how many. Ideally, an instructor regularly uses multiple techniques from all three groups. If an instructor lists only one or two methods, or if most methods are in just one of the three groups, the instructor is using a very small repertoire of teaching methods. These data must be interpreted with care; if an instructor is using a very wide array of techniques, it is quite possible that none of the methods will be used often enough to be scored a 4 or 5. However this is unlikely to happen with TAs or other instructors who are new to inquiry as a teaching modality.

APPENDIX B

Sample Form for Students' Experimental Outline

Guidelines for Instructors

When students conduct experiments as part of an inquiry lab exercise, it is imperative that the instructor review each work group's experimental design before letting the students proceed. The simplest method is to have students outline their experiments using a simple form similar to the one that follows and submit the forms for approval. We recommend that the instructor read and mark up proposals during or immediately after lab, photocopy them for reference purposes, then hand the originals back to the groups shortly before they conduct their experiments. The photocopies provide a record of proposed experiments in case there is any confusion and serve as documentation of all materials requested.

The instructor's goal while reviewing proposals should not be to correct every mistake or clarify specific details. Remember, the central goal of inquiry-based instruction is to facilitate students discovering knowledge for themselves. Resist the temptation to provide them with the answers. Instead, focus on ensuring that

- the group's background thinking is clear and rational,
- the experiment represents a reasonable challenge for the group proposing it, and
- the question and proposed experiment align with the goals of inquiry and the lab unit.

Some students will devise experiments that are relatively minor extensions of procedures used in the first part of the unit (e.g., testing a higher temperature or a broader pH range). This is sufficiently challenging for some students, but less-motivated students may use this tactic to avoid making any mental effort. If the instructor knows that a particular student (or that student's work group) is proposing an experiment that is below the students' current cognitive abilities, the instructor should reject the proposal or ask the group for a better explanation of why their proposed experiment represents a significant mental challenge.

As students become more confident they frequently design experiments that will be too difficult to interpret or that are not feasible given technical or equipment limitations. Experimental outlines let an instructor identify these problems or technical limitations in advance, so they can encourage students to modify their plans accordingly.

Photocopied proposals remain useful after a course ends. Reviewing proposals from several lab sections or semesters right away helps an instructor or coordinator learn to predict what experiments students are likely to propose and purchase materials accordingly. Proposals also can serve as an informal assessment tool to document the raw thinking processes of students as they progress through a particular course or a series of courses in a curriculum.

My Proposed Experiment

Date: ______________ Instructor's Approval: ________________________

Members of my work group:

Background observations:

Hypothesis and rationale:

Experimental Design:
(The strategy I will use to test my hypothesis. How I will interpret potential outcomes.)

Materials Required:
(**Exactly** what live materials I will need, chemicals I plan to test, and equipment I will need that is not available routinely in the lab.)

INDEX

Note: **Boldface** page numbers indicate tables and figures.